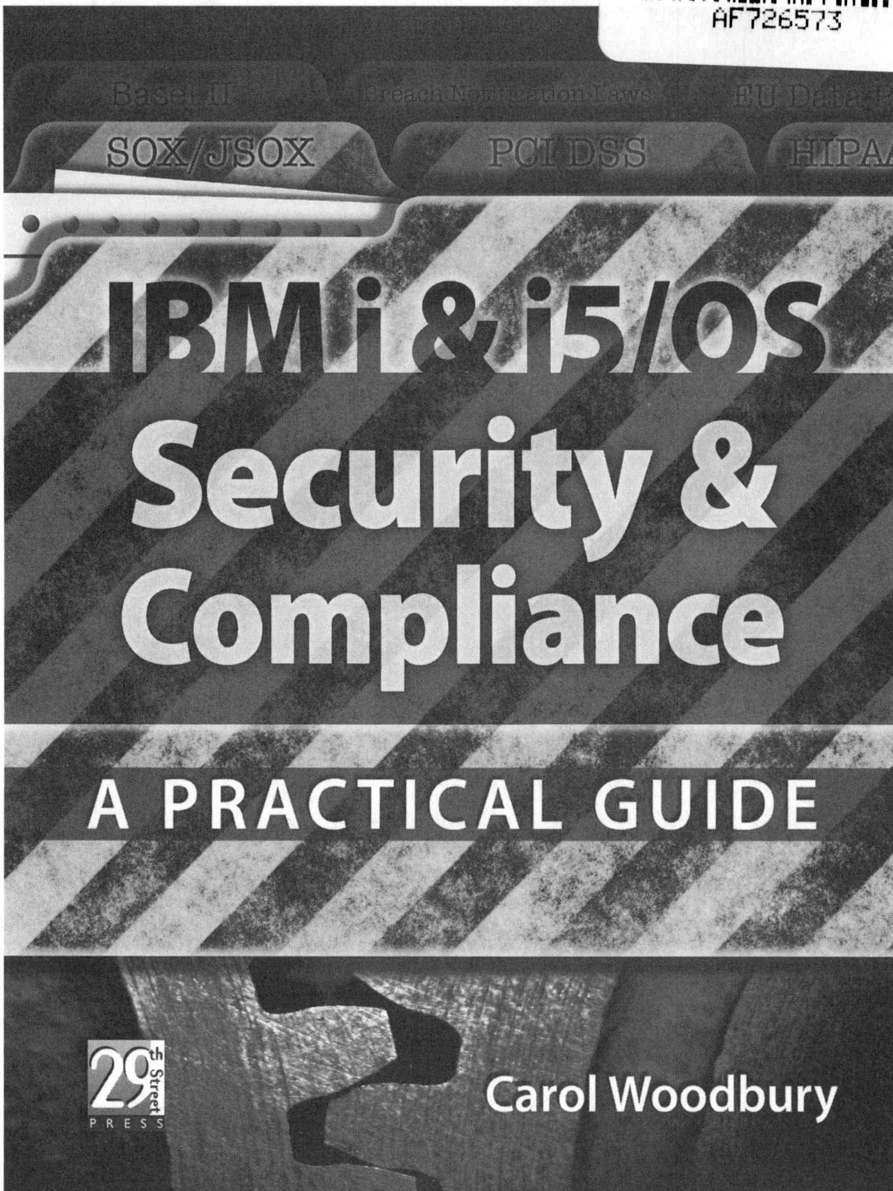

29th Street PRESS®
A Division of **Penton Media, Inc.**
221 E. 29th Street • Loveland, CO 80538 USA
(800) 650-1804 • (970) 663-4700
www.systeminetwork.com • www.left-brain.com

Library of Congress Cataloging-in-Publication Data on file.

29th Street Press® is a division of
Penton Media, Inc.
Loveland, Colorado USA

It is the reader's responsibility to ensure procedures and techniques used from this book are accurate and appropriate for the user's installation. No warranty is implied or expressed.

This book was printed and bound in Canada.

ISBN-10: 1-58304-124-9
ISBN-13: 978-1-58304-124-6

2011 2010 2009 WL 10 9 8 7 6 5 4 3 2 1

HALF PRICE BOOKS®

Half Price Books
1835 Forms Drive
Carrollton, TX 75006
OFS OrderID 39492650

Thank you for your order, Main Address!

Thank you for shopping with Half Price Books! Please contact Support@hpb.com. if you have any questions, comments or concerns about your order (113-1979631-1341038)

Visit our stores to sell your books, music, movies games for cash.

SKU	ISBN/UPC	Title & Author/Artist	Shelf ID	Qty	OrderSKU
S394894641	9781583041246	IBM i & i5/OS Security & Compliance: A Pra... Carol Woodbury	31--15--1	1	

SHIPPED STANDARD TO:
Main Address
800 Avondale Ave
1033837-1-0124
Grandview Heights OH 43212
j1gg094j97htty5@marketplace.amazon.com

ORDER# **113-1979631-1341038**
AmazonMarketplaceUS

To my kids — my nieces and nephews
and great nieces and nephews:
Kelsey and Todd, Abby, Ava Grace and Mia;
Jonathan and Janna, Walker and Grant;
JB, Allison; Christopher; Suz; Michelle and Josh.
I'm so proud of each of you.
Thank you for bringing laughter and love into my life.
I love you more than you can imagine.

I give all glory, praise, and honor
to my Lord and Savior, Jesus Christ —

"My soul finds rest in God alone;
my salvation comes from him.
He alone is my rock and my salvation;
he is my fortress,
I will never be shaken."
Psalm 62: 1 & 2

Acknowledgements

I need to thank several people for their part in helping me complete this book. Katie Tipton lent her editing talents and made this a far better book. I very much appreciate her encouraging comments during the editing process and her patience as I asked for last-minute changes. Jan Fonda provided production support, and Matt Wiebe used his creative talents to design the great new cover. Jan Hazen managed everything, keeping me on schedule and coordinating the process.

Paul Tuohy of ComCon provided the RPG examples. Carsten Flensburg validated the code for V6R1. The IBM i security development team in Rochester, Minnesota, has provided friendship and support throughout my career and has implemented the functions and features that I am privileged to describe in this book. Conference attendees, readers of my books and articles, customers of SkyView Partners, and webinar attendees who have asked the tough questions and pressed for answers to their specific requirements have caused me to dive deeper into the subject of IBM i security and the world of security in general. The practical answers they required are the basis for the information provided in this book.

I would be remiss not to recognize and formally thank previous co-author Wayne Madden for giving me the opportunity to become his co-author, as well as previous co-author Pat Botz for his friendship and encouragement and the lively debates on various aspects of security.

Finally, I'd like to thank my family, my business partner John Vanderwall and his wife Linda, and friends Sam and Tricia Johnson for putting up with me as I went through the writing process.

Thank you all!

About the Author

Carol Woodbury is president of SkyView Partners, Inc., a firm she co-founded in 2002 that specializes in security policy compliance software and remediation services. Carol has worked in the field of security since 1990. She worked for IBM in Rochester, Minnesota, for 16 years, the last 10 of which were spent as team leader of the OS/400 security development team and chief engineering manager for AS/400 security.

Carol is an award-winning speaker who speaks around the world on the topic of security. She is also an award-winning writer and serves as a technical expert on security topics for numerous publications. Her years spent architecting and implementing security solutions have allowed her to amass a wealth of knowledge and experience. It is this knowledge and experience that is the basis for the practical guidelines and examples provided in this book.

Table of Contents at a Glance

Chapter 1: Security — The *Real* Reason You're Reading This Book 1

Chapter 2: Policies and Procedures 9

Chapter 3: Security at the System Level 19

Chapter 4: The Facts About User Profiles 63

Chapter 5: Service Tools Security 93

Chapter 6: Object-Level Security 111

Chapter 7: Security Considerations for the IFS 139

Chapter 8: Securing Your Printed Output 155

Chapter 9: Encryption 163

Chapter 10: Connecting to the System 173

Chapter 11: Internet Security 197

Chapter 12: Evaluating Applications' Current Implementations and Designing New Ones 209

Chapter 13: Role-based Access (RBAC) 223

Chapter 14: Role-based Access for IT 229

Chapter 15: Auditing 239

Chapter 16: Implementing Object-Level Security 267

Chapter 17: Maintaining Compliance 285

Chapter 18: Preparing for the Worst: Creating a Security Incident Response Plan 293

Chapter 19: Creating a Security Awareness Program 299

Index 303

Table of Contents

Chapter 1: Security — The *Real* Reason You're Reading This Book **1**
Evaluating Your Risks 2
Confidentiality 2
Integrity 3
Availability 4
Privacy 4
Evaluating the Threats 5
Managing the Strategic Issues 5
Control Access to Applications, Data, and Systems 5
Review Requirements and Maintain Compliance 6
Getting Started 7
Don't Close the Book 8

Chapter 2: Policies and Procedures **9**
Your Security Policy 9
Physical Security 10
Responsible Parties 10
Data Classification 11
What Classifications Should You Use? 12
Who's Responsible for Classifying Data? 13
Data Ownership 13
Why Not IT? 13
Network Connections 14
Application Design 15
Platform-Specific Issues 15
Employee Guidelines 15
Notification, Enforcement, and Compliance 16
Business Events and Procedures 17
Getting Started with Your Policy 18
Legal Review 18

Chapter 3: Security at the System Level **19**
The System Security Level 19
System Value QSECURITY 19
Security Level 20 20
Security Level 30 21
Security Level 40 22
State and Domain Restrictions 22
Use of Restricted MI Instructions 22
Job Initiation Validation 23

Restoration of Modified Programs 23
Parameter-Passing Validation 24
Why Use Security Level 40? 24
Security Level 50 24
Message Restrictions and Pointer Removal 24
Prevention of Control-Block Modification 25
QTEMP Library Maintenance 25
Why Use Security Level 50? 26
Moving to Security Level 40 or 50 26
Security-Related System Values 27
General Security System Values 27
Password-Related System Values 44
V6R1 Password-Related System Values 52
Audit-Related System Values 56
Locking Down Security-Related System Values 62
A Helpful Tool 62

Chapter 4: The Facts About User Profiles 63
What Are User Profiles? 63
User Profile Attributes 64
USRPRF (User Profile) 65
PASSWORD (User Password) 67
PWDEXP (Set Password to Expired) 68
LCLPWDMGT (Local Password Management) 68
PWDEXPITV (Password Expiration Interval) 69
PWDCHGBLK (Block Password Change) 69
STATUS (Profile Status) 70
USRCLS (User Class) and SPCAUT (Special Authority) 70
Initial Sign-On Options 74
LMTCPB (Limit Capabilities) 75
System Value Overrides 76
Group Profiles 76
Creating Group Profiles 77
Assigning a Group to a User Profile 77
UID (User Identification Number) and GID (Group Identification Number) ... 79
AUT (Authority) 79
Private Authorities and User Profiles 81
Helpful Tools 81
Navigator for i 81
Copying User Profiles 83
Validation List Users 83
Security Implications of Validation List Users 90

Chapter 5: Service Tools Security **93**
Service Tools User IDs 93
Service Tools User ID Passwords 95
Default Password Policy Details 96
Alternate Password Policy Details 96
Choosing a Password Policy 97
Service Tools Functional Privileges 97
Service Tools Features in V6R1 104
Device Profiles 105
The Work with System Security Panel 106
Monitoring Service Tools Use 108
Service Tools Security Recommendations 109

Chapter 6: Object-Level Security **111**
Private Authorities 111
Object Authorities 112
Data Authorities 113
Authority Relationships 114
Authority Groupings 115
Group Profiles 116
Multiple Group Profiles 116
Why Grant Authority to Group Profiles? 117
Public Authority 117
Establishing Public Authority 117
Using Default Public Authority 119
Authorization Lists 121
How IBM i Checks Authority 123
Authority Checking Example: Precedence Between Users and Groups 125
Authority Cache 126
Adopted Authority 127
Adopted Authority Example 128
Authorities and Save/Restore Functions 130
Object Ownership 131
Limit User Function 133
Helpful Tools 136
Navigator for i 136

Chapter 7: Security Considerations for the IFS **139**
IFS Authorities 140
Managing Authorities to IFS Objects 141
File Attributes 144
Adopted Authority and the IFS 145

Auditing Objects in the IFS 146
File Shares: Accessing Objects in the IFS 146
Gotchas and Helpful Hints 148
General Cautions 148
Creating New Objects 149
Copying Objects 149
CPYTOSTMF and CPYTOIMPF 150
Virus Scanning 150
Security Recommendations 151
*PUBLIC Authority for Application and User Directories 151
*PUBLIC Authority for IBM-Supplied Directories 151
Determining Appropriate Authority 152
Home Directory 152
Web Applications 153
QPWFSERVER Authorization List 153
Review (and Remove) File Shares 153
Final Advice 154
Helpful Tools 154

Chapter 8: Securing Your Printed Output 155
Security-Related Output Queue Attributes 155
DSPDTA (Display Data) 156
OPRCTL (Operator Control) 156
AUTCHK (Authority Check) 157
AUT (Authority) 157
*SPLCTL Special Authority 157
Output Queue Ownership 159
Sample Output Queue Security Implementation 159
Helpful Tools 160
Navigator for i 161

Chapter 9: Encryption 163
Encryption Basics 163
Public Key Infrastructure 164
Transmission of Data 165
Encrypting Data in Files 165
Identify the Scope of Your Project 166
Architecting Your Application to Use Encryption 167
The Key Is Key Management 168
Key Generation 168
Separation of Duties 169
Storing the Keys 169

Controlling Access to Keys and Encryption/Decryption Routines 169
Backing Up the Keys 169
Changing the Keys 170
Encrypting Backup Media 170
What IBM Provides 171
Encrypted Auxiliary Storage Pools 171
Disaster Recovery Considerations 172
Success Depends on Planning 172
Helpful Resources 172

Chapter 10: Connecting to the System 173
Physical Security 173
System Values 173
*IOSYSCFG Special Authority 174
Network Security Attributes 174
JOBACN 174
PCSACC 175
DDMACC 175
Security Considerations for TCP/IP 175
Starting TCP/IP Servers 176
Session Time-out Value 176
Securing Ports 177
Internal Addresses 177
IP Packet Filtering 177
NAT 178
FTP 178
SMTP 180
POP 181
DNS 182
REXEC 182
SNMP 182
SNTP 183
DRDA and DDM 183
DDM 183
DRDA 185
Security Considerations for PCs 185
System i Access for Windows 185
Data Transfer and Remote Command Issues 186
Host Servers 186
Limiting What Users See from the Desktop 186
Microsoft Policies 187
Application Administration 187
Selective Install 187

ODBC Security Considerations 187
System i Access for Web 188
Using Exit Points 189
Management Central 190
Secure Communications 191
Digital Certificates 191
Secure Sockets Layer 191
Digital Certificate Manager 193
Virtual Private Networks 193
Secure Shell 194
Wireless Considerations 194
Helpful Tools 195
Navigator for i 195

Chapter 11: Internet Security 197
Determine Your Risk 197
The Process 197
Corporate Security Policy 198
Internet Service Provider 199
Firewalls 200
System Values 201
User Profiles 202
Resource Security 203
Controlling What Goes On 203
Secure Web Applications 204
Exit Programs 206
Monitoring 206
Intrusion Detection 206
Security Configuration 207
Testing and Evaluation 207
Business Contingency Plan 208
The Good News 208

Chapter 12: Evaluating Applications' Current Implementations and Designing New Ones 209
From the Beginning 209
Design Considerations 210
What Roles Will Use the Application? 211
Common Authorization Schemes 212
Menu Access Control 212
*PUBLIC Authority 213
Group Authority 213

*ALLOBJ 214
Authorization Schemes That Are Secure 214
Application-only Access 214
Application-only Access with a Twist 215
Changing the Process 215
Application Ownership 216
Which Profile Runs the Application and Is There Adequate Logging? 217
Does the Application Require a "Powerful" Profile? 217
What Kind of Audit Trail Does the Application Require? 218
Implementation Details 218
Set IBM i Authorities 218
Define *PUBLIC Authority 219
Secure Job Descriptions 219
Manage Your Library List 219
Make Library-Qualified Calls 221
Don't Store Passwords in Clear Text 221
Testing, Testing 221
Moving Forward 222

Chapter 13: Role-based Access (RBAC) 223
Roles 223
Defining the Roles 223
Group Profiles 225
Why Group Profiles? 227
Implementation 227

Chapter 14: Role-based Access for IT 229
Security and Your IT Staff 229
Identify the Roles 230
Define a Secure Environment for Each Business Function 231
Operator 231
Network Administrator for IBM i 233
Help Desk 233
Programmer/Analyst 233
System and Security Administrators 236
Security for Vendors and Consultants 237
Vendor Support 237
Consultant Practices 238
Role-based IT Access 238

Chapter 15: Auditing **239**
The History Log 240
History Log Housekeeping 242
Inside Information 242
The Security Audit Journal 242
The Audit Journal 243
Auditing Controls 243
System-wide Auditing 245
Other Auditing Values 247
User Auditing 247
Object Auditing 249
Object Auditing for New Objects 252
Event-Auditing Recommendations 252
Auditing Controls Security Recommendations 253
System and User Event-Auditing Security Recommendations 253
Object-Auditing Recommendations 253
Working with the Audit Journal 253
Understanding Journal Entry Formats 254
Displaying and Printing Audit Journal Entries 256
Using the DSPAUDJRNE Command to Display Entries 257
Using the DSPJRN Command to Display Entries 259
Using the CPYAUDJRNE Command 262
Reporting on Activities from the Information in the Audit Journal 263
Benefits of the IBM i Architecture 264
Helpful Tools 265
Navigator for i 265

Chapter 16: Implementing Object-Level Security **267**
Determine the Scope of Your Project 267
High-Level Design of the Architecture 268
Building the Big Picture 269
Collecting the Information 271
Dynamic SQL 273
Decision Points 273
Owning All Application Objects Rather Than Being Authorized 274
What Adopts 274
Authorization Lists 275
*PUBLIC Authority 275
How Many Files? 275
*PUBLIC Authority 276
*PUBLIC Authority of Programs 276
Securing Libraries 276
*PUBLIC Authority of Other Application Objects 277

Queries 277
Making Changes to the Application 278
Rolling Out the Changes 280
Changes to Change Management 281
When Something Breaks: Debugging and Recovery Techniques 281
Making Sure the Changes "Stick" 282
Gotchas 282
Summary 283

Chapter 17: Maintaining Compliance 285
Evaluating the Key Areas 285
System Values 286
User Profiles 286
Object Authority 288
An Annual Security Assessment 289
Regular Reviews 290
Group Profile Membership and Special Authority Assignments 290
Authorization Lists 291
Policies and Processes 291
Summary 292

Chapter 18: Preparing for the Worst: Creating a Security Incident Response Plan 293
Be Prepared 293
Assembling the Incident Response Team 293
Responding to an Incident 294
Data Preservation 295
Performing the Investigation 295
Be Proactive 295
Make Sure You're Saving the Right Information 296
Saving Security Data 296
How Often Should You Save Security Data? 296
How Often Should You Save Audit Journal Receivers? 296
Restoring Security Data 297
Order Is Important 297
Security Data Restored with an Object 298
Re-Creating the System After a Breach 298

Chapter 19: Creating a Security Awareness Program 299
What Method Do I Use to Communicate? 300
Getting Started 300

Index 303

Chapter 1

Security — The *Real* Reason You're Reading This Book

When the first version of this book was written by Wayne Madden, the title of this chapter was "Security Is a Business Function." Readers purchased and read the book because they believed in that principle and wanted to learn more. Although that principle is no less significant today, I feel it important to acknowledge — in the very first chapter — why I believe many of you are reading *this* book.

For many of you, it's because your organization is required to be in compliance with some law or regulation. For others, it's because you've been tasked with implementing a "deny by default" security scheme. Some of you may be reading this book because you are auditors assigned to a client running something called an "iSeries" or "IBM i" and you need to understand its security features. Whatever the particular reason, in today's world, implementing a sound security policy is often a forced issue rather than a choice.

So, because you may have picked up this book not so much because you wanted to but because you had no other choice, I've worked hard to make sure reading it will be a good use of your valuable time. To help ensure that it is, I have several goals for this book:

To explain security best practices as they pertain to the IBM i operating system, using clear and understandable terms and concepts. Regardless of the type of organization to which you belong, in most cases you must comply with some law or regulation. If your organization can implement security best practices, not only will you comply with most current laws and regulations, but you will be in a position to comply with future laws and regulations, or at least not have a difficult time coming into compliance. For this reason, it's important to learn what constitutes security best practices on IBM i.

To give you choices. While I recommend that security best practices be every organization's goal, reality shows that most organizations have some area for which they are unable to implement best practices. This book discusses how to evaluate risks so you can determine which exposures you need to remediate and which you're willing to accept.

To provide practical implementation examples. I've probably seen every range of implementation possible, from the unbelievably open system to the system whose users can barely breathe without permission. Most security implementations lie somewhere in between. This book is full of practical and tested implementations.

To help you gain an understanding of the appropriate security scheme for the organization in which you're working. How am I going to do that? By helping you to identify the type of data being stored on your system and to understand the security implications of maintaining this type of data. Not all systems need to be locked down like Fort Knox. Some systems hold data that can be viewed by all users. Others have data that only a few people should be able to see. Understanding the type of data stored on your system is key to understanding the security requirements you need to apply to that system.

Now that you understand the goals of this book, let's look at why part of best practices says that security must be a business function.

Suppose you want to know more about your organization's human resources department, so tomorrow you and the rest of the IT staff go on a "tour" of that department. You find the file cabinets that keep all the employees' employment records, and you start to look through them. In fact, you copy the records of some of the more "interesting" individuals — such as those of your manager, the CIO, and the CEO. Then you take these records and post them on the break-room bulletin boards and on the community bulletin board at your local grocery store.

Although this is an absurd story, it has more basis in reality than most organizations would like to admit. How many members of your IT organization have *ALLOBJ (all object) special authority or access to all production objects on your system? *ALLOBJ authority permits those employees to do electronically what they never would be permitted to do with physical file cabinets. If anyone other than the security administrator has *ALLOBJ special authority, you're permitting that person to violate a written or unwritten guideline that limits access to confidential or private data residing on your system.

Whether you're talking about locks on doors, restrictions on who can access documents in file cabinets, limitations on who can access files on your system, or controls over the information your company makes available on the Internet, your organization's policies — formal or informal, written or unwritten — govern your business. You must take those policies into consideration as you architect and implement your security scheme. If you make compliance with your policies a top concern, you'll go a long way toward being in compliance with many of the laws and regulations that govern your organization.

Evaluating Your Risks

Your security strategy and implementation must take into account the potential for a security breach — whether accidental or intentional — that could result in the disclosure, modification, or deletion of your information assets. What are the risks to those assets, namely the data residing on your systems? The risk is to the confidentiality, integrity, availability, and privacy of the data.

Confidentiality

Some information is obviously confidential because it consists of private information (such as payroll information) or because it could give a competitor an advantage (e.g., early knowledge of an upcoming event, an organization's current financial status, proprietary product information, customer information, vendor lists). You must evaluate your data in terms of its confidentiality and value to the organization to determine how it should be secured. I'm certain you've heard stories about disgruntled employees who've stolen databases and sold the information on the black market. What's scary is how easy it is in most organizations to accomplish this deed. Given the way most systems are configured, most programmers could walk out the door of their organization's office with a copy of virtually any database file. Or they could e-mail the information from the system or carry it home on a memory stick. How often are briefcases examined on the way out of your office? How many e-mails are scanned for inappropriate attachments being sent from your company? When it comes to restricting the access of IT personnel to information assets, an organization's security implementation often fails.

Integrity

Information security also addresses the integrity of your applications and data. Applications and data must be authentic, accurate, and concurrent with your company's other applications and data. Problems such as order entry programs that lose order items, inventory control programs that introduce errors in replenishing stock, or customer invoice records that someone has tampered with can easily bring a business to a crawl or standstill and can endanger profits and even the company's long-term success. You can't afford to risk the integrity of your applications and data because of a nonexistent or weak information security implementation.

For instance, if you let users share user profiles and passwords, you immediately sacrifice your system's ability to authenticate changes to information. In other words, the system can't accurately identify who is doing what on the system because one profile might represent several people. Although limited profile sharing may be acceptable for a specific application, as a rule system integrity requires you to prevent profile sharing.

What if users or management suddenly started doubting the accuracy of the data in your database? Perhaps someone mistakenly uploaded old data from a PC into a file. Or maybe a well-meaning programmer put into production a program that wasn't fully tested, which in turn introduced errors into the database. These two scenarios illustrate the importance of the integrity and accuracy of your programs and data. Your information security plan must minimize the opportunities for users and IT staff to accidentally or intentionally introduce errors into your application repository or database.

Secure information assets are also vital to your company's integrity and reputation. Any security breach, when publicized, can harm a company's reputation. Say your business depends on the cash flow that its Web site produces and you experience a highly publicized security breach in which consumers' credit-card numbers are stolen. You can imagine what this type of breach would do to a company such as Amazon.com or eBay.com. The company's integrity could be so severely damaged as a result of the breach that the company might not be able to continue doing business. A well thought-out and implemented security plan protects a company's integrity and reputation as well as its data.

Availability

You've probably heard the horror stories — or even experienced them yourself — in which production objects were accidentally deleted. Perhaps a developer thought he or she was deleting a test copy of a program, or a user accidentally hit the "upload" icon instead of the "download" icon and overrode the production sales file with data from a spreadsheet containing last month's sales data.

Regardless of the specific scenario, these types of disasters occur simply because users have too much authority to the data. If the user had sufficient object authority to download data but not to upload it, the possibility of overwriting production data wouldn't exist.

Although these disruptions are typically accidental, malicious attacks can also occur. Malicious or accidental, any time data is not available to the organization, the organization is harmed. The degree to which it is harmed depends on what data is no longer available and on the length of the outage. A well-architected and -implemented security scheme can significantly reduce the potential of data being unavailable.

Privacy

Recent years have seen a dramatic rise in concerns about privacy, and numerous laws and regulations have been written to protect it. Consumers are becoming increasingly worried about how organizations use their personal data, whether it's collected over the Web, on an employment application, or in a health insurance form. When data is gathered and inappropriately released, such as information collected from medical tests and credit reports, individuals' lives are often adversely affected — either by sheer embarrassment or through identity theft or fraud.

Security alone doesn't guarantee privacy, but you cannot guarantee privacy *without* implementing good security policies and practices. Don't make the mistake of ignoring this area. Many countries, including the United States, Canada, and most of Europe have legislation that mandates protection for personal information. Financial institutions in the United States must inform their clients about how private data is used and let them opt out of programs that may send that data outside the original institution. In the case of health care, the Health Insurance Portability and Accountability Act (HIPAA) requires physicians to tell their patients how private data is used and to obtain patients' permission before sharing that data with other physicians, health care organizations, or medical research teams.

Even if you aren't in a regulated industry, I strongly recommend that you evaluate *all* data your organization collects. If any of it is personal information, examine how it is used, ensure it is appropriately secured, and inform individuals (through your privacy statements) about how the data will be used so they can choose whether to give your organization their data.

Securing private data is a vital aspect of keeping private data private. However, it's important to realize that this is only one aspect of keeping private information private. Appropriately securing private data provides access control; that is, it determines who can see the data. However, security cannot enforce a policy of acceptable uses of the private data once that data has been collected or viewed. Nor can security enforce the policy of what data is actually collected. So just because you have the private data secured away from public access, don't think you've handled all aspects of dealing with private data.

Evaluating the Threats

What types of problems pose the greatest threat to your information security plan and implementation? A common thread you may have recognized in the scenarios we've discussed is the role accidents can play in breaching confidentiality, accuracy, or availability. A sound security scheme provides significant protection toward reducing threats to your system. If you implement common-sense security measures, such as appropriately securing production data, securing source so that it cannot be updated outside of change management, reducing the number of all-powerful users, auditing the use of critical data and security-relevant actions, and handling data according to the requirements of its data type, you can eliminate many common sources of errors and omissions.

Your security implementation will certainly need to address the threat from disgruntled employees and hackers. But you should first develop a security policy; then concentrate on simple security basics, which can prevent accidents and employees from damaging the system, and a good business contingency plan, which can reduce the effects of a natural disaster or other form of site loss or system outage.

Managing the Strategic Issues

Evaluating the risks and threats to your information assets is the key to getting management's attention. As you expose risks and threats, management begins to see the possible ramifications of inadequate or inappropriate security. Once you have management buy-in, it's important to follow through by implementing a security awareness program throughout your organization and ensuring that the requirements of your organization's security policy are considered with all major changes that occur in the organization. In addition to managing upper management's knowledge and expectations of a security policy, you need to manage the access controls to your applications, data, and systems. And you need to establish and carry out security auditing procedures.

These tasks take time, but once completed they provide the building blocks for the various security implementations you need to undertake on your enterprise's computer systems.

Control Access to Applications, Data, and Systems

Controlling access to your information resources involves physical security, managing user profiles and passwords, defining authorization roles, and implementing resource authorities and specific security programs to enforce and audit access to your system. This particular strategic issue constitutes most of the tasks usually associated with implementing security and consequently occupies the largest portion of this book.

In today's world, access control is a critical part of a security implementation. Perhaps in the past you relied on security measures, such as menu systems, to secure your own application users on the system. That method simply is not sufficient today. With numerous ways to access data that bypass traditional menus — such as System i Access data transfer, File Transfer Protocol (FTP), Open Database Connectivity (ODBC), and Distributed Data Management (DDM) — you must incorporate other approaches to ensure your data is secured appropriately.

One specific subset of access control relates specifically to security within the IT department. IT professionals are often oblivious to the need to secure the system from themselves. Some professionals perceive any form of security aimed at IT as, at best, making IT's job more difficult or, at worst, a personal attack. However, because security is a business function and the IT professional's tasks, like all other users' tasks, require security, IT security is essential.

Some of the keys to a successful IT security implementation are apparent — committing yourself to separate development, test, and production environments, for example. The production environment on your systems must be stable except for planned changes made using some type of change management system. IT professionals shouldn't have authority to production data and applications. However, there must be a contingency plan that provides access in the case of production emergencies.

IT vendors (e.g., consultants, third-party software providers) should fall under the same guidelines as in-house developers. They should have access only to development and test environments. And, as in the case of in-house developers, all actions on production systems should be audited.

Red flags should be thrown and warning bells should go off in your head when third-party software providers require users to have *ALLOBJ special authority to run their software (unless, of course, it is security management software). *ALLOBJ should be the absolute exception and definitely not the rule. Make these vendors justify why users need *ALLOBJ special authority to use the software. There are some legitimate reasons for requiring *ALLOBJ authority, such as when an underlying operating system function requires it. However, no vendor should require *ALLOBJ authority simply to access the application's database.

Regardless of whether you're working with IT or end users, your security scheme should be formed according to the principle of "least privilege access." That is, give users access only to those objects and capabilities that they need to perform their job function — never more.

Review Requirements and Maintain Compliance

In any security implementation, established security requirements and rules become less effective as time passes. Because your security requirements, as well as pieces of the system (e.g., the operating system, application programs, procedures) aren't static, you must continually review and adapt your security plan to stay current with new threats, changing technology, and recently passed laws and regulations.

To maintain a security implementation, you must proactively monitor the compliance of your security implementation as well as review the security-related events that take place on your system each day. That is, you need to regularly check to make sure your implementation complies with your organization's security policy. Many organizations fail to review their policy and its implementation, instead simply trusting that the security implementation works and that "someone" will know when something isn't "right."

Getting Started

The hardest part of implementing a sound security scheme is getting started. As you read the rest of this book, look for simple ways you can get started in your own organization. To do that, you must take these initial steps:

- First, examine any potential risks to your company's information assets — that is, its data. Is there confidential or private information on the system that your company needs to protect? Can you guarantee the integrity of your organization's data? What would the cost be if this data were lost or stolen? To answer these questions, list the risks of your current implementation. Then, list the cost to the organization should the data be lost or stolen.
- Next, obtain management support to begin documenting corporate-wide security policies. After defining the organization's policies, start to define the departmental procedures and establish data ownership. For information about writing a security policy, see Chapter 2.

Only after completing these tasks are you ready to begin implementing a new security scheme. It is in the enforcing that you'll use the technical tools described in this book.

The final step in getting started, from a management point of view, is to commit to maintaining your security implementation and plan. The only way to determine whether your implementation is working is to check the compliance of the current system settings against your security policy and continue to do so on an ongoing basis. When you discover gaps in the configuration, you can fix the implementation to bring it back into compliance with your policy.

The remainder of this book gives you the technical details and approaches for implementing security best practices for i5/OS and IBM i.

Technical Note

The operating system that is the subject of this book has undergone several name changes throughout its history. I use the current name, IBM i, but most of what you'll learn about in this book applies equally to recent versions of i5/OS and OS/400 (with the obvious exception of new functionality provided in V6R1). Similarly, I use the latest names for other IBM products and technology, such as Navigator for i (formerly iSeries Navigator), System i Access (formerly iSeries Access), and so on.

Don't Close the Book

Once you've written your security policy, evaluated your risks, implemented or reworked your security scheme, and tightened down your network, your job is done and you can toss out this book, right? Wrong. Security is a not a one-time event. I like to call it a "lifestyle." New technologies are introduced, new applications are installed, the organization acquires a new company or goes into another line of business. Because security is a business function and the current laws and regulations aren't about to go away, whenever your business changes or grows, you must evaluate those changes against the requirements of your security policy and then implement the appropriate security measures to ensure your organization's data remains secured appropriately. So don't close this book! You'll want to keep it open for future reference.

Chapter 2

Policies and Procedures

Policies and procedures are essential for security. Why? For one thing, many laws and regulations require them. For another, they define the security "rules" against which your organization's decisions should be evaluated. Have you ever played a card game in which the players didn't know all the rules? Play progresses until someone plays a card that someone else claims shouldn't be played. An argument ensues, someone gets upset, and the game ends abruptly. Sounds like fun, doesn't it?

Trying to implement a security scheme without knowing your organization's "rules" — that is, without having a security policy — is a lot like the card game described above. Having an up-to-date and enforceable security policy lets you implement a security scheme confidently and provides a clear pathway for resolving issues.

Your security policy is a vital corporate document. It documents the level of investment required for security resources across your organization, and it spells out the rules and requirements for implementing security. Designing your corporate security policy is a critical business practice — much like laying out your employee compensation plan. For example, once you have a policy in place, you can analyze a proposal for enabling a new technology to see whether it fits within the rules. Thus, a security policy can settle arguments and avert power struggles.

A side benefit of a security policy is that it helps retain "corporate memory." Employee turnover can devastate an IT department if no one has documented the policies and the processes that implement those policies. A security policy also provides legal documentation should your organization ever need to prosecute someone for violating the security policy.

Finally, buy-in from the highest levels of management on your security policy is critical to its successful implementation and enforcement. A member of the IT staff may develop your organization's security policy, but the policy needs to be approved and communicated to the rest of the company through the highest level of management possible.

Your Security Policy

Some security policies tie all areas of the organization into one document. But most organizations use an approach that sets out the policies in a main document and uses appendices to address the implementation of specific areas, such as the operating system, a firewall, or Active Directory. There is no right or wrong format as long as the policy meets your needs and can be communicated effectively.

In addition, while some laws and regulations (e.g., the Payment Card Industry's Data Security Standard) have requirements regarding specific issues that must be addressed in a policy, no law or regulation mandates how many additional areas are addressed. I tend to see organizations that are just creating a policy start out with a very basic policy, addressing just the bare minimum. Then, in subsequent years, when the policy is up for its annual review, new sections are added. Organizations that have addressed security as a business function for many years tend to have "mature" — that is, very comprehensive — security policies.

The point I'm trying to make is that your policy needs to meet your organization's requirements, and those requirements vary from organization to organization. Let's examine the vital parts of a security policy.

Physical Security

Security policies often start with physical security. Perhaps that's because when thinking about security, many people find it easiest to think of items by working from the "outside" toward the "inside." Whatever the reason, the physical security section of the policy describes the physical measures put in place to ensure physical security. These measures can include

- Badge readers
- Cipher locks on critical areas (e.g., machine room doors), as well as requirements and procedures for changing lock combinations
- Security cameras for computer rooms
- Laptop security (laptops are high-theft items, both on-site and when users are traveling)

Responsible Parties

Disagreements occur, new ideas are launched, organizations expand and contract. Any of these circumstances can cause someone to experience a conflict between the job they're trying to do and what the security policy mandates. A clear process for resolving such issues, as well as the person or persons empowered to settle conflicts, must be established and documented. Remember the card game? If someone had been responsible for knowing the rules of the game, the players could have consulted that person when the argument arose and settled the dispute.

A part of doing business is managing and resolving conflict. The topic of security seems to breed conflict. Programmers, for example, always want more function than they're allowed, and some managers seem to have the attitude that they're entitled to see all data on the system. These issues and others must be settled for business to go forward. Your policy will document who is responsible for such decisions so that conflicts can be resolved. The security policy will also identify the person or persons responsible for the security policy itself and the overall security of the corporation.

Data Classification

Let's face it, the reason you're reading this book is because there's data on some computer and that data has value to the organization. The exact value depends on the type of data; therefore, it is critical that data be *classified.* Data classification entails analyzing the data your organization retains, determining its importance and value, and then assigning it to a category. Data that is considered "top secret" (whether contained in a printed report or stored electronically) needs to be classified. Why? So that it can be handled properly. IT administrators and security administrators can guess how long data should be retained and how it should be secured, but unless the organization has taken the time to classify its data, the data may not be secured correctly or retained for the required time period.

When documenting the classification of data in your security policy, you should address the following areas:

- *Who has access to the data* — Define the roles of people who can access the data. For example, information classified as "company restricted" can be viewed only by upper management.
- *How the data is secured* — Determine whether the data is generally available or, by default, off limits. You'll hear this latter approach referred to in the laws and regulations as "deny by default."

Note

Notice that I have not defined what the operating system security setting is to be; I have defined the access in general terms, just as I described who should have access in general terms. Determining who has access and identifying the specific security settings will come when the data custodian (described later) implements this part of the policy. These settings are often documented in the appendices that accompany the policy.

- *How long the data is retained* — Many industries require organizations to retain data for a certain length of time. For example, the finance industry requires a seven-year retention period. Data owners need to know the regulatory requirements for their data; if no requirements exist, they should base the retention period on the needs of the business.
- *What method should be used to dispose of the data* — For some data classifications, the method of disposal won't matter. But some data is so sensitive that data owners will want to dispose of printed reports through cross-shredding or another secure method. Or, they may require employees to use a utility to "scrub" their PCs after erasing files containing sensitive data.

New in V6R1

One concern many of you have is ensuring that no data remains on system disks when you return or resell them. Aside from taking a hammer to your hard drives, there has been no way to ensure your data could not be recovered. IBM Disk Sanitizer for i5/OS is now available via a PRPQ and meets the U.S. Department of Defense requirements for ensuring that data cannot be retrieved from a disk.

- *Whether the data needs to be encrypted* —Data owners must decide whether their data needs to be encrypted. Many laws and regulations make this decision for them. For example, the Payment Card Industry's Data Security Standard (PCI DSS) requires that all cardholder data "at rest" (i.e., in a file) be encrypted.
- *What use of the data is appropriate* — Before data security became such a hot issue for organizations, people in many roles within and outside the company used data in all types of reports. This aspect of the security policy defines whether data is for use within the organization, is restricted for use by only selected roles, or can be made public to anyone outside the organization. In addition, some data has a legal usage definition (for example, the state of California has defined the appropriate use of a Social Security number). Your organization's policy should enumerate these restrictions.

What Classifications Should You Use?

There are no hard and fast rules about the titles and number of classifications. The general guideline is that classification should be defined clearly enough so that it is easy to determine how to classify the data. In other words, little (if any) overlap should exist in the classification definitions. Also, it is helpful to use a term for the title of the classification that indicates the type of data that falls into the particular category.

Here are some examples of categorizing data by title:

- *Private* — Data that is defined as private, such as Social Security numbers (SSNs), bank account numbers, and cardholder data
- *Restricted* — Data that is restricted to a subset of employees
- *Confidential* — Data that can be viewed by all employees but is not for general use outside the organization
- *Public* — Data that can be viewed or used by employees and the general public

Data classifications can also change. For example, IBM will often classify new release information as IBM Confidential Until Announced. The recipients of this information can properly protect and use the information before the announcement and can then more freely use the information after IBM formally announces a new release.

Who's Responsible for Classifying Data?

The individual who owns the data should decide the classification under which the data falls. The committee that wrote the data classification policy can certainly help or provide guidance, but the final determination of the classification should be the data owner's responsibility. The data owner is best qualified to make this decision because he or she has the most knowledge about the use of the data and its value to the organization.

In addition, data owners should review their data's classification at least annually to ensure that the data remains correctly classified. For example, if data owners were reviewing data classifications for the past few years, they probably moved much of their employees' information — especially information such as SSNs — from a "confidential" classification to a "private" classification. SSNs were never considered private until they began being used for identity theft. Since thieves started to steal databases of SSNs, their classification has been upgraded to restrict access and more tightly control their use.

Data Ownership

In addition to classifying data, an organization needs to assign an owner. The owner is not the user profile that owns the database object on the system; rather, it is the person in the organization who owns the data that is stored in the database. The data owner is typically a director, or at least a department head, who has a vested interest in making sure the data is accurately and appropriately secured. The person who is appointed needs to understand the importance and value of the information to the business, the ramifications of inaccurate storage or inappropriate access, and the laws and regulations that may govern the use and retention period of the data.

The data owner's responsibilities include:

- Setting the policy as to who is (what roles are) allowed to access the data
- Determining how the data is to be secured
- Setting the retention period
- Defining the appropriate disposal methods
- Setting the encryption requirements

The data owner may appoint an administrator to perform the daily tasks associated with these responsibilities. For example, the data owner may appoint someone to approve requests for individuals to have access to the data. The appointed person will act under the direct instructions of the data owner.

Why Not IT?

Unfortunately, IT often ends up being the de facto owner of data. Although the IT department can be the custodian of the data, it should not be the owner. Employees in IT generally do not know how important the data is to the business, how the data is to be used, and which people (or roles) should access the data.

Another reason IT should not be the owner as well as the custodian of the data is separation of duties. If IT decides who has access to the data and then administers that access, no checks and balances are in place to ensure that access control policies are being followed or that inappropriate access is not being assigned.

IT's role is usually that of data custodian. The custodian is responsible for implementing the policies set by the data owner. For example, IT is usually responsible for ensuring that the database files access controls (e.g., *PUBLIC authority) are set according to the data owner's requirements. IT is also responsible for backing up the data as well as properly disposing of any electronic copies of the data in the department's possession.

Network Connections

Your security policy needs to address several network-related areas:

- Internal network (or intranet)
- Internet
- Extranet (or requirements between your company and vendors or suppliers)
- Wireless

For each area, you'll want to address the required authentication methods and indicate who has authority to authorize users to various resources. For example, your policy may state that to access your internal network, users must have a valid network user ID and password. Network authorizations will be permitted once the network administrator receives notification from the human resources department that an employee has been hired or a vendor has started a contract. The internal network section of your policy should also define virus-scanning and anti-spam policies as well as discuss the appropriate use of e-mail.

The Internet section will address the hosting of your company's Web site and any connections back into your internal network. This topic will also cover the use of firewalls and the connectivity requirements of employees connecting from outside a firewall into your internal network. For example, many companies require employees to connect from home (or while traveling) using a virtual private network (VPN) client. VPNs enable strong authentication of the client (in other words, a very high assurance that the client is who it says it is) as well as encryption of the entire flow of data.

Extranet issues discussed in your policy will include who determines which vendors or suppliers are allowed to connect to your internal network, what (types of) systems they are permitted to access (e.g., non-production systems, development systems, customer relationship management systems), and the connectivity requirements for connecting to your network.

Wireless issues to be documented include who or what areas are allowed to use wireless technology, who is permitted to add wireless access points, and what security measures are to be taken when using wireless technology. In considering wireless use outside your internal network, you'll want to document whether employees can use wireless home networks or the wireless network at their local Starbucks to access your network over the Internet. Or perhaps you permit this type of access only if the employee is using a VPN client.

Application Design

If your company writes its own applications, your security policy should address issues of change control, use of programming best practices (including practices that ensure a secure implementation, for legacy as well as Web programming), testing procedures to ensure security policy compliance, and emergency procedures for programmer access to production data.

Programmers shouldn't by default have authority to debug an application once it's promoted into production. Nor should programmers be able to update production files. However, as we all know, programs fail, and there are times when programmers must debug a production program. In such cases, temporarily give the programmer a profile with sufficient access to debug and correct the problem. Use an automated procedure to ensure that this profile is deleted or disabled afterward so it can't be used on an ongoing basis. You'll probably also want to audit the activities this profile performs, including all commands entered from a command line.

Platform-Specific Issues

If you take the approach of documenting general principles in the main part of your policy and providing an appendix for each platform or operating system, the platform-specific sections of your security policy will document the specific policy implementation details associated with each platform. For example, for IBM i you may want to document the settings for the security-related system values, the default access (*PUBLIC) authority setting, which type of user (e.g., system operator, security officer, end user) is allowed to have certain special authorities, and the auditing values to be set.

Employee Guidelines

Most likely, your organization has an employee handbook that is updated regularly and given to all employees when they're hired. This book contains policies, guidelines, and information. Often, this handbook documents the employee-relevant security policies. Items covered in this section often document what constitutes appropriate behavior. Let's take a look at some of the issues you'll want to cover in the employee section of your policy:

- All computer equipment, systems, and information residing on the computers is the property of the company and is for business purposes only.
- Activity on computers can and may be logged and tracked.
- Employees can use e-mail facilities during work hours but should have no expectation of privacy regarding any e-mail sent or received.
- Employees must not defame the competition.
- Employees cannot use social network sites or blog on company time.
- Employees cannot blog and attribute the blog contents to the company.
- Employees will wear identification badges at all times while on company property.

The employee section of the policy is often where password rules are communicated. For example, the policy may have a section that includes the following points:

- Individuals will have their own log-on accounts. These accounts and their passwords are not to be shared with other individuals.
- Passwords must be at least seven characters long and contain a digit and a special character.
- Passwords must be changed on all accounts at least every 90 days.
- Password should not be written down.

The items I've listed here are not meant to be an exhaustive list of issues to be addressed in the employee section of your policy. Rather, they are meant to provide some examples so that your organization can think about and ultimately document what employee policies make sense for your business situation.

Notification, Enforcement, and Compliance

A security policy isn't worth more than the paper it's printed on unless the people affected by it are notified about and comply with the policy. So be sure to develop notification procedures. For example, many companies perform security awareness training as part of their new-employee orientation. One aspect of your organization's security awareness program could require managers to annually review the employee's security policy with their employees and then require the employees to sign something to acknowledge that they're aware of and agree to adhere to the policy.

Communicate changes to the policy to the affected parties in a timely manner. Having executive-level management handle notifications about the security policy can be an effective way to reinforce the message that security is important to the business and is the responsibility of all employees.

Enforcement can come in many forms. The compliance and enforcement section of your policy will document how your policy is enforced. Some portions of your policy can be enforced with technology. For example, to ensure appropriate Internet use by employees, all visited Web sites can be logged and reviewed; employees trying to connect to an internal network from outside the firewall will have their connections rejected; and password composition rules will be enforced by setting the appropriate IBM i system values or Microsoft policies.

Realize, however, that not all aspects of a policy can be enforced with technology. Some items rely on the integrity of the individual employee. Obviously, you hope everyone adheres to the policy and guidelines voluntarily. Unfortunately, that doesn't always occur. Non-compliance must be dealt with, and the compliance and enforcement section of your policy should outline the procedures to be followed in the event of non-compliance.

I can't emphasize enough the need to enforce your policy. If a user doesn't comply with the policy and no ramifications occur, there will be no motivation for that employee or any other to follow the policy. It's like a parent telling a child, "Don't do that, or you'll be punished." If the child continues to misbehave and the parent never disciplines the child, what happens? The child learns that it's okay to misbehave.

Effectively communicating your security policy and guidelines sends employees the message that security is important to your business and is part of their job. Top-level management can take this opportunity to show support for information security efforts. Such cooperation ultimately makes your job easier by helping everyone realize — and take to heart — that security is everyone's responsibility.

Business Events and Procedures

Most organizations have established and documented processes for handling changes in employee status — that is, dismissals, leaves of absence, transfers between departments, and so on. The impact on security is often overlooked when these processes are being developed and implemented. Although these processes may not be part of your security policy itself, their security ramifications shouldn't be ignored. For example, the termination of an employee should involve immediate notification of the IT department so that the user's profile (or profiles) can be disabled or deleted. In addition, you might want to gather certain "forensic" information at that time. Perhaps look for recent changes to programs or updates to files made by this user.

Similarly, an extended leave of absence requires IT notification so that the user's profile (or profiles) can be disabled to protect against exploitation during the user's absence. Automated tools can also be updated so that a particular profile isn't flagged as "inactive" and inadvertently deleted while an employee is away.

Although IT is often notified of terminations and leaves of absence, the most overlooked human resource activity is when users transfer between departments. When users transfer, their job responsibilities often shift. They usually receive the ability to perform their new job, but rarely are their old capabilities removed — especially when they have to train their replacement and need to retain their capabilities even though they've moved to their new position. To safeguard security in all these cases, a communications process should be put in place between human resources and IT, and a regular review of all users on the system, as well as their capabilities, should be performed at least once a year. I've never seen a review of user capabilities that didn't find users with capabilities that weren't appropriate for the current job position.

Getting Started with Your Policy

Here are some pointers to remember if you're just starting to write your organization's policy:

- Keep it high-level and generic. That is, don't describe specific settings. Rather, describe the requirement so that it can apply to all operating systems, applications, and so on.
- Don't add a section of a policy that you aren't doing. Just because a concept or a particular action would be a good thing to do, if your organization does not currently have it implemented or isn't in the middle of the implementation, don't add it to the policy. Auditors will audit you against your written policy, and audits will go significantly worse if sections have not been implemented than if you have them on a list somewhere to add in the future.
- Keep it current. An out-of-date policy is almost as worthless as having no policy. Review policies at least annually and more often if you're involved in a major technology shift or organizational change such as a merger.
- Don't try writing a policy from scratch. A number of resources on the Web can at least give you a starting point. For example, the SANS Institute (*http://www.sans.org*) provides free samples of policy topics that you can download and alter to meet your business requirements.

Legal Review

A final word: Because your security policy and, to a certain extent, the business processes that affect security are documents that need to be enforceable and hold up in a court of law, it's imperative that you have your organization's legal counsel review your security policy.

Chapter 3

Security at the System Level

A successful security plan considers all available security methods and tools, combining and tailoring them to implement the requirements of a specific security policy. One of the toolsets you'll want to implement are the system's global settings, or system values. This chapter focuses on the security-relevant system values, discussing recommended settings and the trade-offs you may need to consider with some of the settings.

The System Security Level

I like to say that the system's security level sets the tone for security across the entire system. It is the setting that controls whether you have operating system and data integrity or permit everyone to have access to every object on the system. Security levels now range from level 20, which provides only password security, to level 50, which, in addition to password and resource security, provides system integrity security (the ability to prevent users and applications from executing restricted machine-level instructions) as well as specific security features that comply with the Common Criteria security specifications. (The Common Criteria for Information Technology Security Evaluation is an international standard for computer security certification.) Understanding the security levels and choosing the correct level for your circumstances is the first step in securing your system.

System Value QSECURITY

To control the system security level, you use system value QSECURITY (System Security Level). The possible QSECURITY values of 20, 30, 40, and 50 correspond to the four available security levels. In early OS/400 releases, there existed a security level 10, which provided no password or resource security. IBM removed support for this level in V4R3.

IBM ships all new systems running IBM i or i5/OS with a QSECURITY value of 40. In early releases of the operating system, the shipped default setting was level 10. You can view your system's current QSECURITY value by running the DSPSECA (Display Security Attributes) CL command. This command displays the current and pending values for QSECURITY. If you change QSECURITY, you must perform an Initial Program Load (IPL) of your system to have the new value to take effect. The pending value displayed by the DSPSECA command is the value that QSECURITY will have after the next system IPL.

You can also use the DSPSYSVAL (Display System Value) command to display a QSECURITY value. The value reported by this command is equivalent to the DSPSECA command's pending value. DSPSYSVAL doesn't reveal whether the post-IPL value of QSECURITY will differ from the current one.

You can change the security level by running the CHGSYSVAL (Change System Value) command. For example, to change the QSECURITY value to 40, you would enter

```
CHGSYSVAL  SYSVAL(QSECURITY) VALUE('40')
```

Security Level 20

Security level 20 provides "password security." To access the system at level 20, a user must have a user profile and a valid password, but that's all the security this level provides. By default, all users on a system using security level 20 are created with *ALLOBJ special authority (the equivalent of having root authority in Unix or Administrator rights in Windows).

Unfortunately, there are still systems running at security level 20. These systems have significant exposures, and those responsible for them should put a project in place immediately to move to level 40. The challenge in such a project is planning how users are going to obtain enough authority to access their application's objects once their *ALLOBJ special authority is removed. (Chapter 16 provides details about implementing object-level security.)

Table 3.1 lists the default special authorities for each user class at security level 20.

Table 3.1
Special Authorities for Level 20

User class	Special authorities
*SECOFR (security officer)	*ALLOBJ, *SECADM, *SAVSYS, *JOBCTL, *SERVICE, *SPLCTL, *AUDIT, *IOSYSCFG
*SECADM (security administrator)	*ALLOBJ, *SECADM, *SAVSYS
*PGMR (programmer)	*ALLOBJ, *SAVSYS
*SYSOPR (system operator)	*ALLOBJ, *SAVSYS, *JOBCTL
*USER (user)	*ALLOBJ, *SAVSYS

Technical Note

Resource security (i.e., the system requirement that users have authority to an object to access it) is active at all security levels. The difference between security levels lies in the convenience with which resource security can be implemented. By default at security level 20, all users have *ALLOBJ special authority, which grants all users access to every object on the system. At levels 30, 40, and 50, users are not automatically created with *ALLOBJ. So, to access an object, users must obtain authority from another source, such as a private authority, *PUBLIC, or adopted authority.

Security Level 30

By default, security level 30 provides resource security (users don't receive *ALLOBJ authority by default), permitting objects to be accessed only by users who have authority to the objects. Table 3.2 lists the default special authorities for all user classes at level 30. These special authorities are also the defaults at levels 40 and 50.

Table 3.2

Special Authorities for Levels 30, 40, and 50

User class	Special authorities
*SECOFR	*ALLOBJ, *SECADM, *SAVSYS, *JOBCTL, *SERVICE, *SPLCTL, *AUDIT, *IOSYSCFG
*SECADM	*SECADM
*PGMR	No special authorities
*SYSOPR	*SAVSYS, *JOBCTL
*USER	No special authorities

Notice that *ALLOBJ special authority isn't a default for any user class except *SECOFR, the security officer class. Under level 30, user profiles don't automatically have the authority to create, change, or delete objects on the system. The authority to work with, create, change, or delete objects must be granted either specifically or because it's the default public authority.

Moving from level 20 to the more secure level 30 requires careful thought and planning, as I alluded to earlier. When you move from level 20 to level 30, IBM i removes any special authorities users have that level 30 doesn't allow by default. For planning purposes, compare Table 3.2 with Table 3.1. You can see that levels 30, 40, and 50 grant each user class different default special authorities than level 20 does. So, in addition to planning your object-level security implementation, you need to be aware of the special authorities that are removed when moving off of level 20 to ensure your users will be able to continue to perform their required job functions.

Although you might be tempted to think it's sufficient to let systems remain at security level 30, you cannot ensure data integrity or the completeness of the system's audit journal, or be assured users aren't masquerading as other users, if the system remains at level 30. If you get nothing else out of this book, understand that to effectively safeguard data, programs, and other production objects and to prevent unintentional data loss or modification, all systems must be running at security level 40 or above.

Security Recommendation

Run all your systems at least at security level 40.

Security Level 40

To level 30's resource security, level 40 adds operating system integrity. You saw the default special authorities for level 40 in Table 3.2. System integrity strengthens level 30 security in at least five ways:

- By checking program states and object domains
- By preventing use of restricted machine interface (MI) instructions
- By ensuring users have authority to start jobs under a different profile from their own
- By preventing restoration of invalid or modified programs (this happens at all security levels beginning in V6R1)
- By validating parameters passed to system interfaces

State and Domain Restrictions

Regardless of security level, IBM i classifies all objects (e.g., programs, files, commands) into one of two *domains*: system or user. All programs and service programs also have a *state* assignment — again, either system or user. The concepts of object domain and program state work with level 40 to close the integrity issue that exists at level 30. Under level 40, system-state programs can reference (i.e., directly manipulate through pointers) all objects regardless of their domain, and user-state programs can reference only user-domain objects.

At levels 20 and 30, the system tests programs against these rules when the programs are run. If a program violates the rules, the system simply logs the violation in the audit journal (see Chapter 15 for a discussion of the audit journal). But at level 40, the system won't run a program that violates these rules.

Object domain and program state also prevent users from abusing operating system objects by accessing them directly. Most IBM i objects are in the system domain, and most programs and service programs, other than application programming interfaces (APIs), are in the system domain and have the system state. At security levels 40 and 50, operating system objects can't be accessed directly. Why is this important? Some interfaces that are shipped with the operating system can provide authorized pointers, change jobs to run under QSYS or QSECOFR (the security officer profile), modify internal operating system control objects, or let users who run jobs access objects to which they don't have authority. These are legitimate interfaces that IBM i requires in order to run properly. At security levels 40 and 50, these interfaces can't be abused. Security levels 20 and 30 permit these interfaces to be used to harm your system.

Use of Restricted MI Instructions

At security level 40, the system restricts access to certain MI instructions, including those that create system-domain objects, manage processes or events, and access the database. At all security levels, the system detects all attempts to create a program that uses restricted MI instructions and prohibits the program from being compiled. At levels 40 and 50, an attempt to run an existing program that uses restricted instructions causes the program to fail; at all other security levels, the attempt simply causes an entry to be logged in the audit journal.

Job Initiation Validation

At security levels 40 and 50, an attempt to submit a job fails if the user submitting the job lacks authority to the user profile specified in the job description. At all other security levels, the system logs the attempt in the audit journal but lets the profile submit the job.

One attribute of a job description is the USER (User profile) attribute. You can specify this attribute's value by changing the USER parameter on the CRTJOBD (Create Job Description) or CHGJOBD (Change Job Description) command. The possible values are

- *RQD — The job using the job description must specify a user profile.
- user_profile_name — You can specify a user profile to be associated with the job description. A job can then use this user profile for execution.

Technical Note

Security levels 20 and 30 don't require a user who submits a job to have authority to the job description's default user profile. This means users can use a job description that names a powerful profile and gain authority and capabilities they weren't intended to have.

Restoration of Modified Programs

Whenever a program is created, the system calculates a validation value. This validation value enables the system to determine whether a program being restored is identical to the original. If the program validation fails (e.g., because the program has been modified), the system tries to retranslate the program.

If the retranslation succeeds, the system logs the validation failure and restores the program. If the retranslation is unsuccessful, the system restores the program and logs an entry to the audit journal, regardless of the system's security level. However, when the system is at security level 40 or higher and the ALWOBJDIF (Allow object differences) attribute is specified as *NONE on the RSTOBJ (Restore Object) command, the system changes the program's owner to QDFTOWN (the IBM-supplied default owner) and sets public authority to *EXCLUDE. If the restore command specifies ALWOBJDIF(*ALL), the system restores the program and logs an audit entry but lets the owner and public authorities stay the same.

Security level 40 is your primary defense against Trojan horses. A *Trojan horse* is a program disguised as a legitimate IBM i program, such as QSYS/QSTRUP, the default system start-up program that your system runs each time you IPL. Someone with harmful intent could patch a program on another system, name it QSYS/QSTRUP, and save it. If this person had access to a workstation at which another user was signed on with *SAVSYS authority, and if your system were not at level 40 or 50, the hacker could restore the revised version of QSTRUP to your machine. The next time you IPLed, your system would run the Trojan horse QSTRUP program.

Technical Note

System-state programs have to have a valid IBM digital signature. If they don't, they cannot be restored (as of V6R1).

Parameter-Passing Validation

At level 40, each time a user application calls an IBM i program that runs in system state (e.g., an API, an application that performs any I/O), the operating system checks every parameter passed between the programs. Both the storage domain and the actual values passed are checked to ensure neither will cause IBM i to perform an operation that compromises system integrity.

Why Use Security Level 40?

Simply put, if you want operating system and data integrity, the system must be running at security level 40 or higher.

Security Level 50

Security level 50 adds additional system integrity features as well as the functions and checks necessary to meet the Common Criteria evaluation requirements. For the level 50 default special authorities, see Table 3.2.

When you use level 50, the system

- Imposes certain message restrictions and removes pointers from messages
- Prevents control-block modification
- Maintains the QTEMP library

Message Restrictions and Pointer Removal

As I explained earlier in the chapter, every IBM i object belongs to either the system domain (*SYSTEM) or the user domain (*USER). Most IBM-supplied objects are in the system domain. All user-created objects and some IBM-supplied objects are in the user domain. In addition, every IBM i program is either a system-state (*SYSTEM) program or a user-state (*USER) program. Most IBM-supplied programs are system-state programs, and all user-created programs are user-state programs. User-state programs can access only user-domain objects, but system-state programs can access both user- and system-domain objects.

The rules of object domains and program states mean that you can access system-domain objects only by using the appropriate command or API. Level 50 imposes the following message restrictions:

- Any system-state program can send a message of any type to any user- or system-state program.
- Any user-state program can send a message of any type to any other user-state program.
- A user-state program can send a non-exception message to any system-state program.
- A user-state program can send an exception-type message to a system-state program only if the system-state program is a request program (e.g., QCMD) or if the system-state program called the user-state program.

In addition, at level 50 IBM i removes the pointers from replacement text in messages sent to a user application from another job. Removing pointers prevents the receiving job from using the pointers to compromise security and bypass system auditing.

Prevention of Control-Block Modification

At level 50, no internal control blocks can be modified. A *control block* is an internal structure available to the operating system. IBM i relies on control blocks to remain unmodified. However, some third-party vendors change internal control blocks so that their code runs instead of IBM i code. Such changes often cause unpredictable results when the system runs, and they frequently require a lengthy debugging process to locate and fix the problem. Although level 40 prevents modification of some control blocks, level 50 covers all control blocks, including the open data path (ODP), the spaces for CL commands and programs, and the System/36E job control blocks. This feature may cause some third-party utilities to cease functioning, so you should check with your application and utility vendors before moving to security level 50.

QTEMP Library Maintenance

On the surface, objects in library QTEMP appear to be temporary. The external view of the QTEMP implementation is that a unique QTEMP library exists for every job on the system, that you can access objects in QTEMP only from within its associated job, and that when a job ends normally, the job's QTEMP library as well as all the objects in it are deleted.

The facts of QTEMP implementation, however, are somewhat different. If your system doesn't run at security level 50, every QTEMP library is actually a permanent object on the system. When a job begins, IBM i creates a QTEMP library for the job and assigns an address for the library and each object in it in the operating system's permanent root address system. If a job or the system ends abnormally, IBM i uses these addresses on the next IPL to locate all stranded QTEMP libraries and objects and clear them from the system. So, internally, QTEMP and the objects in it are permanent.

Level 50 prevents hackers from using IBM i's internal permanent address system to access QTEMP library objects. At level 50, the operating system no longer stores those addresses. Instead, the system actually does treat every QTEMP library and its objects as temporary.

That sounds fine, of course, but as a result, if a job or the system ends abnormally, IBM i can't automatically locate and delete the "lost" QTEMP library and its objects when you IPL. Therefore, when you implement level 50, you need to run the RCLSTG (Reclaim Storage) command more frequently to recover objects that get "lost" and then specifically delete them.

Why Use Security Level 50?

Organizations that require the highest level of security and integrity should run at level 50. Level 50 provides the best defense against exploitation of internal interfaces and the introduction of Trojan horses and viruses.

Moving to Security Level 40 or 50

Before moving to security level 40 or 50 from level 30, you'll first want to determine whether any applications will fail at the higher security level. Run auditing long enough to assure yourself that most of your applications have been run — for example, over month end or, perhaps, quarter end. You're looking for two types of violations: applications doing something they shouldn't and the use of job descriptions that name a user profile to which the user doesn't have authority. The following steps tell you what to look for and how. For details about auditing and the audit journal, see Chapter 15.

Step 1. Add the *PGMFAIL value to the QAUDLVL (Auditing Level) system value (discussed in more detail later). This setting lets you start gathering information to determine whether any applications need to change. It's unlikely they do, but you'll want to audit to make sure. Otherwise, if an application is doing something it shouldn't, it will fail at level 40 or 50, and you'll have to restore the system value to its previous setting and re-IPL to allow the application to run.

Step 2. Add the *AUTFAIL value to system value QAUDLVL. This setting lets you gather entries for users who use a job description that names a user profile. An audit entry is generated if a user lacks *USE authority to the user profile named in the job description.

Step 3. When you've audited over (at least) month's end, query the audit journal. Both the *PGMFAIL and the *AUTFAIL parameters produce audit entries of type AF. The first letter of an audit entry's specific data determines which type of AF entry it is. Look for the following entries:

- B — Restricted (blocked) instruction violation (not sent starting with V6R1)
- C — Object validation failure
- D — Unsupported interface (domain) violation
- J — Job description and user profile authorization failure (using a job description that names a user profile)
- R — Attempt to access a protected area of disk (enhanced hardware storage protection)
- S — Default sign-on attempt

The type of the entry will determine what action you need to take to ensure a successful move to security level 40 or 50. For example, if you see an entry that begins with D, you'll know that the program is trying to use an unsupported interface (e.g., trying to directly call a command processing program or other operating system program). You'll need to replace the MI instruction with either a command or an API call. All J entries indicate the use of a job description that names a user profile to which the user of the job description lacks *USE authority. Note that the entry will name the user profile, not the job description. You'll have to look through your job descriptions that name user profiles to determine which job description needs to be changed. It's my experience that failures usually occur with job descriptions that you've created (versus those supplied by a vendor).

There are several ways to query the audit journal to determine whether your system has logged any of the entries listed above. Chapter 15 discusses these methods.

Note that sometimes you'll get audit journal entries that indicate third-party application failures. If you encounter this type of entry, contact the vendor. It's likely that the application takes one code path at security levels 20 and 30 and takes another, which does not cause violations, at security level 40 and 50. It is rare today that a third-party application doesn't run at security level 40 or 50. These days, if there are any audit journal entries, they are often the result of an organization's "home-grown" utility or application calling an operating system program directly rather than using a command or an API. A simple change to call an API or command is usually the only program modification required to clear up the issue.

Security-Related System Values

Once you've determined the security level at which you want to run your system, you can use the other security-related system values to enforce other security policy requirements. These system values fall into three categories: general security system values, system values related specifically to password creation and maintenance, and system values that control auditing. Let's take a look.

General Security System Values

The general security-related system values provide a variety of global settings for security functions. For example, by changing a single system value, you can limit the number of successive incorrect sign-on attempts allowed at a workstation or the number of concurrent sessions a user can be signed on to.

Table 3.3 summarizes the general security system values.

Table 3.3
General Security System Values

System value	Description
QALWOBJRST	Determines whether security-sensitive objects (i.e., system-state objects, programs that adopt authority) can be restored to the system
QALWUSRDMN	Specifies in which libraries you can place *USRSPC (user space), *USRIDX (user index), and *USRQ (user queue) objects
QAUTOCFG	Determines whether new local controllers and devices are configured manually or the system configures them
QAUTOVRT	Determines whether virtual device descriptions are created automatically
QCRTAUT	Specifies the default public authority for newly created objects
QDSCJOBITV	Indicates the length of time (in minutes) that an interactive job can be disconnected before the system ends the job
QDSPSGNINF	Specifies whether the sign-on information display appears when a user signs on
QFRCCVNRST	Determines whether programs will be retranslated when restored
QINACTITV	Determines the interval (in minutes) that a workstation can remain inactive before the system sends a message to a message queue or terminates the job
QINACTMSGQ	Specifies the name of the message queue that receives messages about the workstation when the inactive time interval is reached, or specifies that the job at the workstation is to be terminated or disconnected when the interval is reached
QLMTDEVSSN	Determines whether users can have more than one device session concurrently
QLMTSECOFR	Determines whether users with *ALLOBJ (all object) or *SERVICE (service) special authority can sign on to devices without explicit authority
QMAXSGNACN	Specifies the action taken when a user reaches the maximum number of invalid sign-on attempts
QMAXSIGN	Sets the maximum number of sign-on attempts allowed at a workstation
QRETSRVRSEC	Determines whether decryptable passwords are stored on the system
QRMTSIGN	Specifies whether automatic sign-on from a remote system is allowed
QSCANFS	Determines whether file system scanning takes place
QSCANFSCTL	Provides fine-grained control over the details of file system scans
QSECURITY	Sets the system security level
QSSLCSL	Specifies the cipher algorithms that can be used for Secure Sockets Layer (SSL) connections (V6R1)
QSSLCSLCTL	Specifies whether the operating system controls the list of SSL cipher specifications or a user-defined list is used (V6R1)
QSSLPCL	Specifies the supported protocols available for SSL connections (V6R1)
QUSEADPAUT	Defines which users can create programs that use adopted authority (i.e., set the USEADPAUT(*YES) program attribute)
QVFYOBJRST	Defines the level of object signature verification performed when an object is restored

You've already learned about the QSECURITY system value. Let's look at the details of the other security-relevant system values, including their recommended settings. For more information about any of the security-related system values, visit the IBM i and i5/OS Information Center (*http://publib.boulder.ibm.com/infocenter/systems/scope/i5os/index.jsp*); look under "Security|Security reference|Security system values."

QALWOBJRST — Allow Restoration of Security-Sensitive Objects

System value	Description	Possible values	Default
QALWOBJRST	Determines whether security-sensitive objects (i.e., system-state objects, programs that adopt authority) can be restored to the system	*ALL *NONE *ALWSYSST *ALWPGMADP *ALWPTF *ALWSETUID *ALWSETGID *ALWVLDERR	*ALL

System value QALWOBJRST lets you control whether anyone can restore to your system system-state objects or programs that adopt authority. The default value of *ALL lets a user with the appropriate authority restore any object. You can specify *NONE to prevent anyone from restoring security-sensitive objects. Value *ALWSYSST lets anyone restore system-state objects, and *ALWPGMADP lets anyone restore programs that adopt authority. (In V6R1, system-state objects must be digitally signed by IBM, or else they cannot be restored regardless of the QALWOBJRST setting.)

Value *ALWPTF lets you restore system-state programs and programs that adopt authority when you load PTFs. This option makes system value QALWOBJRST more usable because you don't have to change the system value every time you load PTFs.

The *ALWSETUID and *ALWSETGID values pertain to few organizations. *ALWSETUID enables restoration of Unix programs that have the S_ISUID attribute and run in the IBM i Portable Application Solutions Environment (PASE). *ALWSETGID lets Unix programs that have the S_ISGID attribute and run in the PASE be restored.

The *ALWVLDERR value permits objects to be restored even if they have validation errors. The addition of this value accommodates the capability of the QFRCCVNRST system value (covered later) to re-convert programs if errors are discovered. Because the errors are corrected when you use QFRCCVNRST in this way, the object should be allowed to be restored.

Security Recommendation

For maximum control over what is being restored to your system, set system value QALWOBJRST to *NONE. In all likelihood, however, this setting will be too restrictive for most organizations. If that is the case, consider using value *ALWPTF so you won't have to change QALWOBJRST every time you load PTFs. If you restore frequently or your change management software promotes programs that adopt authority, set QALWOBJRST to *ALWPGMADP.

QALWUSRDMN — Allow User Domain Objects in These Libraries

System value	Description	Possible values	Default
QALWUSRDMN	Specifies in which libraries you can place *USRSPC, *USRIDX, and *USRQ objects	*ALL *DIR library_names	*ALL

System value QALWUSRDMN specifies the names of the libraries that can contain *USRSPC (user space), *USRIDX (user index), and *USRQ (user queue) objects. The default value of *ALL permits these objects to exist in all libraries and is generally acceptable for any system that doesn't need to comply with Common Criteria security specifications. The value *DIR allows user-domain objects in all directories on the system.

You can also name specific libraries into which you want to be able to put *USRSPC, *USRIDX, and *USRQ objects. System administrators who want to be able to easily track these objects might want to specify libraries.

If you need to comply with the Common Criteria requirements, you must use security level 50 and specify the library name QTEMP for system value QALWUSRDMN. (See the next Technical Note, which explains why you must also include library QRCL.) The system can't audit information that moves to and from *USRSPC, *USRIDX, and *USRQ objects. At level 50, library QTEMP is a temporary object, preventing the passing of confidential data between users.

Some people think this system value takes effect at security level 50, but it actually is active at all security levels.

Technical Note

If you specify anything other than *ALL for QALWUSRDMN, you should include library QRCL in the list of libraries. When you run the RCLSTG command, the system might need to move *USRSPC, *USRIDX, or *USRQ objects into this library.

Security Recommendation

Unless you must comply with the Common Criteria level of security, specify *ALL for system value QALWUSRDMN.

QAUTOCFG — Automatic Device Configuration

System value	Description	Possible values	Default
QAUTOCFG	Specifies whether locally attached devices are configured automatically	0 (off) 1 (on)	0 (off)

System value QAUTOCFG determines whether you manually configure new local controllers and devices or let the system configure them for you. QAUTOCFG doesn't affect the creation of virtual devices and thus is becoming less pertinent as virtual devices become more common.

Security Recommendation

Set system value QAUTOCFG to 1 (on) when you install the system and when adding many new devices. Otherwise, leave it at 0 (off), the default.

QAUTOVRT — Automatic Configuration of Virtual Devices

System value	Description	Possible values	Default
QAUTOVRT	Determines whether virtual device descriptions are created automatically	0 (off) 1–32500 *NOMAX *REGFAC	0 (off)

System value QAUTOVRT determines whether incoming Display Station Passthrough (DSPT) and Telnet session requests create virtual devices automatically as needed. A value of 0 tells the system that you'll create the virtual devices to be used and the system shouldn't create them automatically. A numerical value in the range 1–32500 indicates the maximum number of temporary devices that the system can configure automatically. The value *NOMAX specifies that no limit is imposed on the number of automatically created device descriptions. The value *REGFAC specifies that the program registered for the Virtual Device Selection exit point (QIBM_QPA_DEVSEL) is called when a virtual device description is selected or automatically created by the system. If an exit program is not registered or not found, the system behaves as if the system value is set to 0.

Security Recommendation

Set QAUTOVRT to 0 or a number (but not *NOMAX) so you can explicitly control the number and use of virtual devices on your system.

QCRTAUT — Create Authority

System value	Description	Possible values	Default
QCRTAUT	Specifies the default public authority for newly created objects	*ALL *CHANGE *USE *EXCLUDE	*CHANGE

The CRTAUT (Create authority) parameter is available on both the CRTLIB (Create Library) and CHGLIB (Change Library) commands. This parameter defines the default public authority for all objects created in the library and permits the following values:

- *SYSVAL — The value specified in system value QCRTAUT is used.
- *ALL — The default public authority is *ALL.
- *CHANGE — The default public authority is *CHANGE.
- *USE — The default public authority is *USE.
- *EXCLUDE — The default public authority is *EXCLUDE.
- authorization_list_name — The default public authority is *AUTL, using the specified authorization list (Chapter 6 covers authorization lists).

Parameter CRTAUT's default value is *SYSVAL, which references the QCRTAUT system value. Thus, when you create a library and specify no other value for the CRTAUT parameter, the public authority for new objects created in that library defaults to the authority specified in system value QCRTAUT. When you create objects in the library, the AUT (Public authority) parameter on the create (CRT*xxx*) command defaults to the value *LIBCRTAUT, which instructs the system to refer to the library's CRTAUT parameter to determine the default public authority for the new object.

Security Recommendation

You have two choices for changing the default public authority for objects created in a library. One approach is to change system value QCRTAUT from the default *CHANGE to either *USE or *EXCLUDE.

The second approach is to leave QCRTAUT with its default value of *CHANGE and control the public authority at the library level. In other words, change the CRTAUT parameter for user libraries to either *USE or *EXCLUDE, depending on each library's contents. I often find that I use a combination; that is, I often change QCRTAUT to *USE and then may have to modify the create authority setting of some libraries if the objects being created into those libraries need to be something other than *USE. For example, I often set the create authority of the libraries in which queries and their files are being created to *ALL and authorize only the handful of users running the queries to the library itself. I may also set vendor libraries' create authority setting to *CHANGE when I'm not sure whether objects are being created into the library or just being read out of them.

QDSPSGNINF — Display User Sign-on Information

System value	Description	Possible values	Default
QDSPSGNINF	Specifies whether the sign-on information display appears when a user signs on	0 (off) 1 (on)	0 (off)

System value QDSPSGNINF controls the display of user sign-on information. A value of 1 causes the system to display certain information to users when they sign on to the system — specifically, the last date and time when the profile signed on and the number of failed sign-on attempts due to an invalid password.

If you're using System i Access and require your users to sign in to System i Access before a 5250 emulation session is displayed, the last sign-on for which information is displayed will be the sign-on to System i Access. This information isn't at all helpful because that sign-on time is about two seconds before the current sign-on. But the real issue is this: Even when the information is helpful, unless you educate your end users to pay attention to this data and tell them who to call if the information looks suspicious, you might as well not bother turning on this system value. However, I believe it's very appropriate to turn on this attribute in powerful profiles. These profiles constitute a much smaller subset of users to educate, and these users are more likely to pay attention to the information than the general user community might be. If you take this approach, enable this setting in the profiles of these users, not via the system value.

Technical Note

QDSPSGNINF controls the display of sign-on information system-wide. To override this system value at the user profile level, change the DSPSGNINF (Display sign-on information) user profile parameter.

Security Recommendation

Set QDSPSGNINF to 0, and control sign-on information at the user profile level by modifying the DSPSGNINF parameter for individual user profiles.

QFRCCVNRST — Force Conversion on Restore of Object

System value	Description	Possible values	Default
QFRCCVNRST	Determines which objects being restored should be converted before being restored	0–7	1

Understanding the behavior of system value QFRCCVNRST and what setting you should use requires you to know a bit about the internal makeup of an executable object on IBM i. When you compile a source program on IBM i, the system creates a "creation template." The trusted translator that resides below the machine interface translates the creation template and produces executable hardware instructions. The trusted translator is guaranteed to produce a program that does not circumvent the integrity of the system.

Executables not produced by the trusted translator may circumvent system integrity. A program that circumvents system integrity is a large security issue and in most cases should not be trusted by the administrator. As an administrator, you cannot afford to allow these types of programs on your system.

It's impossible to determine with 100 percent certainty whether a program was generated by the trusted translator or whether instructions were altered after the program was translated. Nor can it be proved whether an altered program bypasses system integrity. It is possible, however, to retranslate programs using the trusted translator. This process removes any alterations that may have existed. A retranslated program is 100 percent guaranteed not to bypass system integrity. Retranslating programs is a very good thing for security.

Before V5R1, when observability was removed from a program, the source and the creation templates were removed. Since V5R1, removing observability removes only the source code. Anything compiled at V5R1 or later is therefore retranslatable unless someone has taken specific steps to manually remove the creation template. Any program compiled at V5R1 or later that can't be retranslated should be extremely suspect!

You use system value QFRCCVNRST to tell the system which objects it should try to retranslate, or convert. Valid values range from 0 (do not attempt to convert anything) to 7 (try to convert every executable). For all values, any object that meets the criteria for conversion but cannot successfully be converted will not be restored.

Value 0, as noted, instructs the system not to convert any objects.

Value 1, the default, indicates that only objects with validation errors should be converted. Validation errors are found in executable objects. They signal that the executable itself has been modified in an unsupported way; in other words, it has been manually patched, but the person who patched the program didn't successfully cover his or her tracks. These programs are capable of bypassing system integrity and could contain viruses. Although a patched program may be benign, it is nearly impossible to prove that it is. You most likely do not want programs with validation errors on your system.

Value 2, like value 1, causes objects with validation errors to be converted. In addition, it tells the system to convert any objects that require retranslating to be used on the current version of the operating system.

Value 3 is the same as 2 but adds the precaution of converting objects suspected of having been tampered with. These objects have been patched with Display/Alter/Dump but don't appear to have validation errors. This might mean something was changed and then changed back, or it could mean the person who patched the program was able to do so in such a way that the validation process couldn't detect an error. In general, no programs should ever be patched; rather than patching the executable directly, you should change and recompile the source. In effect, a QFRCCVNRST value of 3 says, "Try to convert any object that appears to have been patched."

The remaining values for QFRCCVNRST cause the system to try to retranslate any program that would be retranslated under value 3 as well as some executables that appear to be unaltered.

Value 4 instructs the system to convert objects that have creation templates and are not digitally signed. Unsigned objects that don't have creation templates will be restored.

Value 5 causes the system to try to convert any object that has a creation template, regardless of whether the object is signed or unsigned. The system will not try to convert objects compiled before V5R1 that do not have creation templates (i.e., have had observability removed). It makes the decision to convert regardless of whether the object is signed or unsigned.

Value 6 tells the system to try to convert all objects that lack a valid digital signature.

Value 7 causes the system to try to convert all objects being restored.

Security Recommendation

Set system value QFRCCVNRST to 3 to require the conversion of objects that have been patched or altered.

System value	Description	Possible values	Default
QINACTITV — Inactivity Time-out Interval			
QINACTITV	Determines the interval (in minutes) that a workstation can be inactive before the system sends a message to a message queue or ends the job	*NONE 5–300	*NONE
QINACTMSGQ — Inactivity Message Queue			
QINACTMSGQ	Specifies either the action to be taken when the inactivity time-out interval is reached or the name of the message queue that will receive messages about the workstation	*ENDJOB *DSCJOB message_queue_name	*DSCJOB
QDSCJOBITV — Disconnect Job Interval			
QDSCJOBITV	Specifies the interval (in minutes) that a job can be disconnected before the system ends the job	*NONE 5–1440	240

The next three system values work together to manage inactive workstations. System value QINACTITV determines the length of time, in minutes, that a workstation is allowed to remain inactive before the system ends the job, disconnects the job, or sends a message to the message queue specified in system value QINACTMSGQ. If QINACTMSGQ is set to *DSCJOB, the system will disconnect, rather than end, a job that times out. The job will remain disconnected for the length of time specified in system value QDSCJOBITV, and then it will end (the system will cancel the job, close all files and programs, and automatically roll back all uncommitted transactions running under commitment control). Users who sign back on to a disconnected job are simply reconnected to the job and can continue processing.

It's important to realize that the term "inactive" refers not to the job itself but to the workstation. Has the user pressed the Enter key? The System request key? Page up or Page down? In other words, a workstation is considered inactive when no user interaction occurs for the specified time. If the workstation job is waiting for a user response, the workstation is inactive.

The QINACTITV system value also monitors the activity of System i Access emulation sessions. These sessions might appear inactive if the user isn't directly interacting with the IBM i session, such as when a user is editing a document or spreadsheet using Microsoft Word or Excel.

Another important point to note is that the QINACTITV value applies to all sessions on the system. That is, you can't have one user group — say tellers — with a time-out value of 5 minutes and another — say loan officers — with a time-out value of 30 minutes. "One size fits all" for this system value. If you want the time to be variable, you'll need to write your own device time-out utility or buy one.

Before you change the QINACTITV system value, consider whether your applications can handle having files closed in the middle of a transaction. For applications that have used commitment control, this shouldn't be an issue. Without commitment control, however, data integrity errors may be encountered when a session is closed with transactions partially completed.

For compliance with the Payment Card Industry Data Security Standard, inactive sessions must be timed-out after 15 minutes of inactivity. If you can't implement a time-out function on IBM i, you may be able to achieve compliance by timing out the entire workstation — not just the IBM i session. This is usually most easily accomplished by setting a group policy and applying it to the computers (PCs) in your network.

Security Recommendation

If possible, set system value QINACTITV to 15, and set system value QINACTMSGQ to *DSCJOB (the default). Be sure to set system value QDSCJOBITV to a reasonable number, such as 30. With QDSCJOBITV set to 30, if the user doesn't sign on again before the job has been disconnected for 30 minutes, the job will end. In addition (or if the IBM i function is too restrictive), consider adding a group policy to time out devices throughout your network.

QLMTDEVSSN — Limit Device Sessions

System value	Description	Possible values	Default
QLMTDEVSSN	Lets you specify whether a user can have concurrent device sessions	0 (off) 2–9 (V6R1 only)	1 (on) 0 (off)

System value QLMTDEVSSN determines whether a user profile can be signed on to more than one workstation at once. A value of 0 lets a user profile be signed on to an unlimited number of device sessions simultaneously. A value of 1 permits a user to be signed on to only one workstation at a time. From a security perspective, this restriction is a good thing because it greatly reduces the ability for users to share their user IDs and passwords.

However, there are times when users may need more than one session to effectively do their jobs. Operators, for example, often use one session to monitor system activity and another to perform work from an operator's menu. Application developers often need two sessions, especially when testing or debugging their code.

Technical Note

QLMTDEVSSN controls concurrent device sessions system-wide, but you can override this system value at the user profile level by changing the user profile parameter LMTDEVSSN (Limit device sessions).

Security Recommendation

Use a value of 1 for QLMTDEVSSN to discourage users from sharing user profiles. Before you do, however, plan for which users require multiple sessions, and update their profiles.

New in V6R1

V6R1 offers the ability to set the number of workstations a user can be signed on to simultaneously. In addition to one device or an unlimited number, you can now specify that users can sign on from two to nine devices at a time. As in previous releases, you can override this setting in the user profile.

QLMTSECOFR — Limit Security Officer Access to Workstations

System value	Description	Possible values	Default
QLMTSECOFR	Limits users with *ALLOBJ or *SERVICE special authority to authorized devices	1 (on) 0 (off)	1 (on)

System value QLMTSECOFR determines which workstations can be accessed by user profiles that have *ALLOBJ or *SERVICE special authority, such as QSECOFR and QSRV. If you set the value at 0, users with *ALLOBJ or *SERVICE special authority can use any workstation attached to the system. When the value is 1, these users are limited to accessing workstations to which they are specifically authorized (by virtue of being granted authority or through ownership).

Quite honestly, system value QLMTSECOFR has lost its usefulness with the advent of TCP/IP and the use of virtual devices. Unless you implement "named devices" by using a Telnet exit program, this system value is totally ineffective. Because you don't know which virtual device you'll be getting when establishing a Telnet session, you would have to grant QSECOFR authority to all virtual devices to ensure you could sign on with an *ALLOBJ profile. The purpose of this system value is to limit where a powerful profile can sign on. But if you have to authorize all devices, the purpose is defeated.

If you're using named devices, you can choose which device is allocated to a particular user and therefore authorize specific users to specific devices. In this case, the QLMTSECOFR system value can be effective.

Security Recommendation

If you're using named devices, set QLMTSECOFR to 1 to limit access to specific devices. Regardless of which value you assign to QLMTSECOFR, the QSECOFR user profile can always sign on at the system console. If you're not using named devices, set this system value to 0.

System value	Description	Possible values	Default
QMAXSIGN — Maximum Number of Sign-on Attempts			
QMAXSIGN	Determines the maximum number of sign-on attempts allowed for a workstation	*NOMAX Any number	3
QMAXSGNACN — Maximum Sign-on Action			
QMAXSGNACN	Specifies the action taken when a user reaches the maximum number of failed sign-on attempts	1 2 3	3

System value QMAXSIGN determines the number of invalid sign-on attempts the system allows before taking action to disable the device or user profile being used. Users cause invalid sign-on attempts by mistyping user IDs or by forgetting their password and entering the wrong one.

When you set a numeric value for QMAXSIGN, the system sends a message to the workstation when only one attempt remains. The message warns the user that the next incorrect sign-on will disable the device or user profile.

Setting QMAXSIGN to *NOMAX will cause your organization to be out of compliance with many laws and regulations. But worse, this setting constitutes a serious security exposure. *NOMAX means a hacker or disgruntled employee can make unlimited tries to guess a password. Obviously, you want to set QMAXSIGN to a number that lets users mistype their password a couple of times on Monday morning without having to get things reset. But you don't want to set the value to such a large number that it allows too many guesses at a profile's password.

System value QMAXSGNACN controls the action the system takes when a user reaches the maximum number of failed sign-on attempts specified in QMAXSIGN. QMAXSGNACN has three possible values:

- 1 — Disable the display device.
- 2 — Disable the user profile.
- 3 — Disable both the display device and the user profile.

As you can see, this system value lets you disable the user profile of a user who exceeds the maximum allowed number of sign-on attempts. After a user profile is disabled, you can restore it to active state only by using the STATUS(*ENABLE) parameter on the CHGUSRPRF (Change User Profile) command.

Security Recommendation

Set QMAXSGNACN to either 2 or 3. Setting this system value to 3 (the default) when using virtual devices carries minimal security advantages. If your QAUTOVRT system value is set to *NOMAX, setting QMAXSGNACN to 1 or 3 can actually cause a denial-of-service attack in which a hacker or disgruntled employee disables all virtual devices so that no one can sign on (except at the console).

QRETSVRSEC — Retain Server Security

System value	Description	Possible values	Default
QRETSVRSEC	Determines whether you allow decryptable passwords to be stored on the system	0 (off) 1 (on)	0 (off)

System value QRETSVRSEC was created to let you control whether IBM i stores decryptable passwords. But rest easy; this option doesn't mean IBM i user profile passwords are stored so that they can be decrypted. (These passwords are stored in encrypted form, with no way for the system to decrypt the password.) System value QRETSVRSEC pertains to passwords that are typically used to sign on to other systems. For example, Serial Line Internet Protocol (SLIP) requires a password that can be decrypted so it can be used to sign on to another system. In addition, you can store decryptable passwords in a validation list. (I discuss validation lists in the next chapter.)

The default QRETSVRSEC value of 0 prevents decryptable passwords from being stored. A value of 1 permits decryptable values.

Technical Note

Changing QRETSVRSEC from 1 to 0 disables the stored passwords but does not remove them from the system. If you want to remove the passwords from the system, you'll have to run the CLRSVRSEC (Clear Server Security Data) command after setting QRETSVRSEC to 0.

Security Recommendation

Unfortunately, many system functions require system value QRETSVRSEC to be on (value 1) — that is, to allow decryptable passwords. Service Director and other IBM service functions won't work if you set QRETSVRSEC to 0. Therefore, most organizations will want to set this value to 1.

QRMTSIGN — Remote Sign-on Value

System value	Description	Possible values	Default
QRMTSIGN	Determines whether automatic sign-on from a remote system is allowed	*FRCSIGNON *SAMEPRF *REJECT *VERIFY program_name	*FRCSIGNON

The STRPASTHR (Start Pass-Through) command lets a user enter a user profile, password, initial program, initial menu, and current library to bypass the target system's typical sign-on procedures. This capability lets you automate passthrough services. You can also configure your System i Access 5250 session to bypass the sign-on screen. System value QRMTSIGN determines how your system handles these requests. The allowed values are

- *FRCSIGNON (the default) — All passthrough and System i Access sessions are presented with a normal sign-on screen. (In other words, the sign-on screen is not bypassed.)
- *SAMEPRF — The session is permitted to bypass the normal sign-on screen but only when the user profile names on the source and target systems are identical.
- *REJECT — All passthrough requests are rejected, and System i Access requests are presented with the sign-on screen.
- *VERIFY — Automatic passthrough requests to bypass the sign-on screen and System i Access requests must have a valid user profile and password.
- program_name — You can designate the name of a program to be run at the beginning and end of every passthrough session. The specified program controls and logs access for every passthrough request. You can qualify this value by library name.

Security Recommendation

If you don't use passthrough, give system value QRMTSIGN the value *REJECT; otherwise, use *VERIFY to require both a user profile and a password or *SAMEPRF to let System i Access bypass the sign-on screen.

System value	Description	Possible values	Default
QSCANFS — Scan File Systems			
QSCANFS	Determines whether file system scanning takes place	*NONE *ROOTOPNUD	*ROOTOPNUD
QSCANFSCTL — Scan File Systems Control			
QSCANFSCTL	Provides fine-grained control of file system scans	*NONE *ERRFAIL *FSVRONLY *NOFAILCLO *NOWRTUPG *NOPOSTRST *USEOCOATR	*NONE

When combined with exit programs, the next two system values can help you prevent or detect PC viruses in objects stored in the integrated file system (IFS).

System value QSCANFS works in conjunction with the QIBM_QP0L_SCAN_OPEN (Scan on Open) and QIBM_QP0L_SCAN_CLOSE (Scan on Close) exit points. The QSCANFS setting tells the system whether objects in the IFS should be scanned when opened, closed, or both. The exit points indicate which programs should be called to perform the scanning. You can set QSCANFS to *NONE (no scanning) or to *ROOTOPNUD, which indicates that every stream file (*STMF) object in the root ('/'), QOpenSys, and user-defined file systems should be passed to the registered exit programs for scanning.

System value QSCANFSCTL provides fine-grained control over matters such as when scans will occur and what should happen when they are attempted. This system value has seven possible values, any or all of which you can specify (with the obvious exception of the default *NONE, which indicates that no controls are being specified).

Setting QSCANFSCTL to *NONE will be sufficient for many organizations. With value *ERRFAIL, if errors occur when the exit program is called (e.g., exit program not found), the system treats the object to be scanned as if it had been scanned and had failed the operation. If you don't specify this value, the system treats the object as if it had not been scanned.

Value *NOWRTUPG instructs the system not to upgrade to write access for the scan exit program, thus preventing the exit program from modifying or repairing the scanned object if possible. If you don't specify this value, the system ensures that the scan exit program has write access to the object that is to be scanned.

For details about the other QSCANFSCTL values, see IBM's Information Center. For information about exit points QIBM_QP0L_SCAN_OPEN and QIBM_QP0L_SCAN_CLOSE, including all the conditions under which scans will occur, see the API reference information (Programming|Application programming interfaces) in the Information Center.

Security Recommendation

Leave QSCANFS at the default setting. You can also leave QSCANFSCTL at the default setting unless you have very high security requirements.

QSECURITY — System Security Level

System value	Description	Possible values	Default
QSECURITY	Determines the system security level	20 30 40 50	40

As you learned early in the chapter, system value QSECURITY sets the security level of the system. Level 40 — which provides password security, resource security, and operating system integrity (protecting the system from undocumented system interfaces) — is the default setting.

QUSEADPAUT — Use Adopted Authority

System value	Description	Possible values	Default
QUSEADPAUT	Specifies which users can create programs that use adopted authority	*NONE authorization_list_name	*NONE

Many people misunderstand the USEADPAUT (Use adopted authority) program attribute. This attribute doesn't control whether a program adopts authority; it controls whether the program can *use* adopted authority that is passed down to it from another program higher in the call stack.

Many applications use programs that adopt a powerful profile (often QSECOFR) and call other programs but don't library-qualify the program call. If a clever employee or hacker can create a program with the same name and put it into a library that precedes the real program's library in the library list, the impostor program will be called instead of the real program and will be able to use the adopted authority that the calling program provides.

System value QUSEADPAUT lets you prevent users from being able to create programs that use adopted authority. To implement this security measure, you create an authorization list (a topic I cover in Chapter 6), and specify it for QUSEADPAUT. Create the authorization list with *PUBLIC(*EXCLUDE), and grant specific authority to only those users who are allowed to create programs that use adopted authority (such as the person allowed to promote programs or perhaps the profile under which your change management software runs).

It's very appropriate to use system value QUSEADPAUT for a production system; however, if you implement it in an application development environment, it can seriously affect how programmers can do their jobs. Some application designs depend on programs that can use adopted authority.

Security Recommendation

For maximum protection in a production environment, use system value QUSEADPAUT to determine which users (if any) are allowed to create programs that use adopted authority.

QVFYOBJRST — Verify Object on Restore

System value	Description	Possible values	Default
QVFYOBJRST	Verifies the digital signatures of objects being restored and whether they are allowed on the system	1–5	3

System value QVFYOBJRST lets you control whether objects must be digitally signed and what should happen to objects that are signed but have invalid signatures. An object can have an invalid signature for two reasons: if the certificate used to sign the object is not in the system's list of trusted object signers, or if the object was changed since it was signed by the sender. You must decide what your policy is with respect to digitally signed objects.

Value 1 is the least restrictive setting. With this value, the system ignores object signatures at restore time. This setting might be a good choice if you're performing a system restore after a disaster and you just want to get the system back up and running the way it was at the last SAVSYS (Save System) operation.

A value of 2 indicates that signatures for signed objects should be checked during restore and a message issued if any signature validation fails. However, all user-state programs will be restored regardless of the result of the verification and whether the objects are signed or not. This setting is a slightly better choice in the disaster recovery scenario, but it still doesn't provide much protection.

Value 3 is the same as 2, except objects with invalid signatures are not restored. Any unsigned objects are restored. This setting provides better protection than the previous values, but there are more restrictive values that give you better control of your system. Value 3 is the default value.

With value 4, unsigned objects are not restored. Signed objects are restored, whether the signatures are valid or not.

Value 5, the most restrictive setting, gives you the greatest control over your system. With this setting, unsigned objects or objects that are signed with an untrusted signature are not restored. A signature is not trusted if it is invalid or if the certificate authority that signed the certificate used to sign the object is not in the system's list of trusted certificate authorities.

Digital signatures are a great way to ensure that an object you're restoring actually came from who you think it came from and that it hasn't changed since that person or company signed it. If everyone signed their products, the world would be a little safer for system administrators. The reality, however, is that I've seen no vendors other than IBM sign their code.

Security Recommendation

Set system value QVFYOBJRST to 3. That way, if the object is signed, the signature must be valid.

New in V6R1

If your organization needs to control the type of encryption as well as the protocol used for Secure Sockets Layer (SSL) connections to your system, you'll be interested in three new system values that debuted in V6R1:

- QSSLCSLCTL (Secure Sockets Layer Cipher Control) — Lets you specify whether the operating system will control the list of cipher specifications for SSL or the list will be user-defined.
- QSSLCSL (Secure Sockets Layer Cipher Specification List) — Lets you specify the cipher algorithms that can be used. The list that comes defaulted in this system value will be used unless you've specified *USRDFN (user-defined) for system value QSSLCSLCTL.
- QSSLPCL (Secure Sockets Layer Protocols) — Specifies the list of SSL protocols that are supported (and usable) on an SSL connection. If your organization has a requirement that only Transport Layer Security (TLS) connections are allowed, for example, you can remove the rest of the protocols and allow only *TLSV1.

Unless your organization has a specific requirement to control the security attributes of SSL connections, leave these system values at their default, which lets the system manage the attributes.

Password-Related System Values

The second group of system values sets the rules regarding passwords on your system. The system checks the password-related system values whenever users use the CHGPWD (Change Password) command or the QSYCHGPW (Change Password) API to change their current password. You can use these system values to control the composition of users' passwords. For example, you can require users to include a digit in their passwords, to not use their previous 32 passwords, and so on. If a user's new password violates any of the rules determined by the password system values, the system sends an error message, and the user must select another password.

Remember that passwords should be secret, hard to guess, and changed regularly. However, sometimes "hard to guess" also translates into "hard to remember." When passwords are hard to remember, users tend to write them down and post them near their workstations! Obviously, this is something you want to avoid. Although you can address this situation in your security policy, the fact is that if you make the composition of passwords too difficult, users will likely write down their passwords regardless of the policy. So while you definitely need to set some of the password system values, I caution you against setting too many.

Table 3.4 lists the password-related system values. I'll start by looking at QPWDLVL, the password level system value.

Table 3.4
Password-Related System Values

System value	Description
QPWDCHGBLK	Specifies the time period following a password change during which the password cannot be changed (V6R1)
QPWDEXPITV	Specifies the maximum number of days a password is valid
QPWDEXPWRN	Determines when users start to receive password expiration warning messages (V6R1)
QPWDLMTAJC	Determines whether passwords can contain adjacent digits
QPWDLMTCHR	Specifies characters that are not allowed in a password
QPWDLMTREP	Determines whether passwords can contain repeating characters
QPWDLVL	Determines the password level of the system
QPWDMINLEN	Specifies the minimum number of characters in a password
QPWDMAXLEN	Specifies the maximum number of characters in a password
QPWDPOSDIF	Determines whether each position in a new password must contain a different character than the corresponding position in the old password
QPWDRQDDGT	Determines whether a digit is required in the password
QPWDRQDDIF	Determines whether the password must be different from previous passwords
QPWDRULES	Specifies the rules used for valid password formation (V6R1)
QPWDVLDPGM	Specifies the name of a user-written password approval program

QPWDLVL — Password Level

System value	Description	Possible values	Default
QPWDLVL	Determines the password level of the system	1–3	0

The value of the QPWDLVL system value determines the possible values of passwords on the system. When QPWDLVL is set at its default value of 0, "traditional" IBM i password values are allowed. Possible values for passwords are uppercase A–Z, the numbers 0–9, and the special characters #, $, and @. In addition, weakly encrypted passwords used by the old Microsoft operating systems are stored on the system.

At password level 1, possible values for IBM i passwords remain the same, but the weakly encrypted Microsoft passwords are removed from the system. At level 2, possible password values are expanded to include upper- and lowercase letters, numbers, special characters, punctuation, and spaces. Passwords may be up to 128 characters long (typically called "passphrases"). At level 3, the strongest password security level, possible values are the same as level 2, and the weakly encrypted Microsoft passwords are removed from the system.

Although you might be tempted to immediately set QPWDLVL to 3 and IPL, you need to do a little planning. Here are the things to consider:

- If you use i5/OS NetServer and the clients connect using Windows 95/98 or ME, these operating systems require the weakly encrypted password, so you'll need to stay at password level 0 or 2. This situation is rarely the case today, so, at the very least, level 1 is possible for most organizations.
- Changing system value QPWDLVL requires an IPL. To display the pending value for QPWDLVL, use the DSPSECA command.

Before moving between password levels, take note of the following considerations.

Moving from level 0 to 1. Your risk here is that you could break the connection between your clients and the Windows Network Neighborhood on the IBM i system. If all your clients are running a current operating system, there is no opportunity for failure.

Moving to level 2. Things start to get a bit more interesting. IBM likes to call this stage the "test level." That's because you can leave your maximum-password-length system value (QPWDMAXLEN, covered later) at 10 and limit the character set for passwords to uppercase A–Z, the numbers 0–9, and special characters and test to see whether all the systems in your network will continue to work with the new algorithm. In case they don't, a password using the old encryption algorithm is kept so you can return to level 0 or 1 and get things up and running again.

Moving to level 3. If you move to level 3, you should plan to stay there because returning to a lower level is difficult from here. Once you move to level 3, the old passwords — both the weakly encrypted passwords used by NetServer and the passwords that use the original IBM i password algorithm — are removed from the system. If you need to return to a lower level, you must move to level 2 first so that the system will start retaining passwords using the original algorithm again.

Security Recommendation

Set system value QPWDLVL to 1 if you want to continue to keep the current IBM i password character set. If you want to use passphrases, migrate this value to 3.

QPWDEXPITV — Password Expiration Interval

System value	Description	Possible values	Default
QPWDEXPITV	Determines the maximum number of days a password is valid	*NOMAX 1–366	*NOMAX

System value QPWDEXPITV determines the maximum number of days a password is valid. When the limit is exceeded, the password expires and the user's next attempt to sign on causes the system to prompt the CHGPWD command. The user must then select a new password to access the system.

Technical Note

System value QPWDEXPITV controls the password expiration interval system-wide, but you can override it using the user profile parameter PWDEXPITV (Password expiration interval).

Security Recommendation

Set QPWDEXPITV to a value of 60 days or 90 days. This period will be adequate for most users. For powerful profiles (e.g., users with *ALLOBJ special authority), override this value in the user profile and set the PWDEXPITV parameter to 30 days.

QPWDLMTAJC — Limit Use of Adjacent Digits

System value	Description	Possible values	Default
QPWDLMTAJC	Limits the use of digits next to each other in a password	0 (off) 1 (on)	0 (off)

System value QPWDLMTAJC restricts the use of adjacent digits in passwords, thus preventing users from choosing passwords such as 11111111, a telephone number, or a social security number.

Security Recommendation

Unless your organization's security policy requires it, leave this value off (set to 0).

QPWDLMTCHR — Limit Use of Specific Characters

System value	Description	Possible values	Default
QPWDLMTCHR	Limits the use of up to 10 specified characters in passwords	*NONE Up to 10 characters	*NONE

You can use system value QPWDLMTCHR to specify up to 10 characters that are not allowed in a valid password. The characters that normally can be used in a password are the uppercase letters A–Z, 0–9, and the special characters #, $, and @.

Technical Note

When the password level system value, QPWDLVL, is 2 or 3, the system ignores system value QPWDLMTCHR.

Security Recommendation

Using system value QPWDLMTCHR to restrict the valid character set allowed for passwords can actually make passwords easier to guess. That's because prohibiting certain characters reduces the number of combinations that a hacker or disgruntled employee must try before hitting upon a valid password. Thus, I don't recommend using this system value.

QPWDLMTREP — Limit Repeating Characters

System value	Description	Possible values	Default
QPWDLMTREP	Limits the use of a repeating character in a password	0 1 2	0

System value QPWDLMTREP lets you prevent more than one occurrence of the same alphabetic character in a password. The default value of 0 lets the same character be repeated in a password.

The value 1 prevents the same character from being repeated anywhere in a password. For instance, the passwords "papaya" and "raspberry" would both be invalid with this QPWDLMTREP setting because a character is repeated in some form in each.

Specifying the value 2 for QPWDLMTREP prohibits only consecutively repeating characters. In that case, the password "raspberry" would not be allowed, but "papaya" would be because its characters aren't repeated consecutively.

Technical Note

When the password level is 2 or 3, this system value's check is case-sensitive. For example, if QPWDLMTREP is set to 1, "Aabcd" would be an allowed password at password level 2 or 3.

Security Recommendation

Set QPWDLMTREP to 2 to prevent users from repeating any character consecutively in a password.

System value	Description	Possible values	Default
QPWDMINLEN — Minimum Number of Characters Required			
QPWDMINLEN	Determines the minimum number of characters in a password	1–10 or 1–128, depending on password level	6
QPWDMAXLEN — Maximum Number of Characters Allowed			
QPWDMAXLEN	Determines the maximum number characters in a password	1–10 or 1–128, depending on password level	8

System values QPWDMINLEN and QPWDMAXLEN determine, respectively, the minimum and maximum number of characters a valid password can contain. Every installation should use these password system values.

The possible values for QPWDMINLEN and QPWDMAXLEN vary depending on your system's password level (as specified in system value QPWDLVL). If the password level is 0 or 1, passwords can be from 1 to 10 characters long. If the password level is 2 or 3, passwords can be from 1 to 128 characters long. You can't specify a maximum password length that is smaller than the specified minimum password length (and vice versa).

Security Recommendation

Set QPWDMINLEN to at least 7 to prevent the use of short names and initials. Set QPWDMAXLEN to at least 10.

QPWDPOSDIF — Require Positions to Be Different

System value	Description	Possible values	Default
QPWDPOSDIF	Determines whether each position in a new password must contain a different character than the same position in the old password	0 (off) 1 (on)	0 (off)

System value QPWDPOSDIF lets you prevent the use of a character in a new password in the same position that the character occupied in the old password. This system value was designed to prevent a succession of passwords such as PASSWORD1, PASSWORD2, PASSWORD3.

Again, the idea behind this system value is good, but in practice it can make it difficult for users to come up with new passwords because a password that has nothing to do with the previous password may not be allowed. For example, if the old password is carolw0od, and the user tries to set the new password to ja85rtez, the new password will be rejected if QPWDPOSDIF is set on because the second letter is the same in both passwords.

Technical Note

When the password level is set to 2 or 3, the comparison made by QPWDPOSDIF is case-sensitive.

Security Recommendation

Because the use of system value QPWDPOSDIF can make it difficult for users to come up with new passwords, I don't recommend turning on this system value.

QPWDRQDDGT — Require a Digit

System value	Description	Possible values	Default
QPWDRQDDGT	Determines whether a digit is required in passwords	0 (off) 1 (on)	0 (off)

System value QPWDRQDDGT determines whether valid passwords must include at least one digit (0–9). Many regulations require passwords to contain a digit, so you may need to set this value to remain in compliance.

Security Recommendation

Set QPWDRQDDGT to 1 to require users to include at least one digit in their passwords.

QPWDRQDDIF — Require Different Password

System value	Description	Possible values	Default
QPWDRQDDIF	Determines whether new passwords must be different from previous passwords	0–8	0

With system value QPWDRQDDIF, you can prevent users from choosing a new password that is the same as a selected number of previous passwords. You can tell the system to let users duplicate a previous password (to do so, specify the default value of 0) or to prevent users from using the previous 4, 6, 8, 10, 12, 18, 24, or 32 passwords, depending on which nonzero value you enter. Here are the meanings of the nonzero values:

- 1 — User can't duplicate previous 32 passwords.
- 2 — User can't duplicate previous 24 passwords.
- 3 — User can't duplicate previous 18 passwords.
- 4 — User can't duplicate previous 12 passwords.
- 5 — User can't duplicate previous 10 passwords.
- 6 — User can't duplicate previous 8 passwords.
- 7 — User can't duplicate previous 6 passwords.
- 8 — User can't duplicate previous 4 passwords.

This system value is a must for every organization. You should prevent users from duplicating at least the previous four passwords, and if you want tighter security, you probably should stretch that guideline to cover the previous 32 passwords.

Security Recommendation

Many laws and regulations allow repeated passwords every fourth password. To comply with these laws and regs, you must set QPWDRQDDIF to 8. However, I prefer to be a little more restrictive and recommend you set QPWDRQDDIF to 1 to prevent users from reusing their past 32 passwords.

QPWDVLDPGM — Password Validation Program

System value	Description	Possible values	Default
QPWDVLDPGM	Specifies the name of a user-written password approval program	*NONE *REGFAC program_name	*NONE

System value QPWDVLDPGM provides a way to enforce more stringent rules on user passwords than those supported by the security system values. You can write your own password validation program, using any rules you want to enforce. The advantage lies in the flexibility such a program provides. For example, using this type of program, you could check a file containing commonly used words and prevent their use, make sure users don't use their own names or initials, verify that a password doesn't contain part of the previous password, or prevent the use of the local sports team names. Specify the program name as the value for QPWDVLDPGM. Or, to have the system retrieve the program name from the registration facility, specify value *REGFAC and register the program name for the Validate Password exit point (QIBM_QSY_VLD_PASSWRD).

Figure 3.1 shows a simple shell for a password validity-checking CL program. This particular program prevents users from changing their password more than once a day. Some users try to bypass the QPWDRQDDIF system value limitation by changing their password again and again until they can reuse the same password they started with. A validation program such as the one shown here at least forces users to work harder and wait longer to keep their old familiar password.

After you've upgraded your system to V6R1, you can use the QPWDCHGBLK system value (described later) to limit how often a password can be changed.

The disadvantage of using a password validation program is that the validation program provides user passwords in clear text. If you create a password validation program, be sure to secure its source to ensure one modifies it to store the clear-text passwords.

Figure 3.1
Sample Password Validity-Checking Program

```
/*=================================================================*/
/*   Author: Paul Tuohy, ComCon  - 2003                            */
/*=================================================================*/
/*=================================================================*/
/* Password Validity-Checking Program                              */
/* Set system value QPWDVLDPGM to *REGFAC.                         */
/* Register this program at exit point QIBM_QSY_VLD_PASSWRD.       */
/* This program will be called as part of CHGPWD command process.  */
/*=================================================================*/

 Pgm Parm( &InData        /* IN  Data in VLDP0100 Format         */ +
           &RtRtn_Code    /* OUT return code (0 = Good, 1 = Bad)  */ +
          )

 Dcl &InData        *char   320
 Dcl &RtRtn_Code    *lgl      1
 Dcl &User          *char    10
 Dcl &SysDate       *char     6
 Dcl &PwdChgDate    *char     6
 Dcl &PwdExpired    *char     4

/* Preset return code                                              */
  ChgVar  &RtRtn_Code '0'

/* Get user profile                                                */
  ChgVar  &User  %SST(&InData 33 10)
/*=================================================================*/
/* Allow only one password change per day.                         */
/*=================================================================*/

/* Get system date                                                 */
  RtvSysVal  SysVal(QDATE) RtnVar(&SysDate)
  CvtDat  Date(&SysDate) FromFmt(*SYSVAL)       +
             ToVar(&SysDate) ToFmt(*YMD) ToSep(*NONE)

/* Retrieve password change date                                   */
  RtvUsrPrf UsrPrf(&User) PwdChgDat(&PwdChgDate) PwdExp(&PwdExpired)

/* If equal and password not expired, reject request               */
  If (&PwdChgDate = &SysDate *AND &PwdExpired = '*NO ') Do
       ChgVar  &RtRtn_Code '1'
       EndDo

 Return
 EndPgm
```

Security Recommendation

If you implement a password validity-checking program, be sure to secure the source and program object to prevent anyone from changing the program. If the program is unsecured, someone could use it to record passwords.

V6R1 Password-Related System Values

V6R1 brings many additional ways to configure and ensure strong passwords. Three new system values fine-tune your ability to implement password security.

New in V6R1

V6R1 provides three new system values related to passwords: QPWDRULES (Password Rules), QPWDCHGBLK (Block Password Change), and QPWDEXPWRN (Password Expiration Warning).

QPWDRULES — Password Rules

System value	Description	Possible values	Default
QPWDRULES	Specifies the rules used to check whether a password is formed correctly	*PWDSYSVAL, *CHRLMTAJC, *CHRLMTREP, *DGTLMTAJC, *DGTLMTFST, *DGTLMTLST, *DGTMAX*n*, *DGTMIN*n*, *LMTSAMPOS, *LMTPRFNAME, *LTRLMTAJC, *LTRLMTFST, *LTRLMTLST, *LTRMAX*n*, *LTRMINn, *MAXLEN*nnn*, *MINLEN*nnn*, *MIXCASE*n*, *REQANY3, *SPCCHRLMTAJC, *SPCCHRLMTFST, *SPCCHRLMTLST, *SPCCHRMAX*n*, *SPCCHRMIN*n*	*PWDSYSVAL

If you've ever turned on all the rules provided by the pre-V6R1 system values, you know how difficult they can make it to come up with a new password. With all the possibilities you can now specify using V6R1's QPWDRULES system value, you must use even more caution when deciding which values to set, or you risk making it too difficult for users to create new passwords.

Valid values for QPWDRULES are as follows:

- *PWDSYSVAL (the default) — This value instructs the system to use the pre-V6R1 password system values.
- *CHRLMTAJC — A password can contain repeated characters, but the characters must be next to each other. This value has the same effect as specifying 2 for system value QPWDLMTREP.
- *CHRLMTREP — Characters cannot be repeated anywhere within a password. This value is the same as specifying 1 for the system value QPWDLMTREP.
- *DGTLMTAJC — A password can contain repeated digits, but the digits must not be next to each other.
- *DGTLMTFST — The first position of a password cannot be a digit.
- *DGTLMTLST — The last position of a password cannot be a digit.
- *DGTMAX*n* — A password can contain at most *n* digits.

- *DGTMIN*n* — A password must contain at least *n* digits. This setting serves the same function as the QPWDRQDDGT system value but lets you require multiple digits in passwords.
- *LMTSAMPOS — A character in a new password cannot be used in the same position it occupied in the previous password. This value is the equivalent of setting on the QPWDPOSDIF system value.
- *LMTPRFNAME — A password cannot contain the user profile name.
- *LTRLMTAJC — A password cannot contain two adjacent letters anywhere in it.
- *LTRLMTFST — The first position of a password cannot be a letter.
- *LTRLMTLST — The last position of a password cannot be a letter.
- *LTRMAX*n* — A password can contain at most *n* letters.
- *LTRMIN*n* — A password must contain at least *n* letters.
- *MAXLEN*nnn* — A password can be a maximum of *nnn* positions long. This setting is equivalent to the QPWDMAXLEN system value.
- *MINLEN*nnn* — A password must be at least *nnn* positions long. This setting is equivalent to the QPWDMINLEN system value.
- *MIXCASE*n* — A password must contain *n* uppercase and *n* lowercase letters.
- *REQANY3 — A password must include three of the following four types of characters: uppercase characters, lowercase characters, special characters, and digits.
- *SPCCHRLMTAJC — A password can repeat special characters, but the repeated characters must not be next to each other.
- *SPCCHRLMTFST — The first position of a password cannot be a special character.
- *SPCCHRLMTLST — The last position of a password cannot be a special character.
- *SPCCHRMAX*n* — A password can contain no more than *n* special characters.
- *SPCCHRMIN*n* — A password must contain at least *n* special characters.

Technical Note

If you leave system value QPWDRULES at its default value of *PWDSYSVAL, the system uses the pre-V6R1 system values for the password rules. If you specify some other value (or multiple values) for QPWDRULES, system values QPWDLMTAJC, QPWDLMTCHR, QPWDLMTREP, QPWDMAXLEN, QPWDMINLEN, QPWDPOSDIF, and QPWDRQDDGT are ignored.

Security Recommendation

If you choose to use system value QPWDRULES, you'll want to require at least one digit (*DGTMIN1), set the minimum length to at least 7 (*MINLEN7), and set the maximum length to at least 10 (*MAXLEN10). In addition, I recommend adding *LMTPRFNAME so that the user profile name cannot be used as part of the password. Last, to be consistent with some network password requirements, you may need to add *MIXCASE1 or *REQANY3.

QPWDCHGBLK — Block Password Change

System value	Description	Possible values	Default
QPWDCHGBLK	Specifies the time period following a password change during which the password cannot be changed	*NONE 1-99	*NONE

System value QPWDCHGBLK lets you specify the number of hours required before a user can change his or her password again. The purpose of this system value is to prevent users from changing their password repeatedly so that they can get back to their original password, thus essentially never having to change their password. I've seen users (usually programmers) write a script that changes their password the number of times required to meet the QPWDRQDDIF system value requirement so that they can always use the same password.

Security Recommendation

Set QPWDCHGBLK to 72 to require at least three days between password changes.

Technical Note

You can override system value QPWDCHGBLK by using the PWDCHGBLK (Block password change) parameter in a user's profile.

Technical Note

A user's password can always be changed using the CHGUSRPRF command or the QSYCHGPW API. System value QPWDCHGBLK is meant to prevent end users from repeatedly changing their password; it does not prevent administrators from changing it.

QPWDEXPWRN — Password Expiration Warning

System value	Description	Possible values	Default
QPWDEXPWRN	Determines when users start to receive password expiration warning messages	1–99	7

System value QPWDEXPWRN lets you set the number of days before the expiration of a user's password when the user will start to receive a "password is expiring" warning message.

Security Recommendation

Leave QPWDEXPWRN at the default of 7, or set it to the value that makes the most sense for your organization. (No, you can't turn off the warnings completely!)

Audit-Related System Values

One area that continues to evolve significantly in IBM i security is the auditing capabilities of the operating system. The audit-related system values are key to implementing IBM i auditing functions. This chapter provides an overview of the audit-related system values, which are listed in Table 3.5. For a complete guide to IBM i auditing, see Chapter 15.

Table 3.5

Auditing System Values

System value	Description
QAUDCTL	Turns IBM i security auditing on or off
QAUDENDACN	Determines the action the system should take if it's unable to continue auditing (i.e., if it can't write another audit record to the journal)
QAUDFRCLVL	Determines when new auditing journal entries are physically written to disk from memory
QAUDLVL	Determines which security-related events are recorded in the security audit journal for all system users
QAUDLVL2	Specifies additional values that, in conjunction with those specified in system value QAUDLVL, determine which security-related events are logged
QCRTOBJAUD	Specifies the default auditing level for any new object

QAUDCTL — Auditing Control

System value	Description	Possible values	Default
QAUDCTL	Turns security auditing on or off	*NONE, *AUDLVL *OBJAUD, *NOQTEMP	*NONE

System value QAUDCTL is the on/off switch for IBM i security auditing. The default value, *NONE, tells the operating system to perform no security auditing. The values *AUDLVL and *OBJAUD tell IBM i to perform auditing. You can specify either or both of these values for QAUDCTL. *AUDLVL activates action auditing across the entire system or by individual user, and *OBJAUD activates object auditing. The other available value, *NOQTEMP, prevents extraneous auditing entries for objects in library QTEMP when you specify auditing with *AUDLVL and/or *OBJAUD.

Security Recommendation

Set QAUDCTL to at least *AUDLVL and *NOQTEMP. If you need to enable object auditing, add *OBJAUD.

QAUDENDACN — Auditing End Action

System value	Description	Possible values	Default
QAUDENDACN	Determines the action the system should take it it's unable to continue auditing (i.e., if the system can't write another audit record to the journal)	*NOTIFY *PWRDWNSYS	*NOTIFY

IBM i can help the security officer audit security by logging data about security-related events to a journal named QAUDJRN. System value QAUDENDACN determines the action the system should take if it can't write another audit record to the journal and thus is unable to continue auditing. The default value, *NOTIFY, causes IBM i to send a message to the system operator and continue processing. The system also sends the message to QSYS/QSYSMSG if that message queue exists. The other choice is *PWRDWNSYS, which instructs the system to immediately power down.

Both *NOTIFY and *PWRDWNSYS trigger an automatic change to *NONE in the QAUDCTL system value, causing auditing to be turned off. You must turn auditing back on manually after correcting the problem that prevented the system from writing to the journal. (For more information about the implications of the *PWRDWNSYS value, see Chapter 15.)

Security Recommendation

Unless you must comply with the Common Criteria security requirements, use the default value for QAUDENDACN, *NOTIFY.

QAUDFRCLVL — Auditing Force Level

System value	Description	Possible values	Default
QAUDFRCLVL	Determines when new auditing journal entries are physically written to disk from memory	*SYS number_of_records (1–100)	*SYS

System value QAUDFRCLVL determines the maximum number of audit records the system can cache before forcing the records to disk. You can specify any number in the range 1–100, or you can set QAUDFRCLVL to *SYS to let the operating system control this number dynamically for performance purposes.

If a system failure occurs, you might lose any audit records not yet written to disk. Systems that require Common Criteria compliance must specify 1 for this system value. For all other systems, however, I recommend the value *SYS, which maximizes performance when you use auditing.

Security Recommendation

Unless you must comply with Common Criteria security requirements, use the QAUDFRCLVL default of *SYS.

QAUDLVL — Auditing Level

System value	Description	Possible values	Default
QAUDLVL	Determines which security-related events are logged to the security audit journal (QAUDJRN)	*NONE *ATNEVT, *AUDLVL2, *AUTFAIL, *CREATE, *DELETE, *JOBDTA, *NETCMN, *OBJMGT, *OFCSRV, *OPTICAL, *PGMADP, *PGMFAIL, *PRTDTA, *SAVRST, *SECURITY, *SERVICE, *SPLFDTA, *SYSMGT	*NONE

System value QAUDLVL determines the type of auditing the system performs. You can specify one or more values (unless one of the values is *NONE, which causes other values to be ignored). The following list describes the types of journal entries made for each QAUDLVL value.

- *NONE — Perform no auditing on the system.
- *ATNEVT — Audit intrusion detection events.
- *AUTFAIL — Audit authority failures (e.g., user doesn't have authority to open a file, user doesn't have authority to run a program).
- *CREATE — Audit object creation operations.
- *DELETE — Audit object deletion operations.

- *JOBDTA — Audit job start and end. Specify *JOBDTA to cause both of the following types of audits to be performed, or specify either or both these subset values:
  - *JOBBAS — Audit actions taken on a job.
  - *JOBCHGUSR — Audit changes to the job's user or group.
- *NETCMN — Audit violations detected by the Advanced Peer-to-Peer Networking (APPN) filter. Specify *NETCMN to audit all of the following, or specify one or more of these subsets:
  - *NETBAS — Basic network events are audited. These include IP rules changes, SSL connections, APPN directory search filter, and APPN endpoint filter.
  - *NETCLU — Events associated with cluster or cluster resource group operations are audited.
  - *NETFAIL — Events associated with network failures are audited (e.g., secure socket port not available).
  - *NETSCK — Tasks associated with sockets are audited. These include accept, connect, Dynamic Host Configuration Protocol (DHCP) address assigned or not assigned, mail filtered, and mail rejected.
- *OBJMGT — Audit object management operations (e.g., move an object, rename an object).
- *OFCSRV — Audit office mail actions and changes to the system distribution directory.
- *OPTICAL — Audit use of optical storage volumes.
- *PGMADP — Audit use of programs that adopt authority.
- *PGMFAIL — Audit system integrity violations (e.g., use or attempted use of restricted machine instructions).
- *PRTDTA — Audit direct and spooled printing.
- *SAVRST — Audit restore operations.
- *SECURITY — Audit changes to security controls (e.g., system value changes, object authority changes, auditing control changes). Specify *SECURITY to audit all of the following, or specify one or more of these subsets:
  - *SECCFG — The security-related configuration is audited. This includes user profile events; changes to system values, environment variables, and network attributes; and more.
  - *SECDIRSRV — Use of directory service functions is audited.
  - *SECIPC — Use of interprocess communications is audited.
  - *SECNAS — Events associated with Network Authentication Service ticket verification are audited.

  - *SECRUN — Runtime functions associated with changes of object ownership, changes to authorization list or object authority, or changes in the primary group of an object are audited.
  - *SECSCKD — Events associated with secure socket descriptors are audited.
  - *SECVFY— Events associated with the use of user profile verification events are audited. This includes events related to profile handles and tokens and more.
  - *SECVLDL — Events associated with use of validation list object entries are audited.
- *SERVICE — Audit use of the service tools (i.e., the STRSST, or Start System Service Tools, command and options).
- *SPLFDTA — Audit spooled file control (e.g., moving a spooled file, releasing a spooled file).
- *SYSMGT — Audit system management activities (e.g., changes to the power on/off schedule, system reply list, backup options, Distributed Relational Database Architecture).
- *AUDLVL2 — As you can see from this list, you can assign a lot of values in the QAUDLVL system value. If you need to specify more values than the system value's limit of 16, include *AUDLVL2 as one of the values in QAUDLVL. This setting instructs the system to look for additional action auditing settings in system value QAUDLVL2 (Auditing Level Extension).

When configuring auditing for the first time, follow the recommendations given below for the initial settings. Then, make sure you spot-check to determine how quickly a new audit journal receiver is being created. On busy systems, journal receivers can fill up quickly, causing a new journal receiver to be generated. Make sure you have a plan for managing journal receivers and a save strategy that accommodates your organization's compliance requirements. For more details and guidance about auditing, see Chapter 15.

Security Recommendation

Start by setting QAUDLVL to *AUTFAIL, *CREATE, *DELETE, *SAVRST, *SECCFG, *SECRUN, and *SERVICE. To prepare for moving the system to security level 40, add *PGMFAIL to determine whether there are any issues to resolve before changing system value QSECURITY and IPLing the system.

Technical Note

If you are using high availability (HA) software, it's likely you may need to specify more values for QAUDLVL. That's because most HA packages rely heavily on the audit journal to know what to replicate to the target system. If you are using an HA package, do not remove values from QAUDLVL without first checking with the vendor.

QAUDLVL2 — Auditing Level Extension

System value	Description	Possible values	Default
QAUDLVL2	Holds additional values that, in conjunction with those specified in system value QAUDLVL, determine which security-related events are logged to the security audit journal (QAUDJRN)	*NONE *ATNEVT, *AUTFAIL, *CREATE, *DELETE, *JOBDTA, *NETCMN, *OBJMGT, *OFCSRV, *OPTICAL, *PGMADP, *PGMFAIL, *PRTDTA, *SAVRST, *SECURITY, *SERVICE, *SPLFDTA, *SYSMGT	*NONE

As noted above, you use system value QAUDLVL2 in conjunction with system value QAUDLVL to indicate the type of auditing to be performed on the system.

QCRTOBJAUD — Auditing for New Objects

System value	Description	Possible values	Default
QCRTOBJAUD	Specifies the default auditing level for new objects	*NONE *USRPRF *CHANGE *ALL	*NONE

System value QCRTOBJAUD and the library attribute CRTOBJAUD (Create object auditing) together determine the default object auditing level for new objects. QCRTOBJAUD works much like the QCRTAUT system value, which determines the default public authorities for new objects. When you create a new object, IBM i checks the CRTOBJAUD attribute for the library in which the new object will reside. If the attribute's value is *SYSVAL, IBM i refers to system value QCRTOBJAUD to determine which of the object auditing values (*NONE, *USRPRF, *CHANGE, or *ALL) to assign to the new object's OBJAUD (Object auditing) attribute.

The default value for the CRTOBJAUD attribute of all libraries is *SYSVAL. As a consequence, you could use the QCRTOBJAUD system value to control the object auditing level for every new object on your system. However, for existing objects, you must use the CHGOBJAUD (Change Object Auditing) command to change the OBJAUD value.

Security Recommendation

Use system value QCRTOBJAUD with caution. Specifying *ALL for this system value has the potential to produce massive numbers of audit journal entries. That's because the system will generate an audit journal entry every time a new object is touched (read) or modified. If you are using objects such as data queues, user queues, or user spaces that are modified or read frequently, an audit journal entry will be generated for each action. Setting the value to *CHANGE produces fewer values, but the number can still be significant.

Technical Note

Some high availability software requires a change to system value QCRTOBJAUD. If that is the case for you, be ready to manage more audit journal receivers as a result.

Locking Down Security-Related System Values

As a safeguard against unintended changes to your system values, it is possible to put a "lock" on system value settings. This practice prevents even system administrators with authority to the CHGSYSVAL command from changing any security-related system value setting. When you choose the lock-down option (accomplished via a service tools menu selection), the only way someone can change these values is by first logging in to the service tools (either System Service Tools or Dedicated Service Tools) and unlocking the security-related system values.

A Helpful Tool

IBM i provides a set of integrated security tools that can help you manage your security environment. One tool that relates specifically to this chapter is the PRTSYSSECA (Print System Security Attributes) command, which is available on the SECTOOLS and SECBATCH menus. This command prints a list of the current settings for security-related system values and network attributes. It also shows IBM's recommended system value and network attribute settings.

Chapter 4

The Facts About User Profiles

When we think about how people access an IBM i system, we normally think of the system object called a user profile. However, IBM i actually provides at least three different kinds of user identities: user profiles, service tools user identities, and validation list users. Although the most used and recognized of these identities are certainly user profiles, all three types are important for security.

User profiles are the entities by which users log on to the system and under which all jobs are run. *Service tools user IDs* identify those users who are allowed to use the low-level operating system tools, including Dedicated Service Tools (DST), System Service Tools (SST), and IBM Navigator for i (formerly System i Navigator) logical partitioning and disk unit management. *Validation list users* are created, managed, and used by applications on the system as a secure way to store user IDs and passwords — especially when an application needs access to a clear-text password. For example, some people will use a validation list to store the user ID and password used in an FTP script. This way, rather than the password being stored in the FTP script, it is retrieved from the validation list and then used to sign on to the FTP server.

In this chapter, we discuss user profiles and validation list users. Chapter 5 covers the service tools user IDs.

What Are User Profiles?

User profiles are the mechanism used to identify users to the system. They provide the means to authenticate a user (proving that the user is who he or she claims to be); they define the user's capabilities and the group profiles to which the user belongs; and they are the identity the system checks to verify whether the user has the authority required to run a particular function or perform a specific task. But more than defining the security information associated with a user, the user profile defines the environment in which the user operates, including the user's initial menu, initial program, and national language preference.

With a profile defined for each user, you have the ability to create a sound security implementation. You can ensure that users have access to only those functions required to perform their jobs. And you can configure the system to log (audit) user actions.

In IBM i terms, a user profile is an object (type *USRPRF). Although the object's name (e.g., CJW, VANDERWALL) is what you normally think of as the user profile, a user profile is much more than a name. The attributes of each user profile object define a particular user to the system, enabling the system to establish a custom initial session (i.e., job) for that user at sign-on. To make the best use of user profiles, you need to understand these attributes and how they can help you implement your security strategy.

Technical Note

IBM ships a number of predefined user profiles with the operating system, including some that I've mentioned previously, such as the powerful QSECOFR profile. These IBM-supplied profiles are used as object owners for various system functions. Some system functions also run under IBM-supplied user profiles. For details about these profiles and their default attributes, see the IBM i and i5/OS Information Center (go to Security|Security reference|IBM-supplied user profiles) or Appendix B of the *System i Security Reference* (available as a PDF file under the InfoCenter's Security topic).

To create and maintain user profiles on your system, you must have *SECADM (security administrator) special authority. The commands you use to accomplish these tasks are CRTUSRPRF (Create User Profile), CHGUSRPRF (Change User Profile), and DLTUSRPRF (Delete User Profile).

User Profile Attributes

You use the CRTUSRPRF command to create a user profile. When you prompt the command and press F10, the system displays the command's parameters, which are shown (with the default values in most cases) in Figure 4.1. In the figure, the parameters that relate directly to security are highlighted. The parameters shown in italics are new or changed in V6R1. In the following sections, I discuss each of these parameters in detail. (Note that the figure reflects the V6R1 user profile attributes. If your system is on an earlier release, you will see fewer parameters.)

Figure 4.1
CRTUSRPRF Parameters

```
                          Create User Profile (CRTUSRPRF)

Type choices, press Enter.

User profile . . . . . . . . . . > NEW_PROF__      Name
User password  . . . . . . . . .   *USRPRF___      Character value, *USRPRF...
Set password to expired  . . . . > *YES            *NO, *YES
Status . . . . . . . . . . . . .   *ENABLED_       *ENABLED, *DISABLED
User class . . . . . . . . . . .   *USER___        *USER, *SYSOPR, *PGMR...
Assistance level . . . . . . . .   *SYSVAL__       *SYSVAL, *BASIC, *INTERMED...
Current library  . . . . . . . .   *CRTDFT___      Name, *CRTDFT
Initial program to call  . . . . > APP_INIT__      Name, *NONE
  Library  . . . . . . . . . . . >   APP_LIB___    Name, *LIBL, *CURLIB
Initial menu . . . . . . . . . . > *SIGNOFF__      Name, *SIGNOFF
  Library  . . . . . . . . . . .     __________    Name, *LIBL, *CURLIB
Limit capabilities . . . . . . . > *YES            *NO, *PARTIAL, *YES
Text 'description' . . . . . . . > 'Make this meaningful!'____________________
_____
```

```
Special authority . . . . . . .   *USRCLS__      *USRCLS, *NONE, *ALLOBJ...
              + for more values   __________
Special environment . . . . . .   *SYSVAL        *SYSVAL, *NONE, *S36
Display sign-on information . .   *SYSVAL        *SYSVAL, *NO, *YES
Password expiration interval . .  *SYSVAL        1-366, *SYSVAL, *NOMAX
Block password change . . . . .   *SYSVAL        1-99, *SYSVAL, *NONE
Local password management . . .   *YES           *YES, *NO
Limit device sessions . . . . .   *SYSVAL        *SYSVAL, *YES, *NO, 0, 1...
Keyboard buffering . . . . . . .  *SYSVAL___     *SYSVAL, *NO, *TYPEAHEAD...
Maximum allowed storage . . . .   *NOMAX_____    Kilobytes, *NOMAX
Highest schedule priority . . .   3              0-9
Job description . . . . . . . .   QDFTJOBD__     Name
  Library . . . . . . . . . . .     *LIBL_____   Name, *LIBL, *CURLIB

Group profile . . . . . . . . . > GROUP1____     Name, *NONE
Owner . . . . . . . . . . . . . > *GRPPRF        *USRPRF, *GRPPRF
Group authority . . . . . . . .   *NONE___       *NONE, *ALL, *CHANGE, *USE...
Group authority type . . . . . .  *PRIVATE       *PRIVATE, *PGP
Supplemental groups . . . . . . > GROUP2____     Name, *NONE
              + for more values > GROUP3____
Accounting code . . . . . . . .   *BLANK__________
Document password . . . . . . .   *NONE___       Name, *NONE
Message queue . . . . . . . . .   *USRPRF___     Name, *USRPRF
  Library . . . . . . . . . . .     __________   Name, *LIBL, *CURLIB
Delivery . . . . . . . . . . . .  *NOTIFY        *NOTIFY, *BREAK, *HOLD, *DFT
Severity code filter . . . . . .  0_____         0-99
Print device . . . . . . . . . .  *WRKSTN___     Name, *WRKSTN, *SYSVAL
Output queue . . . . . . . . . .  *WRKSTN___     Name, *WRKSTN, *DEV
  Library . . . . . . . . . . .     __________   Name, *LIBL, *CURLIB

Attention program . . . . . . .   *SYSVAL___     Name, *NONE, *SYSVAL, *ASSIST
  Library . . . . . . . . . . .     __________   Name, *LIBL, *CURLIB
Sort sequence . . . . . . . . .   *SYSVAL___     Name, *SYSVAL, *HEX...
  Library . . . . . . . . . . .     __________   Name, *LIBL, *CURLIB
Language ID . . . . . . . . . .   *SYSVAL        *SYSVAL...
Country or region ID . . . . . .  *SYSVAL        *SYSVAL...
Coded character set ID . . . . .  *SYSVAL____    *SYSVAL, *HEX...
Character identifier control . .  *SYSVAL__      *SYSVAL, *DEVD, *JOBCCSID
Locale job attributes . . . . .   *SYSVAL        *SYSVAL, *NONE, *CCSID...
              + for more values   _______
Locale . . . . . . . . . . . . .  *SYSVAL_________________________________

_____
User options . . . . . . . . . .  *NONE____      *NONE, *CLKWD, *EXPERT...
              + for more values   _________
User ID number . . . . . . . . .  *GEN_______    1-4294967294, *GEN
Group ID number . . . . . . . .   *NONE______    1-4294967294,*NONE, *GEN

Home directory . . . . . . . . .  *USRPRF_________________________________

_____
EIM association:
  EIM identifier . . . . . . . .  *NOCHG___________________________
  Association type . . . . . . .  _______        *TARGET, *SOURCE, *TGTSRC...
  Association action . . . . . .  ________       *REPLACE, *ADD, *REMOVE
  Create EIM identifier . . . .   __________     *NOCRTEIMID, *CRTEIMID
Authority . . . . . . . . . . .   *EXCLUDE__     *ALL, *CHANGE, *USE, *EXCLUDE
```

USRPRF (User Profile)

The USRPRF parameter specifies the name of the user profile. What's in a user profile name? Although selecting names for user profiles might seem simple enough, developing and following a naming strategy can help immensely in implementing and maintaining your security plan.

One approach to user profile names is to make each one as similar as possible to the name of the person to whom it belongs (e.g., AYOUNG, JBROWNING). This is the approach followed by most organizations, but it has some drawbacks:

- Profiles can be guessed easily. A disgruntled employee has half of the equation and only needs to guess at the password to gain access to the system as a powerful profile.
- When a user changes his or her name (e.g., after getting married), it's difficult to change the profile name. (IBM i provides no method for completely copying one profile to another.) So the user either has to live with the old profile name or make do with not having ownership of all the old profile's objects (such as spooled files).

Another approach holds that user profile names should be completely meaningless (e.g., ABC9876, Q12345, RST0152). Meaningless names make profiles hard to guess and don't link the profile name to a department or location that might change as the employee moves within the company. The correlation between the user's name and his or her profile obviously must be kept somewhere, though — in a file or another user repository, such as Active Directory. Although this naming method is the most secure, it is rarely used because it makes administering the system and maintaining user profiles so difficult.

A third approach is to use a naming strategy that aids system administration. Under this strategy, each user profile name identifies the user's location and perhaps function or department. To implement this strategy, your naming convention should incorporate the user's location or department and a unique identifier for the user's name. For example, if Sven O. Johanson works at your Minnesota distribution center in the procurement department, you might assign one of the following profiles:

User profile name	Naming scheme
MNSJOHANSO	The first two letters of this 10-character profile represent the location (MN for Minnesota). The remainder consists of the first letter of the user's first name (S) followed by as much of the last name as will fit in the remaining seven characters.
MNPRSOJ	This profile name is similar to the first, but the branch (MN) is followed by a department abbreviation (PR for procurement) and then the user's initials (SOJ). This method provides more departmental information while reducing the unique name identifier to initials.
B25PRSOJ	This example is identical to the second, but the Minnesota branch is designated using a letter–number combination (B25) rather than letters.

When profile names provide this type of information, programs in your system that supply user menus or functions can resolve the profiles at run time based on location, department, or group. Auditing can also be more effective if you take the time to analyze departmental trends. However, just like the first method, these types of profile names are less secure than meaningless profile names because someone who understands the naming scheme can easily deduce them, leaving only the password to guess.

A final consideration in choosing a naming convention for user profiles is whether your users will have access to multiple systems, including multiple operating systems. If they will, you might consider using the same name across all systems. To do so, however, you must take into consideration all limitations that the other systems in the network place on user names and apply those limitations when you create the user profiles for your system. For example, another platform in your network might limit the number of characters allowed for user names.

Obviously, there's no perfect way to name user profiles. It's up to you to decide which user-profile naming convention will work best for your organization.

PASSWORD (User Password)

As I stressed in Chapter 3, passwords should be secret, hard to guess, and changed regularly. You can't ensure users keep their passwords secret, but you can help make the passwords hard to guess by controlling password format, and you can require users to change their passwords regularly. Several user profile parameters relate to passwords.

This discussion assumes that you require users to select and maintain their own passwords. No one in IT should ever know user passwords. When administrators ask us why they can't see clear-text passwords on the system, I ask them, "Why would you want to?" If a breach occurred and you, as an administrator, knew every user's password or had the ability to discover it, you would be put in the position of having to prove that you didn't commit the breach. In other words, accountability for who did what on your system would be gone. IBM i does not permit the retrieval of user passwords into a clear-text form, and you as an administrator should not maintain a list of clear-text passwords.

For the PASSWORD parameter, you can specify a value of *NONE, *USRPRF, or the password itself. *NONE means that the user profile can't sign on to the system. Unfortunately, the default parameter value, *USRPRF, creates a password that is the same as the user profile name. This is what is known as a "default" password. Default passwords constitute a huge security exposure. I prefer profiles to be created with the password parameter set to *NONE — meaning, no password. That way, if the person for whom the profile was created never shows up for work, you don't have user profiles on the system that have default passwords. In addition, profiles such as group profiles and profiles that own objects are not meant to represent individuals and, therefore, should not be used for sign-on. Creating those profiles with PASSWORD(*NONE) enforces this policy.

Security Recommendation

Consider changing the CRTUSRPRF command's default value for the PASSWORD parameter from *USRPRF to *NONE. When a profile is created for an individual, have the individual call the help desk to have a password assigned to the profile.

Security Recommendation

IBM-supplied profiles QPGMR, QSRV, QSRVBAS, QSYSOPR, and QUSER can have passwords, but IBM ships these profiles with PASSWORD(*NONE) so that no one can use them to sign on. In early releases of OS/400, these profiles were shipped with default passwords that were the same as the profile name. Those passwords were common knowledge in the hacker community and, if left unchanged, were a very serious security exposure. To ensure that none of these profiles have passwords on your system, analyze the password section of the report produced by the PRTUSRPRF (Print User Profile) command.

The QSECOFR profile is still shipped with a default password, but the first time you sign on as QSECOFR, you must change the password. Use the ANZDFTPWD (Analyze Default Passwords) command to determine whether QSECOFR (as well as any other profile) has a default password.

PWDEXP (Set Password to Expired)

The PWDEXP parameter lets you set the user profile's password to the expired state. When you create new user profiles, if you don't set the password to *NONE, you should specify PWDEXP(*YES) so that new users are prompted to choose a new password the first time they sign on. You should also specify PWDEXP(*YES) when passwords are reset for users who have forgotten theirs. (IBM i doesn't allow a profile to be created or changed to have password *NONE and password expired set to *YES.)

Security Recommendation

When giving a user a password — either for the first time or when the user calls in because he or she has forgotten the password — specify PWDEXP(*YES) to ensure that the user changes the initial password when he or she signs on.

LCLPWDMGT (Local Password Management)

The LCLPWDMGT (Local password management) parameter comes into play for users who exploit the single sign-on capabilities of IBM i. This parameter's value indicates whether the user's password will be stored on the local (IBM i) system. The default value is *YES. A value of *NO indicates that the password will be passed to other IBM products that perform password synchronization (e.g., i5/OS Integration for Windows Server) but will not be stored on the local system.

When LCLPWDMGT is set to *NO, the user profile's PASSWORD value on IBM i is automatically set to *NONE. The user won't be able to sign on directly to the IBM i system through a green-screen interface, but he or she will be able to exploit the system's single sign-on capabilities (e.g., from a Windows workstation) to access IBM i. In addition, use of the CHGPWD (Change Password) command to change the user's password will cause the password to be changed remotely, but the local system will continue to have *NONE as the PASSWORD value.

PWDEXPITV (Password Expiration Interval)

This parameter controls how often the user must change his or her password. The default value, *SYSVAL, means the user must change the password at the frequency specified in the QPWDEXPITV (Password Expiration Interval) system value. However, this value can be overridden in the user's profile. Here are two legitimate reasons for overriding this value:

- You want the user to change the password more frequently than the frequency defined in the system value. For example, a good practice is to have powerful users (such as those with *ALLOBJ special authority) change their passwords more frequently than average users (every 30 days).
- Another time to consider overriding this value is for profiles that I call "application" profiles. Examples of application profiles include profiles used to connect to the system via Open Database Connectivity (ODBC) or remote Structured Query Language (SQL) statements. While these profiles' passwords should be changed on a regular basis, having the system warn that the password is expiring and then expire the password if the password isn't changed can be very disruptive to the application, especially if the process is automated. If there is another process for changing these profiles' passwords on a regular basis, setting this parameter to *NOMAX may be acceptable.

However, is it *not* legitimate to set parameter PWDEXPITV to *NOMAX because you don't want to change your password! I have seen many systems where the administrators' and QSECOFR's password expiration interval was set to *NOMAX. This is not an acceptable security practice. Why? In addition to the fact that it violates many laws and regulations, should someone learn the password for these profiles, that person can use (and abuse) the password forever (or until caught) because the password is never changed. If you are one of those administrators who has changed your expiration interval to *NOMAX, I highly recommend you set it back to *SYSVAL or even to 30 (days) and change your password, as you're asking the rest of your users to do!

PWDCHGBLK (Block Password Change)

V6R1 introduces the PWDCHGBLK (Block password change) parameter, whose purpose is to prevent the user from changing his or her password multiple times in order to re-use the same password all the time. For example, if the system's QPWDLMTREP (Limit Repeating Characters) system value is set to 1, users must come up with 32 unique passwords before they can re-use their current passwords. Some users have realized that they can simply change their password 32 times in rapid succession and be able to re-use the same password. In fact, they have often written scripts to automate this process. The net effect is that the password never changes. The PWDCHGBLK user profile parameter lets you control how much time (in hours) must pass between password changes, defeating the intentions of those users with the automated scripts. (The V6R1 system value QPWDCHGBLK, or Block Password Change, provides similar functionality at the system value level.)

This parameter affects only interfaces that let users change their password themselves, such as the CHGPWD command or the QSYCHGPW (Change Password) API. You can always change user passwords through the CHGUSRPRF command if there is a legitimate need to reset or change a password.

STATUS (Profile Status)

The STATUS parameter specifies whether a user profile is enabled or disabled for sign-on. When the value is *ENABLED (the default), the system lets the user sign on. If the value is *DISABLED, the system doesn't let the user sign on until the profile is re-enabled by a user with authority to do so (i.e., change the value of STATUS to *ENABLED).

This parameter is used with the QMAXSGNACN (Maximum Sign-on Action) system value. When you set QMAXSGNACN to 2 or 3, the system disables a profile that attempts too many invalid sign-ons. (System value QMAXSIGN, or Maximum Number of Sign-on Attempts, determines how many attempts are allowed.) When the system disables a profile, the system changes the value of the STATUS parameter to *DISABLED. Unlike some operating systems, a profile must be reset to STATUS(*ENABLED) by a user with sufficient authority. The profile will not automatically be reset after a period of time.

In addition, many organizations have a policy that requires inactive profiles to be set to STATUS(*DISABLED) after a certain number of days; then, after another period of time, they are to be removed from the system. This practice ensures that profiles that are no longer needed are removed from the system in a timely way.

Technical Note

If the QSECOFR profile is disabled, you can always sign on as QSECOFR at the console and enable the profile (assuming, of course, that you know the password).

USRCLS (User Class) and SPCAUT (Special Authority)

The USRCLS and SPCAUT parameters work together to determine the special authorities granted to the user. Special authorities give the user the right to perform a function or task, such as being able to save or restore objects, control (end or re-prioritize) any job on the system, work with any spooled file on the system, work with user profiles, or configure TCP/IP. I'll discuss the details of the special authorities themselves in a moment. First, let's look at how the USRCLS and SPCAUT parameters work.

The USRCLS parameter lets you classify users by type according to their special authorities. Five user classes are recognized on IBM i:

- *SECOFR (security officer)
- *SECADM (security administrator)
- *PGMR (programmer)
- *SYSOPR (system operator)
- *USER (user)

Technical Note

The user class parameter is used to default the special authorities when a profile is created. It is also used to determine what menu options appear on IBM menus displayed to the user. Last, the user class determines the special authorities that the profile should have when the system moves from QSECURITY level 20 to level 30 or higher. Note that the parameter is *not* used when the system checks to see whether the user is allowed to access an object.

When you assign a profile to a user class, the profile by default is granted the special authorities associated with that class. Table 4.1 lists the default special authorities associated with each user class under security levels 30, 40, and 50. You can override these special authorities using the SPCAUT parameter, but you'll find that the default authorities are usually appropriate.

Table 4.1
User Classes (Levels 30, 40, and 50)

User class	Special authorities
*SECOFR	*ALLOBJ, *SECADM, *SAVSYS, *JOBCTL, *SERVICE, *SPLCTL, *AUDIT, *IOSYSCFG
*SECADM	*SECADM
*PGMR	No special authorities
*SYSOPR	*SAVSYS, *JOBCTL
*USER	No special authorities

Table 4.2 lists the permitted SPCAUT values and the meaning of each. The default SPCAUT value is *USRCLS, which instructs the system to refer to the USRCLS parameter and assign the predetermined set of special authorities.

Table 4.2
SPCAUT Parameter Values

Value	Description
*USRCLS	The user class determines the set of special authorities for the user profile. This value is the default.
*NONE	No special authorities are assigned.
*ALLOBJ	The user is allowed to access any object on the system. A user with *ALLOBJ special authority cannot be denied access to any object. Thus, *ALLOBJ should be given to only a few very trusted individuals.
*AUDIT	The user is allowed to manage all aspects of auditing on the system, including changing the auditing system values and running the auditing commands.
*IOSYSCFG	The user is allowed to configure TCP/IP and all other aspects of communications on the system.

continued...

continued...

*JOBCTL	The user can change, display, hold, release, cancel, and clear all jobs on the system. Spooled files must be in an output queue that specifies OPRCTL(*YES).
*SAVSYS	The user can save, restore, and free storage for all objects — regardless of whether the user has the authority to access or manipulate the objects.
*SECADM	The user can create and change user profiles.
*SERVICE	The user can perform functions in the System Service Tools.
*SPLCTL	The user can delete, display, hold, and release all spooled files on the system.

Technical Note

Navigator for i refers to special authorities as *privileges*. For an example of a privilege in Navigator, see the Navigator discussion later in this chapter.

You can override the default SPCAUT setting by typing from one to eight individual special authorities that you want to assign to the user profile. After sending a message that the special authorities assigned don't match the user class, the system creates the user profile as requested. (Just ignore the message!)

When granting special authorities, you want to do so based on the user's job responsibilities or role and grant only the authority the user requires to do his or her job — not more. Many special authorities give the holder great power and control over all or parts of your system. For example, *ALLOBJ special authority gives a user unlimited access to and control over *all* objects — a user with *ALLOBJ special authority can perform any function on any object on your system. Obviously, you want to give this special authority only to someone whose job responsibility is to administer the system.

To appropriately assign special authorities, you need some understanding of the user's job function or role in the organization. In large organizations, new profiles are usually created with no special authorities unless they fall into a specific job category, such as system operator or user profile administrator. The roles and their special authorities must be predefined; it is not up to the new user or manager requesting the profile to determine which special authorities the user is to be assigned.

You should design your security implementation so users are assigned only the special authority that they need to perform their job functions. Avoid — at all costs — an implementation where all users require *ALLOBJ special authority. *ALLOBJ should be assigned only to system administrators — not programmers, not operators, and definitely not end users. In later chapters, I discuss how to provide these users with the capabilities they need to perform their job responsibilities without giving them *ALLOBJ.

Security Recommendation

Applications running on your system shouldn't require a profile with *ALLOBJ authority to manage the application or its objects. For example, vendors should not have to have an *ALLOBJ profile on your system that they use to sign on and maintain their application. If they insist on this, it is because they are lazy and just want to make sure they don't run into any security issues when signing on your system. Do not let them talk you into having a profile with *ALLOBJ unless they are administering the system itself. We'll discuss other ways to manage access for application maintenance in later chapters.

*SECADM special authority enables a user profile to create and maintain user profiles without requiring the profile to be in the *SECOFR user class or giving it *ALLOBJ authority.

The *SAVSYS special authority lets a user save off or restore any object to the system regardless of whether he or she has authority to access the object. Although *SAVSYS may seem powerful, your alternative is to give your operations staff *ALLOBJ special authority so they have sufficient authority to save all the objects on the system. The better option is obviously to give them *SAVSYS instead. However, *SAVSYS authority is not without risks. If you secure a library (set its *PUBLIC authority to *EXCLUDE) but don't secure a file within it that contains confidential information, a user with *SAVSYS could save the file out of the secured library and restore it into a library to which they have authority and gain access to the confidential data.

*IOSYSCFG special authority lets the user create, delete, and manage devices, lines, and controllers and configure and start any TCP/IP server, such as a Web server. *IOSYSCFG has become a powerful authority that you should assign with care.

*SPLCTL special authority is the "*ALLOBJ" of spooled files and should be granted sparingly. With this authority, a user can delete, display, hold, and release all spooled files on the system. There is no way to prevent a user with *SPLCTL from accessing a spooled file. I usually see *SPLCTL given to users because one person has created a report that another person needs to see. Without *SPLCTL authority, IBM i does not allow this. Unfortunately, because *SPLCTL lets users see *any* spooled file on the system, granting this authority when you need to keep some spooled files confidential or private is not appropriate. In Chapter 8, I describe other methods for sharing spooled files without giving users *SPLCTL.

*SERVICE is another special authority you should guard. *SERVICE lets a user perform a communications trace (which includes clear-text passwords from a 5250 emulation sign-on screen), debug a program, and configure and manage logical partitions (LPARs). Fortunately, you now need a service tools ID as well as *SERVICE special authority to be able to access the service tools. This restriction limits the power of *SERVICE special authority. In addition, one of the host functions within Application

Administration (a component of Navigator for i that we discuss in Chapter 6) lets you selectively permit certain users to perform these functions without having to give them *SERVICE special authority. Consider taking this approach rather than giving a user the full capabilities associated with *SERVICE special authority.

Security Recommendation

One of the IBM-supplied profiles, QSRV, has *SERVICE special authority. I often see this profile with a password and find out, after questioning the administrator, that every IBM service person knows this password. A more secure method is to set the profile's password to *NONE and assign a password only when IBM comes on site. When the service person leaves, set the password back to *NONE.

Initial Sign-On Options

Three user profile parameters work together to determine the user's initial sign-on options. The CURLIB (Current library), INLPGM (Initial program to call), and INLMNU (Initial menu) parameters determine the user profile's current library, initial program, and initial menu, respectively. Together, these parameters establish the user's environment when the user signs on to the system. The menu or program that is run at sign-on determines which menus and programs are available to the user. Let's look at two examples.

First, consider the user profile USER, which has the following values:

```
Current library . . . . . . . . . CURLIB     ICLIB_____
Initial program to call . . . . . INLPGM     *NONE_____
  Library . . . . . . . . . . .              __________
Initial menu  . . . . . . . . . . INLMNU     ICMENU____
  Library . . . . . . . . . . .                ICLIB_____
```

When USER signs on, the user's current library is set to ICLIB, and the display shows menu ICMENU from library ICLIB.

Now, let's say USER instead has these values:

```
Current library . . . . . . . . . CURLIB     ICLIB_____
Initial program to call . . . . . INLPGM     ICUSERON__
  Library . . . . . . . . . . .                SYSLIB____
Initial menu  . . . . . . . . . . INLMNU     *SIGNOFF__
  Library . . . . . . . . . . .              __________
```

When USER signs on in this scenario, ICLIB is again the user's current library, and the system runs program ICUSERON from library SYSLIB. But in this case, the initial menu is set to *SIGNOFF. When a user signs on, the initial program is run first, and then the initial menu is presented. If the initial menu is set to *SIGNOFF, the user's session is signed off once the initial program is completed.

LMTCPB (Limit Capabilities)

Parameters CURLIB, INLPGM, and INLMNU are significant to security because the sign-on screen as shipped with the system permits users to change the values of these parameters at sign-on. Users can also run commands when presented with a command line. Obviously, permitting all users to make these changes and run commands isn't a good security practice. The LMTCPB (Limit capabilities) user profile parameter lets you control these user functions. Specifically, it controls users' ability to

- Define (using the CHGPRF, or Change Profile, command) or change (at sign-on) their own initial program
- Define (using CHGPRF) or change (at sign-on) their own initial menu
- Define (using CHGPRF) or change (at sign-on) their own current library
- Define (using CHGPRF) or change (at sign-on) their own attention key program
- Run IBM i or user-defined commands from a command line

Table 4.3 shows the effects of the possible values for LMTCPB.

Table 4.3
Effects of the LMTCPB Parameter

	User can define, change, and execute				
LMTCPB	*Initial program*	*Initial menu*	*Current library*	*Attention program*	*Commands*
*NO	Yes	Yes	Yes	Yes	Yes
*PARTIAL	No	Yes	No	No	Yes
*YES	No	No	No	No	Limited[1]

[1] *A user with LMTCPB(*YES) is limited to the SIGNOFF, SNDMSG (Send a Message), DSPMSG (Display Messages), DSPJOB (Display Job), DSPJOBLOG (Display Job Log), and STRPCO (Start PC Organizer) commands and to other commands that you create or change to specify ALWLMTUSR(*YES).*

Technical Note

If you want to permit users with LMTCPB(*YES) to use a particular command, you can change the command using the CHGCMD (Change Command) command, specifying ALWLMTCPB(*YES).

Most users on production systems should have their LMTCPB parameter set to *YES. The profiles that need LMTCPB(*NO) are typically in a role that needs to run many commands from the command line. These may include users in the administrator, database administrator, or system analyst roles. In addition, the programmers that need to debug production issues may also need LMTCPB(*NO).

Technical Note

Although the LMTCPB parameter possesses value for green-screen menu environments and should continue to be used, current technology has limited its effectiveness. Many interfaces — such as SQL, ODBC, remote command, and Distributed Data Management (DDM) — bypass menus altogether. The limited-capabilities setting is ignored in most cases when the user doesn't actually sign on at a menu. Don't make the mistake of thinking that limited capabilities provide more protection than they do.

Security Recommendation

Change the LMTCPB parameter to *YES unless the user specifically needs the additional capabilities provided by LMTCPB(*NO). Don't bother using the value of *PARTIAL. Very little benefit is obtained by this setting, and I — as well as most administrators — find it more confusing than helpful.

System Value Overrides

As I mentioned in Chapter 3, you can override system values QDSPSGNINF (Display User Sign-on Information), QPWDEXPITV (Password Expiration Interval), and QLMTDEVSSN (Limit Device Sessions) by using the user profile parameters DSPSGNINF (Display sign-on information), PWDEXPITV (Password expiration interval), and LMTDEVSSN (Limit device sessions), respectively. From Figure 4.1, you can see that these three parameters all have the default value *SYSVAL, which lets the respective system value control these functions. To override a system value, specify the desired value in the associated user profile parameter. The available choices are the same as those for the system values themselves.

Group Profiles

All the parameters we've discussed so far are used to define profiles for individual users. The next few parameters — GRPPRF (Group profile), OWNER (Owner), GRPAUT (Group authority), GRPAUTTYP (Group authority type), and SUPGRPPRF (Supplemental groups) — let you associate an individual user profile with one or more *group profiles.*

IBM i provides group profiles as a tool to make security administration easier. It's also the method you'll use to implement "role-based" access. When you have a set of user profiles that all need the same capabilities or require access to the same objects, you'll want to define a group profile for those users (role). Examples of roles include the operations staff, all users in the payroll department, and programmers. When you assign a special authority to a group profile or give a group profile access to an object, everyone in that role has the special authority or object access. Therefore, it is much simpler to assign the capability or access to the group rather than to each individual in the group. User profile creation and maintenance are also easier. For example, if a new operator is hired, all you have to do is put that user in the Operator group.

Doing so gives that person all the capabilities and access he or she needs to perform the responsibilities of the operator role, and it is much easier than assigning them individually. If a user changes position within the company, you can simply add that user to the group associated with the new role and remove him or her from the group associated with the old role.

Creating Group Profiles

Before you can make a user a member of a group, you first must create the group profile. The confusing part of this step is that you create a group profile using the same command you use to create a user profile — that is, CRTUSRPRF. The following list shows the CRTUSRPRF parameters that are especially meaningful for group profiles.

- USRPRF — Carefully consider the name of the group profile. I like to use a naming convention that makes it very obvious that the profile is a group profile as well as what role it represents (e.g., GRP_OPER, GRP_PGMR, GRP_HR).
- PASSWORD — The next parameter to pay attention to is PASSWORD. Group profiles are created to represent a role or a set of users, not an individual user; therefore, you don't want this profile used for sign-on. To enforce this rule, set the profile to have no password — in other words, set PASSWORD to *NONE.
- PWDEXPITV — Leave the password expiration interval at its default of *SYSVAL. Some people think this setting will cause a problem when the password is *NONE, but that's not the case. When the password is *NONE, the system never checks the expiration interval.
- LMTCPB — You can set the LMTCPB parameter to *YES, but because the profile shouldn't be used for sign-on, it shouldn't matter what the setting is. However, if you are generating reports that list the users with LMTCPB(*NO), setting a group profile's LMTCPB parameter to *YES will mean one less profile on the list (and, therefore, one less profile to justify/explain to an auditor).
- TEXT — Use the Text field to further explain the purpose of the group profile.
- SPCAUT — Assign the special authorities required to accomplish the tasks of the role. These special authorities may or may not match the user class in which you place the group. The important thing is to assign the exact special authorities required by the role — no more and no fewer.
- GRPPRF — You cannot nest groups. In other words, a group profile cannot be a member of another group. Therefore, when you create a group profile, the GRPPRF parameter must be set to *NONE.

Assigning a Group to a User Profile

Now that you've created the group profile, you'll want to assign users to the group. In other words, you are making particular users members of the group.

Use the CRTUSRPRF command's GRPPRF parameter to specify the name of the group that represents the user's primary role. For users who perform more than one role in your organization, you can use the SUPGRPPRF parameter to specify up to 15 additional groups. (You may be wondering which access the system checks first, the user's individual access or the group's. I discuss this topic in detail in Chapter 6.)

Three other user profile parameters work with the GRPPRF parameter: OWNER, GRPAUT, and GRPAUTTYP. Let's take a look at how these parameters interact.

By default, when a user creates an object, the user owns the object. However, you may choose to have the user's group own all objects that are created by the members of the group. To set up this configuration, specify *GRPPRF for the user profile's OWNER parameter.

There are times when this approach is an efficient and desirable one. For example, when you have a set of developers all working on the same project, you may want to have their group profile (e.g., GRP_PGMRHR) own all the objects that each developer is currently working on. Another example is when everyone in a department needs to access and manipulate the same queries and their work files. However, this configuration can be a very dangerous security scenario. That's because when the group owns an object, all members of the group also own the object, which means that all group members have *ALL authority to those objects. The danger comes when a group profile owns a production application. That is, each user is a member of the profile that owns the application, and, therefore, all application users have *ALL authority to all application objects — including the production application's data files. This is a configuration you want to avoid.

Technical Note

The OWNER, GRPAUT, and GRPAUTTYP parameters apply only to the group specified in the GRPPRF parameter, not to any group profiles specified in the SUPGRPPRF parameter.

Technical Note

Objects created in the integrated file system (IFS) are owned by the user profile regardless of the value of the OWNER parameter.

If you don't want the group owning the objects, but other group members need authority to the objects, you can have the system grant a private authority to the group. The GRPAUT parameter specifies the authority granted to the group profile when the value for OWNER is *USRPRF. Valid values for GRPAUT are *ALL, *CHANGE, *USE, *EXCLUDE, and *NONE. The first four values are authorities, which we'll discuss in detail in Chapter 6. If you specify one of these four values for the GRPAUT parameter, the individual user profile that creates an object owns it, and the group profile has the specified authority. (Note that you must leave the GRPAUT parameter set to *NONE if you specify OWNER(*GRPPRF). In other words, you can have the group either own the objects or be authorized to them, but you can't do both.)

The GRPAUTTYP parameter determines the type of authority (i.e., private authority or primary group authority) that the user's group profile has to objects the user creates. If you specify GRPAUTTYP(*PRIVATE), IBM i grants the group profile the private authority specified in the GRPAUT parameter. If you specify GRPAUTTYP(*PGP), IBM i assigns the group profile specified in the GRPPRF parameter as the primary group for the new object, and the primary group authorities will be those designated by the GRPAUT parameter's value.

Assigning private authorities or primary group authorities yields the same results in terms of authorization; however, the performance implications differ slightly. For more information about using primary group authority, see Chapter 6.

UID (User Identification Number) and GID (Group Identification Number)

The way the IFS was implemented, when a user accesses an object, the system uses the user's User Identification (UID) number rather than the profile name to verify the user's authority to access objects in the IFS. IBM chose this implementation to comply with the POSIX standards and make it easier to port applications written for Unix systems to run on IBM i. When you create a user profile, IBM i automatically assigns a UID to the profile. Each user must have a unique UID. If you choose to assign users a UID rather than have the system generate one, the UID must be a number between 1 and 4294967294, and no other profile can already have that number.

The Group Identification Number (GID) is identical in purpose to the UID except that it is assigned only to group profiles.

AUT (Authority)

The AUT parameter is the *PUBLIC authority setting on the user profile itself. By default, IBM i sets the user profile to *PUBLIC *EXCLUDE. I do not recommend changing this value.

I've seen numerous situations where the *PUBLIC authority of some or all profiles was set to *USE. With this authority setting, anyone with a sign-on to the system can submit a job as one of these profiles or use a job description that names one of these profiles. This capability lets users run as a more powerful user than themselves. To ensure that one user cannot abuse the power of another profile's capabilities and access, it's vital that you leave the *PUBLIC authority of profiles at *EXCLUDE.

In some cases, however, certain users need to see profiles' attributes. This is often the case when help desk personnel or programmers are debugging production issues and need to view users' group profiles, special authorities, and text descriptions. In this case, you can grant the group profiles of those users (e.g., GRP_HELP, GRP_PGMR) *READ authority to all profiles. This authority lets the group members see the attributes of a profile but not submit a job or otherwise use the profile. While some organizations choose to set the *PUBLIC authority of profiles to *READ, most organizations don't want to expose all profiles' attributes to being viewed by all other users on the system.

Table 4.4
Tools for Managing User Profiles

Security tool	Description
ANZDFTPWD (Analyze Default Passwords)	Prints a report listing each user profile whose password is equal to the user profile name. Such profiles aren't secure, and they require attention.
ANZPRFACT (Analyze Profile Activity)	Checks user profiles' last sign-on date and last-used date every day and disables any profiles that haven't been used in a specified number of days (you supply the number). However, some applications that use profiles don't update the last sign-on or last-used date, making the profile look as though it's been inactive. To exempt a profile from ANZPRFACT processing, use the CHGACTPRFL command to add the profile to the exemption list.
CHGACTPRFL (Change Active Profiles List)	Lets you add and remove user profiles from the ANZPRFACT command's active profile (i.e., exemption) list. ANZPRFACT ignores any profiles that are on the exemption list.
DSPACTPRFL (Display Active Profiles List)	Lists the user profiles that are on the ANZPRFACT exemption list.
DSPACTSCD (Display Activation Schedule)	Lists the schedule information that controls enabling and disabling user profiles by date and time. This information includes the user profile and the valid inactive day and time you've specified for that user on the CHGACTSCDE command.
CHGACTSCDE (Change Activation Schedule Entry)	Creates and maintains user profile entries and their activation day and time periods. For each user profile, you can schedule days and times for which the user profile is inactive, thus controlling that user's system use.
DSPEXPSCD (Display Expiration Schedule)	Lists user profiles that are scheduled to be removed or disabled.
CHGEXPSCDE (Change Expiration Schedule Entry)	Lets you create and maintain a list of user profiles that you want to remove or disable at some point in the future. This tool makes managing temporary profiles easier by scheduling their removal so you don't have to remember to remove them. Each day at one minute after midnight, the system runs a job that examines this list and takes the appropriate action.
PRTUSRPRF (Print User Profile)	Lets you print information about user profiles based on selection criteria that you supply (e.g., special authorities, user class, mismatch between user class and special authorities). You can print authority information, environment information, and password information.
PRTPRFINT (Print Profile Internals)	Lets you print information about how "full" a user profile is. User profiles contain an entry for each object owned, each private authority the user has, each private authority another user has to objects this user owns, and each primary group authority assigned. If there are too many of these entries, the user profile becomes full. (If a user profile is full, it can't own more objects, be authorized to additional objects, allow any objects it owns to be authorized to another user, or become the primary group of an object. This command produces a report indicating how full the profile is and the percentage of each type of entry.

Private Authorities and User Profiles

When you create a user profile, IBM i gives the profile *CHANGE plus *OBJMGT (object management) authority to itself. When you make a profile a member of a group profile, IBM i grants the member profile *OBJOPR (object operation), *OBJMGT, *READ, *ADD, *UPDATE, and *DLT authorities to the group profile. In both cases, these private authorities are required for normal system operations and should not be removed. For example, if you remove the authority to a user's group profile, the user won't be able to sign on. Likewise, you shouldn't add authorities. I've seen some organizations add *EXECUTE to the authorities that users have been given to their group profile. This authority lets each group member submit a job that runs as the group profile rather than being restricted to running jobs only under their own profile.

You'll learn more about private authorities in Chapter 6.

Helpful Tools

Several security tools can help you manage user profiles on your system. Table 4.4 lists and describes each of these commands. You can find these tools on the SECTOOLS and SECBATCH menus.

Navigator for i

You can use IBM Navigator for i instead of the CRTUSRPRF command to create user profiles. To do so, expand the directory for the system on which you want to create the user, and select Users and Groups; then right-click All Users, and select New User. Figure 4.2 shows the New User configuration window.

Figure 4.2
New User Configuration Window

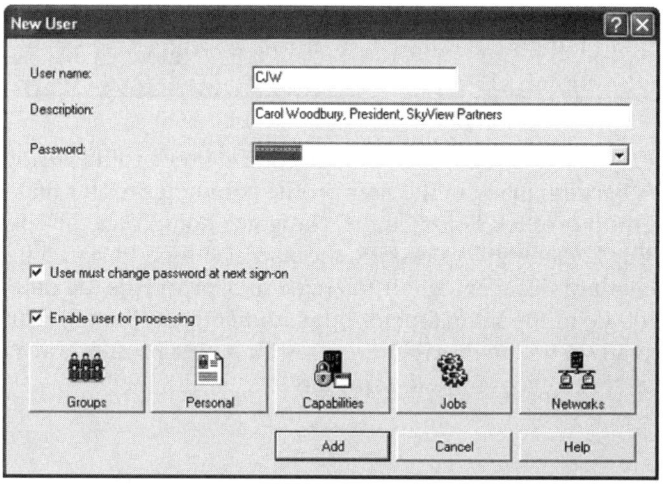

When you use this interface, you'll notice that Navigator's terminology for user profile attributes differs from that of the CRTUSRPRF command. For example, the option labeled "User must change password at next sign-on" is equivalent to the CRTUSRPRF command's PWDEXP (Set password to expired) parameter. You might find this difference a bit confusing if you're moving between Navigator for i and green-screen commands, but IBM's intent with Navigator is to better describe the attribute.

In contrast to the way you create a group profile through the command-line ("green-screen") interface, Navigator provides a separate interface for creating group profiles. To create a new group profile, expand the directory for the system on which you want to create the new group profile; then select Users and Groups, right-click Groups, and select New Group. Figure 4.3 shows the New Group configuration window.

Figure 4.3
New Group Configuration Window

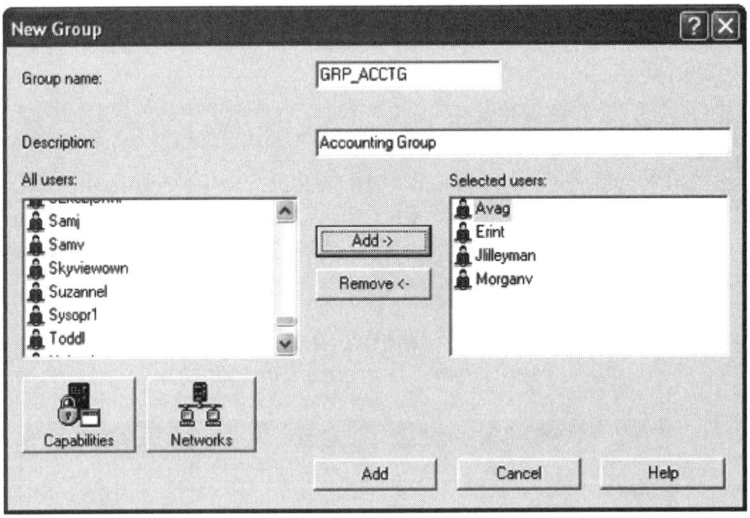

Notice that Navigator doesn't include most of the user profile attributes in this interface. That's because many of the user profile parameters don't make sense for or don't apply to group profiles. For example, Navigator doesn't let you specify a password for a group profile because group profiles shouldn't be used to sign on.

Another benefit of using Navigator to create user profiles is the other things Navigator lets you do at the same time or does automatically for you. For example, it automatically creates a system distribution entry for a user profile. You can also choose to create a Lotus Notes client for a user profile.

If you need to create the same user on multiple systems, you can use Management Central (a component of Navigator for i that lets you manage tasks across multiple servers simultaneously) to get that profile created on all systems being managed by Management Central.

Copying User Profiles

Although most of this chapter has discussed creating user profiles "from scratch," in reality I have found very few organizations that create user profiles by using the CRTUSRPRF command or Navigator's New User interface. Instead, most organizations copy an existing user profile or base the new profile on an existing profile. Often, the profile being copied is the profile of another user in the same area of the organization. Although convenient, this method often propagates special authorities and data access to the new profile even when they are not required for the new user to complete his or her job requirements. That's because many existing profiles were created with more special authorities than necessary. And in most cases, organizations don't go back and remove the excess special authorities. So when an existing profile is copied, the excess capabilities are propagated to the new profile.

A slightly different method that avoids this propagation issue copies a profile that you've created as a "template." The template profile has only the necessary special authorities (if any) and accesses. I recommend you create a template profile for each role on your system. This approach to creating user profiles ensures that each profile gets created with only the capability and access required for that role — and no more.

Validation List Users

Let's turn our attention now to one of the other types of user identifies I listed at the beginning of this chapter: validation list users. Validation list users are defined and managed in validation list (*VLDL) objects by applications and the Apache Web server. The Apache server refers to these users as "Internet users." Validation list users can be authenticated by applications, but they cannot be used as the profile under which an IBM i job is run. The purpose of a validation list user is to give applications a way to define application users or functions without requiring the user to have an IBM i user profile.

Validation list users are useful for certain applications, especially Internet applications, where you need to authenticate users (e.g., customers, members, or frequent Web site visitors) who don't work for your organization and therefore have no need for a full IBM i user profile. Because validation list users can't be used to run jobs on your system, it is more secure for applications to use validation list users rather than user profiles.

A validation list user is an entry in a validation list object. The entry consists of an identifier, data that is encrypted when stored in a validation list entry (encrypted data), and other unformatted data (entry data). Any of these fields can be defined in any coded character set ID (CCSID) in the range of 1 to 65535 and can use any available character. An identifier can be up to 100 characters long. Encrypted data can be up to 600 characters. The entry data field can be up to 1,000 characters. The encrypted data field is intended for applications to use to store passwords. This field is sometimes used to store other kinds of sensitive information.

Applications are responsible for managing all aspects of validation list users, including password composition rules, aging, and expiration and invalid sign-on attempts. Applications can choose whether passwords in validation list entries can be used only for password verification (i.e., one-way encrypted) or can also be returned in unencrypted form (two-way encrypted).

No commands exist to add, remove, change, or verify encrypted data; only APIs are provided. Figures 4.4A through 4.4D show the code for command PRTVLDL, which uses the validation list APIs to retrieve the user attributes. Figure 4.4E shows the entry screen presented by the command, and Figure 4.4F shows the results of running the command against a sample validation list.

Figure 4.4A

CL Command PRTVLDL

```
/*==================================================================*/
/*   Author: Paul Tuohy, ComCon  - 2003                            */
/*==================================================================*/

Cmd  'Print Validation List'

   Parm  VLDL  Q1                                              +
         Min( 1 )                                              +
         Prompt( 'Validation List')

Q1:                                                            +
       Qual  *NAME  10                                         +
             Min( 1 )                                          +
             Expr( *YES )
       Qual  *NAME  10                                         +
             Dft( *LIBL )                                      +
             SpcVal( ( *LIBL ) )                               +
             Expr( *YES )                                      +
             Prompt( 'Library' )
```

Figure 4.4B
DDS for Command PRTVLDL

```
A*----------------------------------------------------------------------
A*    Author: Paul Tuohy, ComCon  - 2003
A*----------------------------------------------------------------------
A          R HEAD
A                                         SKIPB(5)
A                                       13DATE
A                                         EDTCDE(Y)
A                                       53'List Validation List Entries'
A                                      116'Page'
A                                       +1PAGNBR
A                                         SPACEA(2)
A                                        3'Validation List:'
A            QUALNAME      21    0      21SPACEA(2)
A                                        3'Create'
A                                       15'Last'
A                                         SPACEA(1)
A                                        3'Date'
A                                       15'Used'
A                                       28'User Id'
A          R DETAIL
A                                         SPACEB(1)
A            PRTCREATE        L  0       3DATFMT(*USA)
A N90        PRTUSED          L  0      15DATFMT(*USA)
A  90                                   15'Not Used'
A            USERID       100    0      28SPACEA(1)
A          R FOOTER
A                                         SPACEB(1)
A                                       54'***** End of Report *****'
A                                         SPACEA(1)
```

Figure 4.4C
RPG IV Prototypes for Command PRTVLDL

```
  //--------------------------------------------------------------------
  //   Author: Paul Tuohy, ComCon - 2003
  //--------------------------------------------------------------------

  // Prototypes and structures for PRTVLDL
  //
  // Standard API Error Data Structure

D APIError        DS
D  EBytesP                      10I 0 INZ(40)
D  EBytesA                      10I 0
D  EMsgID                        7A
D  EReserved                     1A
D  EData                        40A
  // Prototype for Call to System API
  //     QWCCVTDT - Convert Date and Time Format
  //     Refer to API Documentation for Parameter Descriptions.

D ConvertDate     PR                  ExtPgm('QWCCVTDT')
D  InputFormat                  10    Const
D  InputVar                      8    Const
D  OutputFormat                 10    Const
D  OutPutVar                    17
D  Error                              Like(APIError)
```

continued...

continued...

```
  //------------------------------------------------------------------
  //
  // The following are the structures, parameters, and prototypes for
  // the validation list APIs,
  //          QsyFindFirstValidationLstEntry
  //          QsyFindNextValidationLstEntry
  //          QsyFindValidationLstEntryAttrs
  //
  //     Refer to API Documentation for Parameter Descriptions.
  //     Refer to QSYSINC/H(QSYVLDL) for C equivalents.

  // Parameter Fields and Structures

D Validation_Lst  S             20

 // Qsy_Entry_Id_Info
D Entry_Id        DS                  Qualified Align
D  Len                          10I 0
D  CCSID                        10U 0
D  EntryId                     100

D Entry_Id_Info   DS                  LikeDS(Entry_Id)

 // Qsy_Entry_Encr_Data_Info_T
D Encr_Data_Info  DS                  Qualified Align
D  Len                          10I 0
D  CCSID                        10U 0
D  Data                        600

 // Qsy_Entry_Data_Info_T
D Entry_Data_Info...
D                 DS                  Qualified Align
D  Len                          10I 0
D  CCSID                        10U 0
D  Data                       1000

 // Qsy_Rtn_Vld_Lst_Ent_T
D Rtn_Entry       DS                  Qualified Align
D  EntryIdInfo                        Like(Entry_Id)
D  EncrDataInfo                       Like(Encr_Data_Info)
D  EntryDataInfo                      Like(Entry_Data_Info)
D                                4
D  MoreInfo                       *

 // Qsy_Attr_Info_T
D Rtn_Attributes  DS                  Qualified
D  NumberAttrs                  10I 0 Inz(1)
D                               12
D  AttrDesc                           Like(Attr_Desc)

 // Qsy_Attr_Descr_T
D pAttr_Desc      S               *
D Attr_Desc       DS                  Qualified
D                                     Based(pAttr_Desc)
D  Location                     10I 0
D  Type                         10I 0
D  Res1                          8
D  AttrId                         *
D  OtherDescr                   32
D  DataInfo                     96
D  InVLDL                             Like(In_VLDL)
D                                     Overlay(DataInfo)
D  OtherData                    32
```

continued...

continued...

```
 // Qsy_In_VLDL
D pIn_VLDL         S               *
D In_VLDL          DS                  Qualified
D                                      Based(pIn_VLDL)
D  CCSID                         10I 0
D  Len                           10I 0
D  Res1                           8
D  AttrValue                       *

 // Qsy_Rtn_VLDL_Attr_T
D Rtn_VLDL_Attr    DS                  Qualified Align
D  BytesReturn                   10I 0
D  BytesAvail                    10I 0
D  AttrLen                       10I 0
D  AttrCCSID                     10I 0
D  AttrData                      28

 // Qsy_Rtn_Entry_Usage_Attr_T
D Rtn_Rtn_Entry_Usage...
D                  DS                  Qualified Align
D  CreateDate                     8
D  LastUsedDate                   8
D  EncrDataChgDt                  8
D  NotVldVerify                  10I 0


 //
 // Prototypes
D FirstLstEntry    PR            10I 0 ExtProc('QsyFindFirst-
D                                      ValidationLstEntry')
D  Qsy_Qual_Name_T...
D                                20    Const
D  Qsy_Rtn_Vld_Lst_Ent_T...
D                                      Like(Entry_Id)

D NextLstEntry     PR            10I 0 ExtProc('QsyFindNextValidation-
D                                      LstEntry')
D  Qsy_Qual_Name_T...
D                                20    Const
D  Qsy_Entry_ID_Info_T...
D                                      Like(Entry_Id_Info) Const
D  Qsy_Rtn_Vld_Lst_Ent_T...
D                                      Like(Entry_Id)

D GetVldLstEntry   PR            10I 0 ExtProc('QsyFindValidationLstEnt-
D                                      ryAttrs')
D  Qsy_Qual_Name_T...
D                                20    Const
D  Qsy_Entry_ID_Info_T...
D                                      Like(Entry_Id) Const
D  Qsy_Rtn_Vld_Lst_Ent_T...
D                                      Like(Rtn_Entry)
D  Qsy_Attr_Info_T...
D                                      Like(Rtn_Attributes)

D errno            PR              *   EXTPROC('__errno')

D errno_value      S             10I 0 based(errno_ptr)
D errno_ptr        S               *
```

Figure 4.4D
PRTVLDL Command Processing Program

```
H DftActGrp(*NO) Option(*SrcStmt:*NoDebugIO) BNDDIR('QC2LE') ACTGRP(*NEW)

 //-----------------------------------------------------------------------
 //   Author: Paul Tuohy  - 2003
 //-----------------------------------------------------------------------
 //
 //   PRTVLDL - Print Validation List
 //
 //   With many thanks to Simon Coulter for help with
 //   deciphering the C structures.
 //
 //-----------------------------------------------------------------------
FPrtVLDLP  O    E             Printer OFLInd(OverFlow)

 /COPY PRTVLDLPR

 //________________________________
 // Program Fields.
D  rc               S             10I 0
D  QsyEntryUsage    S             20     Inz('QsyEntryUsage')
D  DateVar          S             17

 //*********************************************************************
 // Mainline.                                                          *
 //*********************************************************************
C      *ENTRY        Plist
C                    Parm                    Validation_Lst

 /Free

    QualName = %TrimR(%SubSt(Validation_Lst:11:10)) + '/' +
               %SubSt(Validation_Lst:1:10);
    Write Head;
   // Set the parameters for the validation list APIs
    ExSR SetParms;

   // Get the first entry in the list then
    rc = FirstLstEntry(Validation_Lst:Rtn_Entry);
    Entry_Id = Rtn_Entry.EntryIdInfo;
    DoW  rc = 0;

      // Print Details
       ExSR PrintDetail;

      // Get the next entry in the list
       rc = NextLstEntry(Validation_Lst:Entry_Id:Rtn_Entry);
       Entry_Id = Rtn_Entry.EntryIdInfo;
    EndDo;

    *InLR = *On;

  //______________________________________________________
  // Set parameters for Validation List APIs
  BegSR SetParms;

     QsyEntryUsage = %Trim(QsyEntryUsage) + x'00';
```

continued...

continued...

```
    pAttr_Desc = %Addr(Rtn_Attributes.AttrDesc);
    Attr_Desc.Location = 0;
    Attr_Desc.Type = 0;
    Attr_Desc.Res1 = *Allx'00';
    Attr_Desc.AttrId = %Addr(QsyEntryUsage);
    Attr_Desc.OtherDescr = *Allx'00';
    Attr_Desc.DataInfo = *Allx'00';
    Attr_Desc.OtherData = *Allx'00';

    pIn_VLDL = %Addr(Attr_Desc.InVLDL);
    In_VLDL.CCSID = -1;
    In_VLDL.Len = %Len(Rtn_VLDL_Attr);
    In_VLDL.Res1 = *Allx'00';
    In_VLDL.AttrValue = %Addr(Rtn_VLDL_Attr);

  EndSR;


  //_____________________________________________________________
  // Print Details for every entry
  BegSR PrintDetail;

    // Get the entry attributes
     rc = GetVldLstEntry(Validation_Lst:Entry_Id:
                         Rtn_Entry:Rtn_Attributes);

    // Store the attributes returned
     Rtn_Rtn_Entry_Usage = Rtn_VLDL_Attr.AttrData;

    // Convert dates from system format (*DTS) to printable format
     ConvertDate('*DTS':Rtn_Rtn_Entry_Usage.CreateDate:
                 '*YYMD':DateVar:APIError);
     PrtCreate = %Date(%SubSt(DateVar:1:8):*ISO0);
     ConvertDate('*DTS':Rtn_Rtn_Entry_Usage.LastUsedDate:
                 '*YYMD':DateVar:APIError);
     PrtUsed = %Date(%SubSt(DateVar:1:8):*ISO0);

    // Check for not used
     *In90 = (PrtUsed = D'1928-08-23');

     UserId = %SubSt(Entry_Id.EntryId:1:Entry_Id.Len);

     Write Detail;
     If OverFlow;
         Write Head;
         OverFlow = *Off;
     EndIf;

  EndSR;

 /END-FREE
```

Figure 4.4E
Print Validation List Command Screen

```
Session A - [24 x 80]
                              Print Validation List (PRTVLDL)

Type choices, press Enter.

Validation List . . . . . . . .                 Name
  Library . . . . . . . . . . .     *LIBL       Name, *LIBL










                                                                     Bottom
F3=Exit   F4=Prompt   F5=Refresh   F12=Cancel   F13=How to use this display
F24=More keys
Parameter VLDL required.
MA     a                 MW
```

Figure 4.4F
Validation List Entries Report

```
      3/08/03              List Validation List Entries               Page 0001
Validation List:  *LIBL/TEST
Create       Last
Date         Used           User Id
06/25/2003   Not Used       joan
06/25/2003   Not Used       mary
06/21/2003   Not Used       paul
```

Security Implications of Validation List Users

As an administrator, you need to be aware of validation list users. Find out whether any applications use them and, if so, whether the applications manage those users according to your security policy requirements.

Administrators can control whether authentication information stored in validation lists can be decrypted. If you have any reason to be concerned about the use or security of validation list users (e.g., if you think one or more may have been compromised), the QRETSVRSEC (Retain Server Security) system value gives you the control you need.

When QRETSVRSEC is set to 0 (the default), the authentication information cannot be decrypted. A value of 1 permits the information to be stored and decrypted. A value of 0 still lets the data be stored, but it cannot be decrypted.

In releases earlier than V5R2, changing QRETSVRSEC from 1 to 0 caused all data in the encrypted data field for all existing entries to be removed, and the data couldn't be recovered. Since V5R2, when you change QRETSVRSEC to 0, existing validation list user information cannot be decrypted, but the information isn't removed from the entries. If QRETSVRSEC is set back to 1, the entries can once again be decrypted.

To remove the encrypted data from all validation list entries in all validation list objects, set system value QRETSVRSEC to 0 and then run the CLRSVRSEC (Clear Server Security Data) command. Doing so will leave the entries but remove the decryptable data.

Technical Note

Data stored in validation lists as non-decryptable is not cleared by the CLRSVRSEC command and is not affected when system value QRETSVRSEC is changed.

Chapter 5

Service Tools Security

The third type of user identity I mentioned at the beginning of Chapter 4, service tools user identities, helps you secure a critical part of the system known as *service tools*. The service tools are a set of low-level tools provided by IBM i through the Dedicated Service Tools (DST), System Service Tools (SST), and IBM Navigator for i logical partitioning and disk unit management functions. Among other things, the functions available in the service tools include the capability to see or change nearly everything stored on disk!

Service functions are a set of capabilities within the service tools that are normally used for problem determination and problem solving, usually with the guidance of IBM support. Examples of service functions include the Licensed Internal Code (LIC) trace, LIC log, and display/alter/dump functions.

DST is available only from the console and can even be used when only the LIC is available and not IBM i. SST lets you access service tools and functions from IBM i by running the STRSST (Start System Service Tools) CL command. Most, but not all of the service tool features are available through SST.

Service tools are very powerful, and improper use of them can damage your server. Controlling access to the service tools is therefore vital to protecting your system and the security of your data. Service tools security consists of three primary elements: service tools user IDs, functional privileges, and device IDs. Let's start with service tools user IDs.

Service Tools User IDs

Service tools user IDs are separate, and different, from IBM i user profiles. Created through DST or SST, service tools user IDs have their own password and can be assigned service tools *functional privileges* that permit a user to perform selected service tool functions. You can define up to 100 service tools user IDs. Included in this total are four IBM-supplied user IDs that are built in to the service tools: 11111111, 22222222, QSECOFR, and QSRV.

Technical Note

The QSECOFR user profile and the QSECOFR service tools user ID are completely different entities. This fact is often (and understandably so) a point of confusion. For example, changing the IBM i QSECOFR user profile password has absolutely no effect on the QSECOFR service tools user ID password.

You create service tools user IDs for users who need to perform service functions for the same reason you create a unique IBM i user profile for each user — to provide accountability and separation of duties. For example, you can create a service tools user ID that lets one person administer the configuration of one LPAR and create another ID that lets a different user administer a different LPAR. Just as you should create user profiles with only the capabilities needed to perform the required job functions, you should create service tools user IDs and grant only the functional privileges required to perform the services or tasks appropriate for each user's job function.

Figures 5.1 and 5.2 show the SST screens used to work with and create service tools user IDs. To use the STRSST command, you must have a valid service tools user ID and password.

Figure 5.1

Work with Service Tools User IDs Panel

```
                    Work with Service Tools User IDs
                                                         System:   SKYVIEW
Type option, press Enter.
  1=Create                 2=Change password       3=Delete
  4=Display                5=Enable                6=Disable
  7=Change privileges      8=Change description

Opt  User ID      Description                              Status
 1   carol_____
 _   QSECOFR      QSECOFR                                  Enabled
 _   QSRV         QSRV                                     Enabled
 _   11111111     11111111                                 Enabled
 _   22222222     22222222                                 Enabled









F3=Exit   F5=Refresh   F12=Cancel
```

Figure 5.2
Create Service Tools User ID Panel

```
                           Create Service Tools User ID
                                                               System:   SKYVIEW
Service tools user ID name . . . . . . . . :  CAROL

Type choices, press Enter
  Password . . . . . . . . . . . . . . . .


  Allow ID access before
  storage management recovery . . . . . . .  1   1=Yes, 2=No
  Set password to expire . . . . . . . . . .  1   1=Yes, 2=No
  Description . . . . . . . . . . . . . . .  Carol Woodbury, President, SkyVie
w Partners, Inc_










F3=Exit   F5=Change privilege   F12=Cancel
```

Service Tools User ID Passwords

Two sets of password policies are available for service tools user IDs. The default password policy uses Data Encryption Standard (DES) encryption. You must use DES encryption if you use pre-V5R1 Navigator clients to connect to service functions.

The other password policy uses Secure Hash Algorithm (SHA) encryption, supports a much longer password length, and significantly expands the password character set. This password policy enables significantly stronger passwords than the default password policy.

Service tools user ID passwords are initially set to expired unless they are explicitly set otherwise. Whenever an administrator sets or changes a password for another user, by default the password is set to expired. The administrator can override this setting by specifying option 2 (No) for the "Set password to expire" option when creating a service tools user ID. This option is also available on the Change Service Tools User Password for Another User panel, reached by selecting option 2 (Change password) on the Work with Service Tools User IDs panel.

Passwords for the IBM-supplied service tools user IDs 22222222, QSECOFR, and QSRV are also initially set to default and expired, which means you're required to change the password before you can sign on using one of these IDs. You can always sign on to DST using the QSECOFR service tools user ID and password even if that password is expired.

Security Recommendation

Of the four predefined service tools user IDs that are shipped with IBM i — 11111111, 22222222, QSECOFR, and QSRV — each one except 11111111 comes with a default and expired password. 11111111 is shipped with a default and unexpired password. It should go without saying that you should immediately change the 11111111 password to something other than the default.

Default Password Policy Details

The default password policy has the following attributes:

- Uppercase user IDs, maximum 10 characters in length.
- *Case-sensitive* passwords, maximum eight characters in length. At user ID creation time, the password must be at least one character long. When a password is changed, the new password must be at least six characters long.
- Passwords for user IDs do not expire due to the passage of time; however, they can be set to expired by an administrator. (By default, passwords are expired when the profile is created.)
- With the exception of 11111111, initial passwords for IBM-shipped service tools user IDs default to being the same as the name of the service tools user ID and expired.
- By default, service tools user IDs cannot repeat their past 18 passwords. You can change this interval by selecting option 10 (Duplicate password control) on the Work with Service Tools Security Data menu. The value can be any number in the range 0–32.
- By default, user IDs are disabled after three incorrect passwords are entered. You can change the maximum number of failed sign-on attempts using option 9 (Maximum sign-on attempts allowed) on the Work with Service Tools Security Data menu. The value can be set to 2–99.

Alternate Password Policy Details

The alternate password policy has the following attributes:

- Uppercase user IDs, maximum 10 characters in length.
- 128-digit, case-sensitive passwords. At user ID creation time, passwords must be at least one character long. When passwords are changed, they must be at least six characters.
- By default, passwords for user IDs expire after 180 days; however, you can change the expiration interval using option 8 (Password expiration interval in days) on the Work with Service Tools Security Data menu. The value can be 0 for *NOMAX or up to 999 days.
- By default, passwords cannot be repeated for 18 passwords. You can change this interval using option 10 (Duplicate password control) on the Work with Service Tools Security Data menu. The value can be set to 0–32.

- By default, user IDs are disabled after three incorrect passwords are entered. You can change the maximum number of failed sign-on attempts using option 9 (Maximum sign-on attempts allowed) on the Work with Service Tools Security Data menu. The value can be set to 2–99.

Be aware that if you change to the alternate password policy, you cannot change back to the default policy! To change to use the alternate password policy, use DST to perform the following steps:

1. Sign on to DST using your service tools user ID.
2. On the Use Dedicated Service Tools (DST) menu, select option 5 (Work with DST environment).
3. On the Work with DST Environment menu, select option 6 (Service tools security data).
4. On the Work with Service Tools Security Data menu, select option 6 (Password level).
5. A message asking whether you want to change to the new password policy will appear. Press Enter.

Technical Note

To change the password policy (or any of the other password rules), a service tools user ID must have the "Dedicated Service Tools (DST) environment" privilege and the "Service tool security" privilege. Only QSECOFR is shipped with both of these privileges (which are described in the next section).

Choosing a Password Policy

Which password policy should you use? The alternate password policy provides much stronger password protection than the default policy. Its strength comes primarily from the larger set of characters that can be used in your passwords. I recommend using the alternate password policy unless clients running a pre-V5R1 version of Navigator need to access your system. (These clients support only the weaker algorithm.)

Service Tools Functional Privileges

Service tools functional privileges let you control which panels of SST and DST an individual service tools user ID can access. By controlling which panels a service tools user is permitted to see, you can control which functions within SST and DST the user is allowed to run. IBM provides a large number of privileges, enabling you to create service tools user IDs that are limited to just the specific tasks you want a user to be able to accomplish. Although no auditor would ever know to ask about service tools user ID settings and there are no laws or regulations requiring this, it is simply a good security practice to create user IDs with only the privileges required to perform their job responsibilities. Functional privileges let you do this.

Each of the four predefined service tools user IDs that IBM ships with IBM i comes with a default set of functional privileges. We cover those defaults later. First, let's take a look at the full set of privileges available and how you access them.

In the panels used to work with and create service tools user IDs (Figures 5.1 and 5.2), you may have noticed the option to "Change privilege(s)" (by typing option 7 or pressing F5, respectively). Choosing this option produces a series of screens similar to those shown in Figures 5.3A through 5.3C.

Figure 5.3A

Change Service Tools User Privileges — Screen 1

```
                         Change Service Tools User Privileges
                                                          System:   SKYVIEW
Service tools user ID name  . . . . . . . . . :  CAROL

Type option, press Enter.
  1=Revoke   2=Grant

Option   Functions                                                  Status
  _      Disk units - operations                                    Revoked
  _      Disk units - administration                                Revoked
  _      Disk units - read only                                     Revoked
  _      System partitions - operations                             Revoked
  _      System partitions - administration                         Revoked
  _      Partition remote panel key                                 Revoked
  _      Operator panel functions                                   Revoked
  _      Operating system initial program load(IPL)                 Revoked
  _      Install                                                    Revoked
  _      Performance data collector                                 Granted
  _      Hardware service manager                                   Revoked
  _      Display/Alter/Dump                                         Revoked
  _      Main storage dump                                          Granted
                                                                        More...
F3=Exit    F5=Reset    F9=Defaults    F12=Cancel
```

Figure 5.3B
Change Service Tools User Privileges — Screen 2

```
                         Change Service Tools User Privileges
                                                           System:   SKYVIEW
Service tools user ID name  . . . . . . . . :  CAROL

Type option, press Enter.
  1=Revoke   2=Grant

Option   Functions                                                 Status
  _      Product activity log                                      Revoked
  _      Licensed Internal Code log                                Granted
  _      Licensed Internal Code fixes                              Granted
  _      Trace privilege                                           Revoked
  _      Dedicated Service Tools (DST) environment                 Revoked
  _      Remote service support                                    Revoked
  _      Service tool security                                     Revoked
  _      Service tool save and restore                             Revoked
  _      Debug                                                     Revoked
  _      System capacity - operations                              Revoked
  _      System capacity - administrator                           Revoked
  _      System security                                           Revoked
  _      Start Service Tools                                       Revoked
                                                                         More...
F3=Exit   F5=Reset   F9=Defaults   F12=Cancel
```

Figure 5.3C
Change Service Tools User Privileges — Screen 3

```
                         Change Service Tools User Privileges
                                                           System:   SKYVIEW
Service tools user ID name  . . . . . . . . :  CAROL

Type option, press Enter.
  1=Revoke   2=Grant

Option   Functions                                                 Status
  _      Take over console                                         Revoked










                                                                         Bottom
F3=Exit   F5=Reset   F9=Defaults   F12=Cancel
```

The Change Service Tools User Privileges screens list the functional privileges that can be granted or revoked:

- Disk units – operations
- Disk units – administration
- Disk units – read only
- System partitions – operations — This privilege lets the user access panels that control most attributes of existing partitions. These attributes are
  - Change a default electronic customer support resource
  - Change a partition name
  - Change communication options with resources not in use
  - Change I/O configuration of primary and secondary partitions
  - Change operating mode for a logical partition
  - Change the alternate restart device for a logical partition
  - Change the guest partition host information
  - Change the restart source command line parameters of a guest partition
  - Change the load source for a logical partition
  - Display available hardware resources
  - Display the communication options of a logical partition
  - Display the console for a partition
  - Display the logical partition operating system release level
  - Display the operating system type for a partition
  - Display the partition ID
  - Display the PCI [Peripheral Component Interface] information
  - Display the remote control panel for a logical partition
  - Display the system reference code history for logical partitions
  - Display system resources
  - Dynamic movement of interactive performance
  - Dynamic movement of I/O processors
  - Dynamic movement of memory
  - Dynamic movement of processors
  - Enable the virtual Ethernet communication for a logical partition
  - Find a logical address for a resource
  - Move a dedicated processor
  - Move a dedicated processor to the shared processor pool
  - Move an I/O processor with resources in use (forced)
  - Move an I/O processor with resources not in use
  - Move interactive performance, memory, or shared processing power
  - Move an I/O adapter assigned to an I/O processor to a guest partition
  - Move an I/O adapter to a guest partition
  - Move an I/O adapter assigned to an i5/OS partition

  - Perform main storage dumps on servers with logical partitions
  - Prevent a secondary logical partition from restarting during a system restart
  - Print system configuration for logical partitions
  - Reset a disk unit I/O processor with logical partitions
  - Restart a secondary logical partition during a system restart
  - Restart a system with logical partitions
  - Restore all logical partition configuration data
  - Save all logical partition configuration data
  - Schedule a dedicated processor move
  - Schedule an interactive performance move
  - Schedule an I/O processor move
  - Schedule a memory move
  - Schedule a shared processor move
  - Use remote service with logical partitions
  - View the status of a logical partition
- System partitions – administration — This privilege lets the user access all functions related to managing a partition. The administration privilege includes all the privileges of the operations privilege plus the following:
  - Accept a disk unit as load source for a logical partition
  - Change a bus ownership type
  - Change communication options with resources in use (forced)
  - Change the partition type
  - Clear nonreporting resources on logical partitions
  - Clear partition configuration from nonconfigured disk units
  - Copy partition configuration data between load sources
  - Create a new logical partition
  - Create a guest partition
  - Delete a logical partition
  - Delete all logical partition configuration data
  - Move an I/O processor with resources in use (forced)
  - Recover logical partition configuration data
  - Restart a secondary logical partition during a system restart
  - Update partition configuration data on all logical partitions
- Partition remote panel key *XXXXXXXX* (0–7) — This privilege requires some explanation. Although it does not appear in the figure, if you are on a system that is partitioned, you'll see a privilege for each partition on the system. The purpose of this privilege is to enable the configuration of a service tools user ID that has the ability to manage the virtual control panel of each (or only one) LPAR. A user must possess the "Partition remote panel key" privilege to perform Operations Console functions. In addition, the PC on which the user is working must have the appropriate privilege — something accomplished through device profiles, which I describe later in the chapter.

- Operator panel functions
- Operating system initial program load (IPL)
- Install
- Performance data collector — New service tools users are granted this privilege by default.
- Hardware service manager
- Display/Alter/Dump
- Main storage dump — New service tools users are granted this privilege by default.
- Product activity log
- Licensed Internal Code log — New service tools users are granted this privilege by default.
- Licensed Internal Code fixes — New service tools users are granted this privilege by default.
- Trace privilege — Communications trace
- Dedicated Service Tools (DST) environment — This privilege controls access to panels in DST from which users can access security functions related to DST. To be able to perform security-related functions in DST, a user must possess this privilege as well as the "Service tool security" privilege.
- Remote service support
- Service tool security — This privilege is required to manage service tools security, which includes managing service tools users, privileges, and device profiles.
- Service tool save and restore
- Debug
- System capacity – operations
- System capacity – administrator
- System security — This privilege controls access to the Work with System Security panel (described later). By tightly controlling access to this panel, you can tightly control your overall security configuration (e.g., security-related system value settings) and who is allowed to change it.
- Start Service Tools — You can create user IDs that are allowed to run DST from a console but not from SST. Because the DST panels contain more function than SST, you can use this privilege to split responsibility among different people. An individual with an IBM i user profile that has *SERVICE special authority and a valid service tools user ID will not be allowed to successfully run the STRSST CL command unless the user's service tools user ID has this privilege.
- Take over console

Table 5.1 shows the default functional privileges that are shipped with each service tools user ID.

Table 5.1
Default Functional Privileges

Privilege	Service tools user IDs				
	11111111	*22222222*	*QSECOFR*	*QSRV*	*New**
System partitions – operations		X	X	X	
System partitions – administration		X	X	X	
Disk units – operations		X	X	X	
Disk units – administration		X	X	X	
Disk units – read only		X	X		
Trace privilege	X	X	X	X	
Display/Alter/Dump		X	X		
Licensed Internal Code log		X	X	X	X
Hardware service manager		X	X	X	
Product activity log	X	X	X	X	
Main storage dump	X	X	X	X	X
Install		X	X	X	
Performance data collector		X	X	X	X
Operating system initial program load (IPL)	X	X	X	X	
Licensed Internal Code fixes	X	X	X	X	X
Dedicated Service Tools (DST) environment	X	X	X	X	
Debug		X	X	X	
Service tool save and restore		X	X	X	
Remote service support		X	X	X	
Partition remote panel key 0	X	X	X	X	
Operator panel functions	X	X	X	X	
Service tool security			X		
System capacity – operations		X	X	X	
System capacity – administrator		X	X	X	
System security			X		
Start Service Tools		X	X	X	
Take over console			X	X	

** These privileges are granted by default to newly created service tools user IDs.*

Technical Note

Because service tools user ID 11111111 is granted the "Dedicated Service Tools (DST) environment" privilege by default, it can be used to change the security environment of the service tools. For example, it can be used to change the expired passwords of the other predefined service tools IDs. Coupled with the fact that 11111111 is shipped with a default and unexpired password, this ability constitutes a security exposure until you change the ID's password.

Service Tools Features in V6R1

V6R1 introduced the concept of "linking" a service tools user ID with an IBM i user profile. This capability is provided to enable functions that require one or more of the service privileges but are actually run from a command line rather than from SST or DST. These functions are typically run when you're on the phone with IBM debugging a problem and not something you'd run as a normal part of your daily activities.

Technical Note

The IBM i QSECOFR profile is linked to the QSECOFR service tools user ID, and this link cannot be removed or changed.

You can link an individual's service tools user ID to the individual's IBM i profile or, if you have several IBM i profiles that need the service privilege, you can create a service tools user ID, grant it the required privilege, and then link it to group profile of the IBM i user profiles. When you link a service tools user ID to an IBM i group profile, every profile in the group is linked to the service tools user ID and has that privilege when running the IBM i functions that require it.

Technical Note

Because the IBM i QSECOFR profile is linked to the QSECOFR service tools user ID, every IBM i profile that is a member of QSECOFR is linked to the QSECOFR service tools ID.

V6R1 also provides a long-awaited method of displaying the attributes of service tools user IDs from IBM i. In prior releases, you had to go into SST or DST to view the attributes of individual service tools user IDs. The DSPSSTUSR (Display Service Tools User ID) command lets you display or print the attributes of one or all service tools user IDs. Figure 5.4 shows the information provided for every service tools user ID.

Figure 5.4
Display Service Tools User ID Panel

```
5761SS1 V6R1M0  080215
                                                  Display Service Tools User ID
 Service tools user ID . . . . . . . . . . . :   QSECOFR
 Previous sign-on . . . . . . . . . . . . . :   02/09/08  16:34:54
 Password verification not valid  . . . . . :   0
 Status . . . . . . . . . . . . . . . . . . :   *ENABLED
 Linked user profile  . . . . . . . . . . . :   QSECOFR
   Linked user profile exists . . . . . . . :     Yes
 Date password last changed . . . . . . . . :   13/03/08
 Date password expires  . . . . . . . . . . :   None
 Password is set expired  . . . . . . . . . :   No
 Description  . . . . . . . . . . . . . . . :   QSECOFR
 Privileges:
   Disk units - operations  . . . . . . . . :   *GRANTED
   Disk units - administration  . . . . . . :   *GRANTED
   Disk units - read only . . . . . . . . . :   *GRANTED
   System partitions - operations . . . . . :   *GRANTED
   System partitions - administration . . . :   *GRANTED
   Partition remote panel key . . . . . . . :   *GRANTED
   Operator panel functions . . . . . . . . :   *GRANTED
   Operating system initial program load
     (IPL)  . . . . . . . . . . . . . . . . :   *GRANTED
   Install  . . . . . . . . . . . . . . . . :   *GRANTED
   Performance data collector . . . . . . . :   *GRANTED
   Hardware service manager . . . . . . . . :   *GRANTED
   Display/Alter/Dump . . . . . . . . . . . :   *GRANTED
   Main storage dump  . . . . . . . . . . . :   *GRANTED
   Product activity log . . . . . . . . . . :   *GRANTED
   Licensed Internal Code log . . . . . . . :   *GRANTED
   Licensed Internal Code fixes . . . . . . :   *GRANTED
   Trace  . . . . . . . . . . . . . . . . . :   *GRANTED
   Dedicated service tools (DST) environment:   *GRANTED
   Remote service support . . . . . . . . . :   *GRANTED
   Service tools security . . . . . . . . . :   *GRANTED
   Service tools save and restore . . . . . :   *GRANTED
   Debug  . . . . . . . . . . . . . . . . . :   *GRANTED
   System capacity - operations . . . . . . :   *GRANTED
   System capacity - administrator  . . . . :   *GRANTED
   System security  . . . . . . . . . . . . :   *GRANTED
   Start service tools  . . . . . . . . . . :   *GRANTED
   Take over console  . . . . . . . . . . . :   *GRANTED
```

Device Profiles

To identify and control which PCs are allowed to act as the console for a system or a partition using LAN console, you use service tools *device profiles.* Device profiles provide a way to authenticate a device that is trying to act as the console. Once the PC is authenticated, the user trying to run Operations Console at the PC must provide his or her service tools user ID and password and must have the appropriate privileges to perform Operations Console functions.

Operations Console with LAN connectivity uses a version of the Secure Sockets Layer (SSL) protocol but does so without using digital certificates. Instead, device authentication is accomplished via service tools device profiles. These profiles, which are stored on the IBM i system, contain a cryptographic key (i.e., password) that the PC trying to act as the console must provide to authenticate itself to the IBM i service tools server.

Information about the device profile associated with a particular PC is stored on the PC. This information consists of the device profile name and the same key that is stored in the device profile itself back on the IBM i system. The information stored on the PC is encrypted with a different password, which is provided during configuration. In other words, using device profiles involves two passwords. One password is used to protect the device profile information on the PC, and the other is contained in the device profile itself and is used to authenticate the device to the IBM i system. IBM i servers are shipped with a default service tools device profile named QCONSOLE and an associated default password of QCONSOLE.

To configure a PC to be allowed to run Operations Console, you use the Operations Console setup wizard. The wizard adds the necessary device profile information to the PC. It asks for the service tools device profile name, the service tools device profile password (i.e., the one stored in the device profile), and a password to protect the service tools device profile information. The first two pieces of data must match an existing IBM i device profile. Operations Console with LAN connectivity will change the password stored in the device profile during each successful connection.

In V6R1, you can choose to allow automatic creation of device profiles. To auto-create device profiles, go to DST, and on the Work with Service Tools Security Data menu, specify a value (1–49) in the "Autocreate service tools device IDs" field.

Technical Note

If no cryptographic products are installed, data (i.e., the device profile passwords) flowing over the LAN between the Operations Console device and the IBM i system is not protected.

You can configure SST to permit the management of Operations Console device profiles from SST; however, as shipped, this capability is disabled. To enable this option, go to DST and display the Work with DST Environment menu; select the option to display the Work with Service Tools Security Data menu, and choose option 7 (Work with lock for device IDs from SST).

Unlocking the SST device IDs menu item introduces some additional security exposure because it allows some SST users to create device ID profiles for their own PCs. This is one of those trade-off you'll have to consider — a bit more risk versus the convenience of managing device profiles from SST.

The Work with System Security Panel

You reach the Work with System Security panel (Figure 5.5) by selecting option 7 (Work with system security) from DST or SST. This panel provides three important options that can help you manage the security of your entire system:

- Allow system value security changes
- Allow new digital certificates
- Allow a service tools user ID with a default and expired password to change its own password

Figure 5.5
Work with System Security Panel

```
                    Work with System Security
                                                  System:   SKYVIEW
Type choices, press Enter.

  Allow system value security changes . . . . .  1  1=Yes, 2=No
  Allow new digital certificates . . . . . . . .  1  1=Yes, 2=No
  Allow a service tools user ID with a
   default and expired password to change
   its own password . . . . . . . . . . . . . .  1  1=Yes, 2=No











F3=Exit   F12=Cancel
```

I briefly mentioned this panel's "Allow system value security changes" function at the end of Chapter 3. Setting this option to 2 (No) prohibits many system values from being changed — period. Unless you set this value back to 1 (Yes), these system values cannot be changed, even by a user signed on as QSECOFR. This option protects the system values from being changed by other administrators (as long as you haven't given them a service tools user ID with the ability to change this setting). It also prevents vendors from changing these system values during their installation process. The list of system values affected by this setting can change with each release. To see the current list for your system, press F1=Help while your cursor is on the "Allow system value security changes" parameter on the Work with System Security display.

The "Allow new digital certificates" option lets you control whether new trusted certificate authority (CA) certificates can be added to your system. When this option is set to 2 (No), no program or person can add a new trusted CA to the system. This option helps prevent someone from adding a trusted CA to your system without your explicit consent. You might turn on this option if you were concerned about where digital certificates were created for use in client-side SSL authentication or if you were running applications that are digitally signed.

The "Allow a service tools user ID with a default and expired password to change its own password" option controls whether such service tools user IDs with are permitted to change their own passwords. When you leave this option at its default value, 2 (No), the password can be changed only by another service tools user ID with the appropriate functional privileges or from DST. Leaving this value at the default prevents someone from gaining access to SST by using the well-known service tools user IDs. Such an attempt will be successful if you haven't changed these user IDs from their default passwords and you change this value to 1 (Yes).

Monitoring Service Tools Use

You can use the service tools security log or the IBM i audit journal to monitor use of the service tools. Any use of the service tools via SST is added to the audit journal.

To work with the service tools security log, you use DST with a service tools user ID that has the "Dedicated Service Tools (DST) environment" privilege. Follow these steps:

1. From the Use DST menu, select option 5 (Work with DST environment).
2. From the Work with DST Environment menu, select option 6 (Work with service tools security data).
3. From the Work with Service Tools Security Data menu, select option 3 (Work with service tools security log) to display the Work with Service Tools Security Log screen. This screen shows security-related activity by date and time.

From the security log display, you can print the log by pressing F6. Enter 5 in the Option field of a particular entry to display additional details about that entry. If the activity noted in the entry is related to a granted or revoked privilege, the Display Service Tools Security Log Details display appears, showing the following information:

- Time of activity
- Activity description
- User ID of the person who made the change
- Affected user ID
- Privilege description

If the activity is related to enabling or disabling a user ID, the details screen shows the following information:

- Time of activity
- Activity description
- User ID of the person who made the change
- Affected user ID

If the activity is related to any other type of event, the details screen shows the following information:

- Time of activity
- Activity description
- Affected user ID

Unfortunately, there is no way to get at this log without going into DST, so you're going to have to note this activity on your calendar so you remember to review this log. In addition to using the service tools log, you can use the security audit journal to record service tools actions. To audit the use of service tools, make sure the value *SERVICE is specified for the QAUDLVL system value.

You can view audit entries related to service tools just as you can any other audit entry. On a command line, enter the CPYAUDJRNE (Copy Audit Journal Entries) CL command, specifying ENTTYP(ST). Then run a query or an SQL statement to view the information. The audit entries include attempts to use SST without a service tools user ID, entering an invalid service tools user ID password, tracing a job, and more.

Service Tools Security Recommendations

In case you find the contents of this chapter a bit overwhelming, here are some suggestions for managing service tools security. This advice can be divided into two parts: recommendations that help you retain control over your system's security configuration, and recommendations that can help protect you against people trying to compromise your system by distributing rogue programs to the system.

Here are the minimum recommendations to help you control your security configuration:

- Define the minimum number of service tools user IDs that are absolutely necessary; however, each user that accesses the service tools should have his or her own user ID so that there can be accountability for user actions in the log and audit journal.

    - Monitor the log and audit journal for appropriate use of service tools.
    - Create user IDs with only the functional privileges needed to perform the required job functions.
    - Create user IDs with expired passwords so users are required to change their password upon first use.

- Give only one or two service tools user IDs the ability to manage service tools security.
- For users who need only occasional access to service tools, consider disabling their service tools user ID after each use and re-enabling it the next time access is needed.
- Change all the IBM-supplied service tools user ID passwords to something other than the shipped (default) value.
- If you permit service tools user IDs with expired and defaulted passwords to change their own passwords, make sure no user IDs with default passwords exist.

For organizations with a very restrictive security policy, take the following steps in addition to those listed above:

- Create your own service tools user ID with the same functional privileges as the QSECOFR service tools user ID. Disable the QSECOFR service tools user ID, and use the one you created instead.
- Lock down the security-related system values.
- Use the alternate service tools password policy.
- Do not permit service tools user IDs with expired and defaulted passwords to change their own passwords.
- Do not enable the automatic creation of device profiles.

Frankly, I have rarely seen these service tools features used. Perhaps it's because they are hardly ever written about or promoted, or maybe it's the thought that if someone is using service tools, they must need all the authority available. Or it could be because organizations are under the impression that this type of control is overkill. Regardless of the reason, I encourage you to consider these features and at least use the function that lets you create service tools IDs with only the privileges needed for the tasks you want users to perform.

For more details about using service tools user IDs, see the IBM publication *System i: Security Service Tools,* available in the Information Center under "Security | Service tools."

Chapter 6

Object-Level Security

Once again, I want to acknowledge the reason why you are reading this book. It is because there is data, either on systems you've been tasked to administer or on systems you are auditing, that needs to be protected. How tightly that data should be protected depends on its classification. Some data will require "deny by default" protection so that only users with a specific job requirement can access it. Other data might be classified such that all users can read the data but not update it. This chapter introduces you to the features available in IBM i for protecting data. Later in the book, I provide some practical examples that use these features to implement security schemes such as deny by default and read-only access.

Before a user can access an object, the user's profile must have authority to the object. Authority can come from several places: an object's *PUBLIC authority, private authority given to the user's individual or group profile, a group profile, an authorization list securing the object, adopted authority, or (as you learned in Chapter 4) *ALLOBJ special authority. In this chapter, I discuss each source of authority, how to manage the various options, and why you might use each one. The examples I'll discuss use the QSYS file system (which uses the traditional library/object structure). Authority considerations for the integrated file system (IFS) are discussed in Chapter 7.

Private Authorities

Authorities granted to individual or group profiles — often called *private authorities* — permit users to access and use specific objects. You can grant (give) or revoke (take away) specific authorities by using one of the following commands:

- GRTOBJAUT (Grant Object Authority)
- RVKOBJAUT (Revoke Object Authority)
- EDTOBJAUT (Edit Object Authority)

As the diagram in Figure 6.1 illustrates, IBM i provides two categories of authorities: object authorities and data authorities. *Object authorities* authorize actions that affect an entire object, such as renaming, changing, saving, and deleting the object, as well as a few functions for specific types of objects, such as defining referential integrity rules for database files. *Data authorities* apply at the content level for objects (e.g., database files, data queues, libraries, output queues) and authorize very specific actions, such as adding, updating, reading, and deleting object contents (e.g., records, entries).

Figure 6.1
Authority Categories

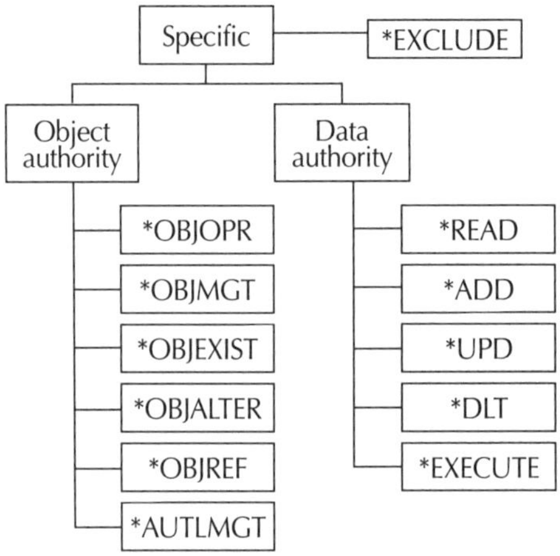

Object Authorities

IBM i provides six object authorities, which I'll group into three categories for purposes of explaining their functions. The first category consists of three object authorities that apply to all the objects on your system: *OBJOPR, *OBJMGT, and *OBJEXIST.

Object authority	Description
*OBJOPR	*Object operation* authority authorizes the use of an object.
*OBJMGT	*Object management* authority authorizes the user to move and rename the object, change its attributes, and grant and revoke the authority of other user profiles to the object.
*OBJEXIST	*Object existence* authority lets the user delete, save, and restore the object and transfer ownership of the object.

Users who have *OBJOPR authority to an object can use the object. What exactly does it mean to be able to "use" an object? *OBJOPR authority lets the user view the object's description, regardless of the object's type. If the object contains data or executable code, the user can access the object's description but cannot access the data or run the program without having the appropriate data authority.

*OBJMGT authority lets the user change the object's attributes, including the authorities that other users have to the object. *OBJMGT is a superset of the object authorities *OBJALTER and *OBJREF (which I discuss momentarily). If you don't want a user to have *OBJALTER or *OBJREF authority, you must revoke not only *OBJMGT authority but also the *OBJALTER and *OBJREF authorities specifically.

*OBJEXIST authority lets the user perform save/restore operations on the object, transfer object ownership, and delete the object.

To reduce the risk of an object being improperly used or inadvertently deleted, you should restrict *OBJMGT and *OBJEXIST authorities to the object's owner. Unfortunately, this recommendation can make some operations a challenge, such as when a file is deleted and re-created when a query is run. In later chapters, I'll discuss techniques for managing these scenarios.

The second category of object authorities consists of those that apply primarily to database files. These authorities are the aforementioned *OBJALTER and *OBJREF authorities.

Object authority	Description
*OBJALTER	*Object alter* authority authorizes the user to add, clear, reorganize, and initialize database file members and to alter the attributes of database files (SQL tables or physical and logical files). Authorized actions include adding and removing triggers and changing the attributes of tables, files, and SQL packages.
*OBJREF	*Object reference* authority lets the user specify a database file or table as the parent file in a referential constraint.

A user with *OBJALTER authority to a database file can change the file's attributes and can add and remove database file triggers. If you want to avoid granting a database administrator or programmer *OBJMGT authority, you can grant only *OBJALTER authority to limit the user to database-specific functions.

You should grant *OBJREF object authority to only the database administrator or the programmer responsible for enforcing referential integrity constraints. This authority is necessary only for the parent file in a referential constraint. For example, say you want to ensure that every new order record in the ORD_MAST file references an existing customer in the CUST_MAST file. You can do so by adding a referential constraint that links the two files by the customer number that is common to both files. In this case, you must have *OBJREF authority to the CUST_MAST file to add the constraint.

The third category of object authorities consists of just one authority, *AUTLMGT (authorization list management), and it applies only to authorization lists. *AUTLMGT authority lets you add and remove users' authorities to an authorization list. I cover authorization lists in more detail later in this chapter.

Data Authorities

Data authorities apply to all object types, but they are most easily explained by using objects that have contents. For example, a physical file (or table) has an object description, and a physical file has contents: records. Other examples of objects that have contents include libraries (object type *LIB), data queues (object type *DTAQ), data areas (object type *DTAARA), and output queues (object type *OUTQ).

There are five data authorities:

Data authority	Description
*READ	Lets the user read and retrieve data and entries.
*ADD	Lets the user insert a new data record or entry.
*UPD	Lets the user modify existing data and entries.
*DLT	Lets the user delete and remove data and entries. (Remember that *OBJEXIST is required to delete the object itself.)
*EXECUTE	Lets the user locate an object in a library or directory and execute a program, a service program, or an SQL package.

The first four data authorities let you read and manipulate the contents of an object. You would typically grant users who need to manipulate data all four of these data authorities. However, if you prefer, you can grant users only a subset of the four.

*EXECUTE authority lets you search the contents of a library or directory. To access an object, you need not only authority to the object itself but also at least *EXECUTE authority to the library or directory in which the object resides. *EXECUTE authority is also one of the authorities you need to run a program, service program, or SQL package (you also need *OBJOPR authority).

As I mentioned earlier, data authorities also apply to objects that don't have contents. For example, to submit a job that uses a job description, you need *OBJOPR, *READ, and *EXECUTE authorities to that job description.

Authority Relationships

One point of confusion for many people when they first encounter IBM i's object security implementation is the difference between having authority to a library and having authority to objects within the library. To access or use an object, you must have authority both to the library in which the object resides and to the object itself. If you don't have authority to the library, you won't be able to access the objects in the library. However, having authority to a library does not give you authority to the objects within it. You need authority to both in order to access an object in a library.

Think of library-level authority as a gate to what goes on with the objects that reside in the library. To access the objects, you must first open the gate; that is, you must have appropriate authority to the library for the request being made. For example, to look for objects in the library, you must have *OBJOPR authority to the library. To read the description of objects in the library, you must also have *READ authority to the library. To add objects to the library, you must have *ADD authority. And so on.

*DLT authority can also be confusing. You would think that if you had *DLT authority to a library, you could delete objects from the library. However, to delete an object from a library, you need only *USE authority to the library. Whether you can delete an object is determined by the authority you have to the object itself. To delete an object, you must have *OBJEXIST authority to that object. To delete a library, you need *OBJEXIST authority to the library.

Now for a quick preview of the next chapter: If you work with directories on the system — that is, if you work with file systems other than the QSYS file system (i.e., libraries) — you need to know that the relationship between directory and object authorities is nearly identical to the relationship I just described between library and object authorities. The only difference is that to access an object in the file system, you must have *EXECUTE authority to all directories in the directory path. For example, to access the file SkyView_Partners/Policy_Minder/Reports/Compliance_report.pdf, you must have authority to every directory in the path that leads to the file (i.e., to the SkyView_Partners, Policy_Minder, and Reports directories) as well as proper authority to the Compliance_report.pdf file itself.

Authority Groupings

To make object and data authorities easier to manage, IBM i defines several authority groupings. You can think of these groupings as shorthand: They let you specify commonly used combinations of object and data authorities using a single value. Except when working with database files, most security administrators use one of these authority groups rather than individual object and data authorities. Table 6.1 lists the authority groupings you can use with libraries and their objects.

Table 6.1

Authority Groupings for Use with Authority Commands for Libraries and Objects

	Authority classes			
Object and data authorities	**ALL*	**CHANGE*	**USE*	**EXCLUDE*
*OBJOPR	X	X	X	None
*OBJMGT	X			
*OBJEXIST	X			
*OBJALTER	X			
*OBJREF	X			
*READ	X	X	X	
*ADD	X	X		
*UPD	X	X		
*DLT	X	X		
*EXECUTE	X	X	X	

To use authority groupings to grant user profile CYOUNG object operational and all data rights to physical file ACCTPAY_PF in the QSYS file system, you could use this GRTOBJAUT command:

```
GRTOBJAUT  OBJ(MYLIB/ACCTPAY_PF) OBJTYPE(*FILE) USER(CYOUNG) +
           AUT(*OBJOPR *READ *ADD *UPD *DLT *EXECUTE)
```

Or, you could use

```
GRTOBJAUT  OBJ(MYLIB/ACCTPAY_PF) OBJTYPE(*FILE) USER(CYOUNG) +
           AUT(*CHANGE)
```

Both commands have the same effect.

*EXCLUDE authority is a bit odd in that users who have this authority are actually prevented from accessing the object rather than being given some type of access rights. Some operating systems, such as Unix, prevent users from accessing a file simply by not granting them any authority to the file. In other words, the absence of authority prevents the users from accessing the file. IBM i works differently. With IBM i, users always have some authority to the object. If they don't have authority through another means, they will, by default, have authority to the object's *PUBLIC authority. (I explain *PUBLIC authority a bit later in this chapter.) To prevent a user from accessing an object, you must grant that user *EXCLUDE authority to the object. To set the object's access to "deny by default," you would set the object's *PUBLIC authority to *EXCLUDE.

Group Profiles

Not only can you grant or revoke private authorities to or from individual users, but you can do the same for group profiles. Group profiles, as you learned in Chapter 4, are a management tool that provides an efficient way to allow all the members of a group (or role) to have the same access to an object. You use the same commands to grant or revoke authority to group profiles. Simply specify the name of the group profile instead of an individual user profile name on the commands.

Multiple Group Profiles

Recall from Chapter 4 that you can specify a user profile as a member of up to 16 groups. To do so, you first specify a group in the user profile's GRPPRF (Group profile) parameter. Then you use the SUPGRPPRF (Supplemental groups) parameter to specify any other group profiles you want the user to be a member of.

When a member of multiple groups creates an object, IBM i refers to the user profile's OWNER (Owner), GRPAUT (Group authority), and GRPAUTTYP (Group authority type) parameters to determine the authority, if any, that the first group (i.e., the group profile named in the GRPPRF parameter) has to the new object. When you assign multiple groups to users, you should assign the most frequently used groups first. The operating system searches each group profile listed in the user profile's GRPPRF and SUPGRPPRF parameters until it finds enough authority to perform the task (or determines that the user lacks sufficient authority). The fewer groups the system has to search to accumulate sufficient authority, the better the performance.

Why Grant Authority to Group Profiles?

Group profiles simplify security management by letting you grant similar authorities to multiple user profiles that perform similar functions (i.e., to a role). Many user profiles in your organization probably have similar basic authorities. Combining similar user profiles into a single group profile lets you administer authorities for the entire group instead of for each user individually. Because many users — especially in IT — have many roles, the ability to make a user profile a member of more than one group is particularly helpful.

When you grant authority to a group profile, an option you have is to use *primary group authority* rather than a private authority. The effect of granting primary group authority is the same as granting a private authority to the group. Only a couple of differences exist between the two authorities. One difference is that primary group authority is stored with the object itself, while private authorities are stored in the user's user profile object. This point has two ramifications. First, primary group authority goes with the object when it is saved and restored. To associate a private authority with an object, you must save the object by running the SAVSECDTA (Save Security Data) command and then running the RSTAUT (Restore Authority) command after the object is restored. The second ramification is that retrieving the authority from the object's header is ever so slightly faster than retrieving the private authority from the user's profile. You can grant primary group authority for an object to only one group. Rather than give the group profile private authority to the object, you can grant the group primary group authority by using the CHGOBJPGP (Change Object Primary Group) or WRKOBJPGP (Work with Objects by Primary Group) command.

From a performance standpoint, primary group authority is a *slightly* faster version of private authority for group profiles. Primary group authority isn't considered a private authority. Rather, it is stored with the object itself, thus shortening the time the system needs to find the proper authorities during an authorization search. However, the performance gain is negligible, so if you find the concept of primary group authority too confusing, go ahead and grant the group private authority if it makes management easier.

Public Authority

The *PUBLIC authority of an object is assigned when a user creates the object. This is the authority users have to an object if they don't obtain authority from any other source. To view an object's authorities, including *PUBLIC authority, you can use the EDTOBJAUT command or the DSPOBJAUT (Display Object Authority) command.

Establishing Public Authority

All objects on the system are created either by using one of the many create (CRT*xxx*) commands or by restoring the object from a save file or save media. The initial public authority for an object is determined when the object is created. (You can always change it later if the original setting isn't appropriate.) When you use a create command to create an object, the command's AUT (Authority) parameter establishes the object's public authorities. The AUT parameter has the following possible values:

AUT value	Description
*LIBCRTAUT	Uses the library's CRTAUT (Create authority) attribute to determine the object's *PUBLIC authorities.
*ALL	Grants all object and data rights.
*CHANGE	Grants *OBJOPR object authority and all data authorities.
*USE	Grants *OBJOPR object authority and *READ and *EXECUTE data authorities.
*EXCLUDE	Grants *PUBLIC *EXCLUDE authority.
Authorization_list_name	Secures the object with an authorization list. *PUBLIC authority for the object comes from the authorization list's *PUBLIC authority. (You can't specify an authorization list for the AUT parameter of a user profile or of another authorization list.)

For most create commands, the default value for the AUT parameter is *LIBCRTAUT, which causes the system to use the object library's CRTAUT (Create authority) attribute to determine the object's public authority. Thus, the CRTAUT library attribute specifies the default public authority that all new objects in the library will have unless you override the default authority on the create command.

You can use the library's CRTAUT parameter to define the default public authority for all objects subsequently created in that library (existing objects aren't affected). The CRTAUT parameter allows the following values:

CRTAUT value	Description
*SYSVAL	Uses the value specified in the QCRTAUT (Create Authority) system value
*ALL	Grants all object and data rights
*CHANGE	Grants *OBJOPR and all data rights
*USE	Grants *OBJOPR, *READ, and *EXECUTE rights
*EXCLUDE	Grants *PUBLIC *EXCLUDE authority
Authorization_list_name	Uses the specified authorization list to determine public authorities

The default value for the CRTAUT parameter is *SYSVAL, which instructs the operating system to reference the QCRTAUT system value to determine the default public authorities for the objects in the library. When you create a library and don't specify a CRTAUT value, the *SYSVAL default value causes the library to use the value found in the QCRTAUT system value.

Technical Note
System value QCRTAUT has no effect on objects created in directories.

Technical Note
Changing the QCRTAUT system value has no effect on objects that already exist. A change to QCRTAUT affects only objects that are created after the system value is changed.

Figure 6.2 illustrates the trickle effect of the default values for the library's CRTAUT attribute and the object's AUT attribute. From bottom to top, the values shown for AUT, CRTAUT, and QCRTAUT are the defaults shipped with IBM i. As you can see, if you don't change these defaults, the system automatically gives the public *CHANGE authority to almost every object created on the system.

Figure 6.2

*How *PUBLIC Authority Is Determined*

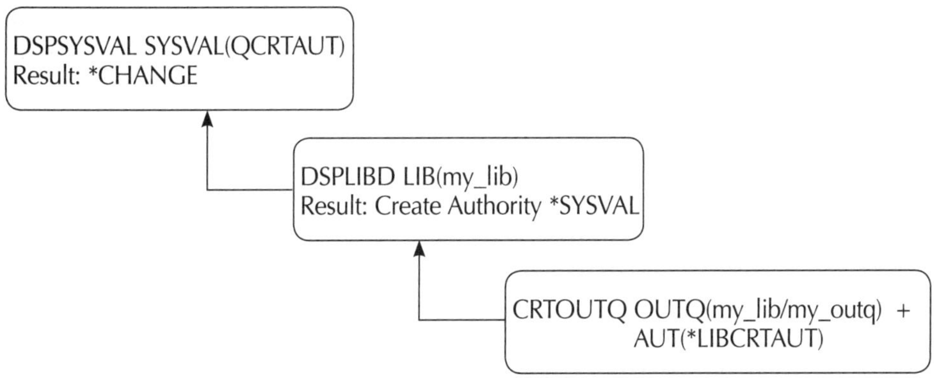

The only tricky part of how *PUBLIC authority is determined comes when you create a library. As implemented, IBM i actually creates all libraries in library QSYS. You can think of QSYS as the "master" library. So, when you create a new library — say, PAYLIB — and specify *LIBCRTAUT for the library's AUT parameter, PAYLIB's public authority comes from the QSYS library's CRTAUT value.

Technical Note

A library's CRTAUT parameter has no effect when you move, duplicate, or restore an object to a library. Instead, the object's public authority is the public authority that the object had before you moved, duplicated, or restored it.

Technical Note

When you create an object and specify REPLACE(*YES), the new object will have the public authority of the replaced object instead of the authority specified in the library's CRTAUT parameter.

Using Default Public Authority

Because IBM ships the QCRTAUT system value set to *CHANGE, the default public authority for most objects created in any library is *CHANGE. In today's environment, where users have access to tools such as File Transfer Protocol (FTP), Open Database Connectivity (ODBC), and SQL and where applications can use sockets or the HTTP server to connect to the system, public authority *CHANGE isn't appropriate — especially for database files or tables. You'll probably want to set the QCRTAUT system value to *USE or even *EXCLUDE to limit public authority to new libraries and objects.

*USE is the most appropriate public authority in today's environment. If you want to change QCRTAUT to *EXCLUDE, you'll have many more considerations. The exact procedure will depend on your applications and system configuration.

Although *ALL is also a valid QCRTAUT value, there is no reason you should ever set this system value to *ALL. This is *not* a wise choice. The obvious reason is because, with this setting, all newly created objects on the system would be created with *PUBLIC *ALL, giving all users the ability to modify or even delete the objects. The other reason to avoid setting QCRTAUT to *ALL is because it is very difficult to know what's going to break if you change this value from *ALL to another value. Avoid setting QCRTAUT to *ALL at all costs.

An alternative is to leave QCRTAUT set to *CHANGE and then change each library's CRTAUT attribute to determine the public authority to new objects created in the library. This approach allows much more flexibility because you can set some libraries to *USE while setting others that contain highly sensitive data to *EXCLUDE.

To ensure that public authorities enhance rather than hinder security on your system, determine the appropriate public authority for each non-system library. Check each one to see whether the currently defined public authority is right for the objects in the library as well as for the library itself. Then modify the CRTAUT attribute of each library to reflect the appropriate default public authority for objects created in that library.

Perhaps one library contains only utility program objects (e.g., date conversion routines, a binary-to-decimal conversion program, a check object/authority program) that various applications use. Because all the programs should be available for execution, it makes sense to set this library's CRTAUT attribute to *USE so that new objects created in the library will have *USE default public authority.

Suppose another library contains the payroll and employee data files. You probably want to restrict public access to this library and secure it by granting only selected users or groups authority. Any new objects created in this library should probably also have *EXCLUDE public authority unless the program or person creating the object specifically selects another public authority using the object's AUT attribute. This library should have a CRTAUT attribute value of *EXCLUDE.

Last, if all the objects in the library are secured with an authorization list (explained in the next section), you can specify the name of the authorization list for the CRTAUT attribute. All objects created into the library will then automatically be secured with the authorization list, and the *PUBLIC authority of the object will come from the authorization list.

If you follow these suggestions for the QCRTAUT system value and the CRTAUT library attribute, an AUT value of *LIBCRTAUT works well as a default. In many cases, public authority at the object level coincides with the public authorities that the CRTAUT attribute establishes at the library level. However, be sure you make this a planned decision instead of a default decision.

Understanding IBM i's implementation of object security can help you plan your authorization scheme. You can secure many or most objects on your system simply through the objects' *PUBLIC authority. You can then protect more sensitive objects by specifying *EXCLUDE authority for *PUBLIC and authorizing only certain user or group profiles. Understanding IBM i's authority search order will also help you make these decisions. You'll learn about the search order later in this chapter.

Authorization Lists

An *authorization list* is a management tool that helps you secure objects that require the same access control settings or authorities. The objects associated with an application, for example, usually all need the same access control or authority settings. Figure 6.3 shows an accounts receivable application that creates a new file each month. The same users need to have the same authority to each monthly file. The best way to secure these files is to use an authorization list.

Figure 6.3

Securing Similar Objects with an Authorization List

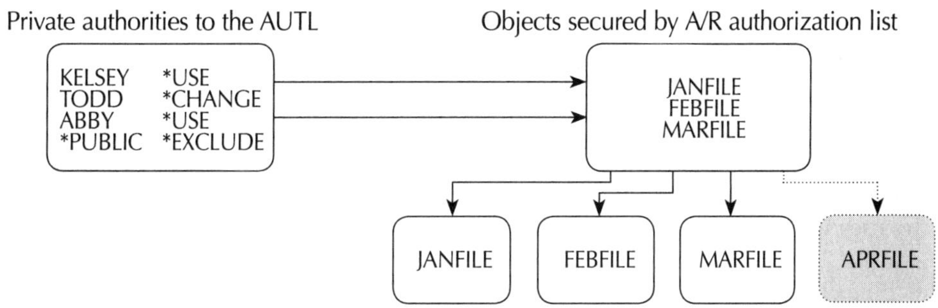

To create an authorization list to secure an object, you run the CRTAUTL (Create Authorization List) command. Then, run the EDTAUTL (Edit Authorization List) command to give selected user or group profiles authority to the authorization list. Next, you associate the authorization list with all the application's files by specifying the authorization list name on the AUTL (Authorization list) parameter on the GRTOBJAUT command. Last, you can set each file's *PUBLIC authority to come from the authorization list by specifying *AUTL in the AUT parameter on GRTOBJAUT.

By virtue of having authority to the authorization list, the users have authority to the objects. The type of authority that each user has to the object is the same as the authority you grant the user to the authorization list. For example, in Figure 6.3, Todd has *CHANGE authority to the authorization list; therefore, he has *CHANGE authority to all objects associated with the authorization list.

Authorization lists have several advantages. They are a wonderful tool for securing files. Many organizations now run a 24x7 shop. For these organizations, managing authorities to files that are always in use can be a challenge because you can't modify the access control settings on a file that's open (in use). By associating an authorization list with those files, you can manage the authorities on open files by managing the authorities on the authorization list, which you can do even when a file is open. Another advantage is that you only have to manage the access controls on the authorization list (i.e., you only manage authorities once) rather than managing access to all the individual files (objects).

Security Recommendation

If you have trouble maintaining authorities for a file because the file is always open, secure the file with an authorization list during the next downtime (when the file is not in use). You can grant and revoke authority to the authorization list even when the file is open.

Another advantage of authorization lists is that they save room in the internal structure of user profiles. In other words, they reduce the number of private authorities and, therefore, reduce the size of the user profile object itself. Rather than putting into a user profile an entry for each object to which the profile has private authority, you put only an entry for the authorization list. In addition to reducing the size of the profile, having fewer entries in a user's profile saves time when you run the SAVSECDTA and RSTAUT commands.

Authorization lists don't preclude using private authorities. You can still grant and revoke private authorities to an object using the GRTOBJAUT, RVKOBJAUT, and EDTOBJAUT commands. A user profile's private authorities supersede any authority the user has to an authorization list. However, as a matter of practice, I try to avoid granting both private authorities and securing the object with an authorization list. While technically possible, using both of these methods tends to add complexity to a security scheme's implementation. This complexity then often leads to mismanagement of the authorities. If I ever grant a private authority in addition to securing an object with an authorization list, it's an exceptional case and not part of a typical implementation.

I often am asked whether it's better to use group profiles or authorization lists. Neither method is inherently better than the other. You should use the option that makes the most sense for what you're trying to do. I often end up using a combination of both methods — granting a group profile authority to an authorization list, for example. Table 6.2 compares the use of authorization lists and group profiles.

Table 6.2
Comparison of Authorization Lists and Group Profiles

Authorization list	Group profile
Secures and helps you manage access to multiple *objects* that need the same authority settings.	Helps you manage multiple *user profiles* that need the same access and capabilities.
Lets each profile have a different authority to the authorization list.	Gives all users in the group the same authorities.
An object can be secured by only one authorization list.	A profile can be a member of multiple group profiles.

How IBM i Checks Authority

We've discussed the fact that before you can access an object, you need to have sufficient authority, and you've learned where the authority can come from. The question many of you probably have at this point is, "In what order are the authority sources checked?"

The order in which authority is checked is a mystery to many people. But unless you understand how it works, securing your system will be a change-something-and-hope-it-works process that will be prone to error and impossible to explain or maintain.

The authorization search order is really quite simple once you see it. Figure 6.4 (page 124) illustrates the progression of IBM i's authority search algorithm.

As you can see, the system first determines whether the profile has *ALLOBJ special authority. If it does, the system lets the user access the object because *ALLOBJ gives the user profile *ALL authority to every object on the system.

If the profile does not have *ALLOBJ authority, the system performs an authority lookup to determine whether the user profile has a private authority to the object. If the user has a private authority, even if it is *EXCLUDE, the system uses that authority, and the check stops. It doesn't go on to check group or *PUBLIC authorities.

If the user has no private authority, the system then determines whether the object is secured with an authorization list and whether the user has authority to the authorization list. If not, the system checks the user's first group profile, if there is one, for *ALLOBJ special authority, then for primary group authority to the object, and then for a private authority to the object. If none of these authorities is available, the system checks whether the group has authority to the authorization list if one has been used to secure the object.

If the first group profile doesn't have sufficient authority, the system checks the user's next group and so on until either sufficient authority is found or all groups have been checked. If a user has multiple group profiles, IBM i sums the groups' authorities, if necessary, as it checks the groups. For example, if GROUP_TWO has *OBJOPR private authority and *READ authority to the file, GROUP_THREE has *ADD, *UPDATE, and *DLT private authority, and GROUP_SIX has *EXECUTE authority to the file, the system accumulates the authorities and determines that the user has *CHANGE authority to the file.

Figure 6.4
Authorization Search Progression

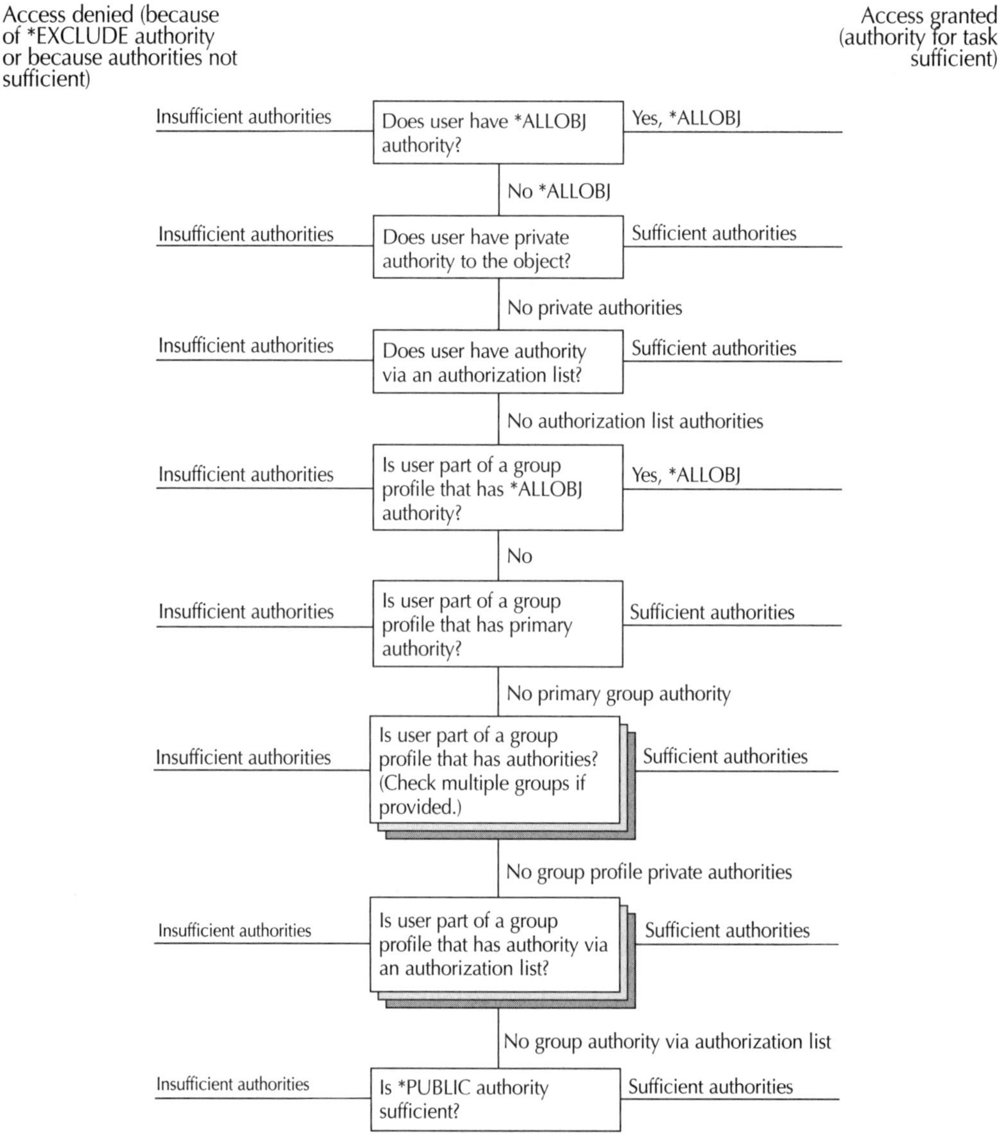

Finally, if neither the user nor his or her group (or groups) has sufficient authority, the system checks *PUBLIC authority. *PUBLIC authority can come either from the object itself or from an authorization list that secures the object.

During each step of the check at the user level and after all group profiles have been checked, when IBM i finds some authority, sufficient or not, the system discontinues the search unless a program in the call stack adopts authority. (I explain adopted authority in detail later in this chapter.) In that case, the system checks the adopting program's owner for authority to the object. The system checks the program owner's authority for *ALLOBJ, private, and authorization list authority (the first three boxes in the figure).

This is how the authorization search works conceptually. However, IBM has made some enhancements to refine the search performance. First, the system determines whether any private authorities exist for the object (i.e., via a private authority granted to a user profile, to a group profile, or to an authorization list securing the object). If no private authorities are present (determined simply by checking a flag in the object's header), the system immediately examines the *PUBLIC authorities. Every object has an internal part called a *header* that IBM i maintains. The header contains basic object attributes, such as the owner and the owner's authorities, the name of the primary group profile and its authorities, and the *PUBLIC authority. Another flag in the object header indicates whether existing private authorities are less restrictive than the public authorities. If existing private authorities are no less restrictive than the public authorities, the system bypasses the rest of the authority search algorithm and immediately checks whether public authority is sufficient. These modifications to the search algorithm speed search performance significantly.

Authority Checking Example: Precedence Between Users and Groups

The following example shows the net effect of users being in a group profile and the precedence that applies when both the user and one of the user's groups has authority to the same object. Take a look at Figure 6.5.

Figure 6.5

Sample Group Profile Implementation

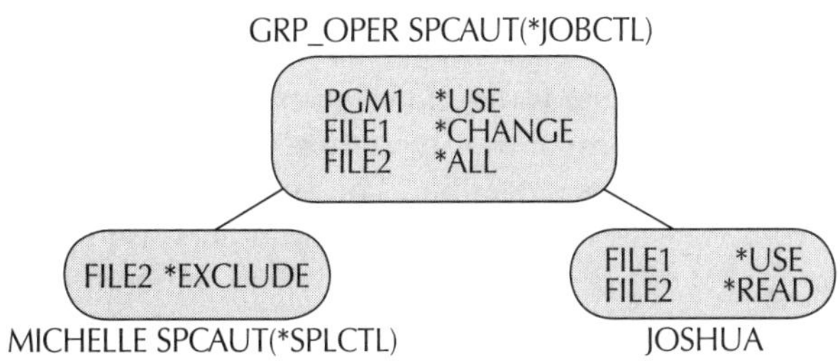

Michelle's authorities	Source	Joshua's authorities
*JOBCTL	Group	*JOBCTL
*SPLCTL	Self	
PGM1 *USE	Group	PGM1 *USE
FILE1 *CHANGE	Group	
	Self	FILE1 *USE
FILE2 *EXCLUDE	Self	FILE2 *READ

The figure illustrates an implementation of a sample group profile, GRP_OPER. Two users, MICHELLE and JOSHUA, are members of this group. When MICHELLE performs IBM i functions, she will do so with both *SPLCTL and *JOBCTL special authorities. MICHELLE has *SPLCTL authority and inherits *JOBCTL from the group. JOSHUA has no special authorities himself but inherits *JOBCTL from the group.

When either MICHELLE or JOSHUA calls program PGM1, the system will determine that the user has *USE authority to the program through the GRP_OPER group.

When MICHELLE accesses FILE1, she will be able to do so with *CHANGE authority from the group. However, because JOSHUA was granted a private authority of *USE to FILE1, he will access the file with *USE authority. The authority that a user has takes precedence over any authority that the group has.

When MICHELLE tries to access FILE2, her access will be denied because she was granted private authority *EXCLUDE. JOSHUA, on the other hand, will be able to access the file with *READ authority. Once again, his user authority takes precedence over the group authority.

Group profiles are a great tool to help deal with the complexities of managing objects' access controls. Many organizations have implemented group profiles, and users within those groups commonly need the same authority to the same objects. Granting authority to the group profile makes this task very easy. Chapter 16 discusses how to use group profiles to implement the security strategy you design for your system.

Authority Cache

IBM i uses an *authority cache* to look up private authorities to objects and authorization lists. The first time the system performs an authority lookup to determine whether a user has specific authority to an object, IBM i creates a cache for the user and adds the specific authority to the cache. The next time the system performs a lookup, it checks the user's cache before checking the profile. This procedure results in a much faster check than looking in the user profile. The cache contains up to 32 private authorities to objects and 32 authorities to authorization lists. Whenever a user's authority changes, the system automatically updates the user's cache. All users' authority caches are cleared at IPL.

The effect of the authority cache is that if users generally access the same objects and have private or authorization list authority to those objects, the performance impact of checking for private authorities is negligible. The authority cache gives you more flexibility in securing your objects, but it doesn't give you license to go wild with private authorities. Massive numbers of private authorities are still difficult to manage and maintain, still cause user profile objects to grow very large, and still affect performance when you save and restore your system.

Adopted Authority

Another way a user can gain authority to an object is through *adopted authority*. You can use adopted authority to temporarily give a user sufficient authority to accomplish a sensitive task. For example, many people write a CL program that adopts and is owned by a profile with *ALLOBJ and *SECADM authority that lets their help desk personnel reset user's passwords. This technique gives help desk personnel the capability to perform the task but doesn't require them to be permanently assigned *ALLOBJ and *SECADM.

When you compile a program on the system (using the CRT*xxx*PGM commands), you can supply a value of either *USER or *OWNER for the USRPRF (User profile) parameter. The default value, *USER, instructs the system to run the program under the authority of the user who is running the program. A value of *OWNER instructs the system to run the program under the authority of the program's owner if the user's profile doesn't have enough authority to perform the program's tasks. Adopted authority remains in effect until the program ends — in other words, for as long as the program is in the program call stack. Any programs called by the program that adopts authority also use the adopted authority unless the called programs specifically use the USEADPAUT(*NO) parameter to block propagation of the adopted authority.

Figure 6.6 illustrates the concepts and implementation of adopted authority.

Figure 6.6
Adopted Authority Concepts

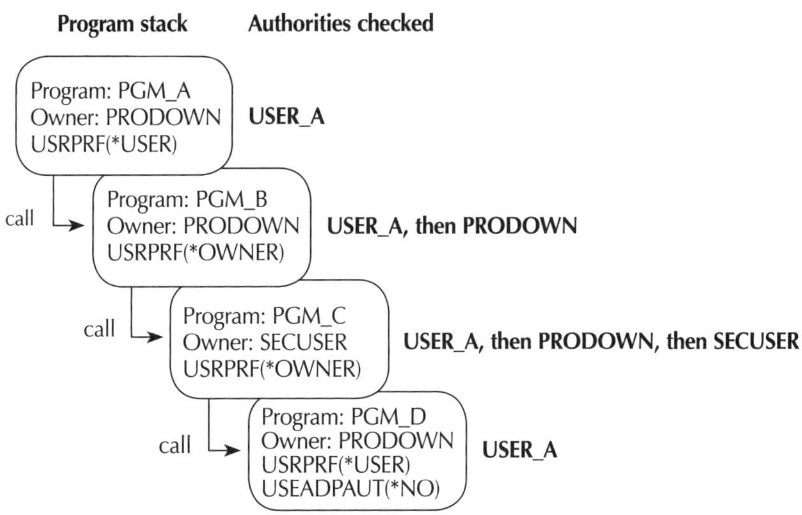

In this scenario, because PGM_A doesn't adopt authority (notice USRPRF(*USER)), IBM i can use only USER_A's authority when the program is run. When PGM_A calls PGM_B, the system uses USER_A's authority. If USER_A's authority is not sufficient, the program owner's (PRODOWN's) authority is used to complete the tasks. When PGM_B calls PGM_C, the authorities available to complete the tasks continue to add up. PGM_C specifies USRPRF(*OWNER), and the owner of PGM_C is SECUSER. The system first determines whether USER_A has sufficient authority to perform the tasks in PGM_C. Then, if necessary, it checks the authority of PRODOWN and SECUSER, in turn, to obtain the needed authorities. Finally, when PGM_C calls PGM_D, two things change. First, PGM_D ignores previously adopted authorities (due to USEADPAUT(*NO)). Second, because program PGM_D doesn't specify USRPRF(*OWNER), PGM_D does not adopt the program owner's authorities. As a result, PGM_D uses only USER_A's authorities.

Technical Note

Many people confuse the purpose of the two program attributes that control adopted authority. It is the URSPRF (User profile) parameter that determines whether the program adopts. The USEADPAUT (Use adopted authority) parameter determines whether the program will "consume," or use, the adopted authority being passed to it from a program higher in the call stack.

Adopted Authority Example

Let's consider another example. A help desk needs to be able to reset users' forgotten passwords and reset the status of the disabled profiles to *ENABLED. To perform these functions, each person on the help desk would need *OBJMGT and *USE authority to the profiles as well as *SECADM special authority. Profiles, by default, are created as *PUBLIC *EXCLUDE, so to set this up you'd have to grant each help desk user or the group these authorities to every user profile. This proposition is problematic on a couple of fronts. First, how do you keep up with granting these authorities as new profiles are created? Second, granting users *USE authority to other individuals' profiles would permit help desk personnel to submit jobs and run as another profile — not a good security practice. In addition, giving them *SECADM special authority would let them create user profiles — again, not a task this role should be performing. How do you allow the help desk to reset passwords and profile status yet not grant the authorities permanently?

First, write a CL program. Let's call it PWDRESET. This program would prompt a help desk member to enter the name of the profile whose password is being changed, along with the new password. The program would then run the CHGUSRPRF (Change User Profile) command, making sure the password is set to *EXPIRED (so the user will have to change the password upon first use) and setting the profile's status back to *ENABLED. If desired, you could add logic to prevent the help desk from changing specific profiles, such as QSECOFR or other profiles with *ALLOBJ.

Next, create the program, specifying user profile *OWNER on the CRTCLPGM (Create CL Program) command and making sure the owning profile has *ALLOBJ and *SECADM special authorities. Set the program's *PUBLIC authority to *EXCLUDE, and grant the help desk group (e.g., GRP_HDESK) *USE authority to the program. Then, assign this program to an option on a help desk menu if one exists. When a member of the help desk group selected this menu option, he or she would be able to change a user's password but would not be able to perform any other tasks as this profile. Nor would this person have *ALLOBJ or *SECADM specify authorities assigned to his or her profile. This is just one example of the type of utilities you can write that permit users to perform tasks while avoiding permanent assignment of special authorities or granting authority to objects on the system.

Here's a summary of some things to keep in mind when you use programs that adopt authority:

- Adopted authority is checked only if existing authority is inadequate.
- The owner's special authorities (e.g., *ALLOBJ, *SECADM) are adopted.
- The owner's private authorities are adopted.
- The owner's group authorities are not adopted.
- Adopted authority is active as long as
    - The program that adopts authority is in the program stack. (You can use the USEADPAUT program attribute to block a program from using adopted authority from a program higher in the call stack.)
    - The program is interrupted by the System request or Attention key, transfer to a group job, a break-message-handling program, or debug functions. (In any of these events, adopted authority is not passed on.)
- Adopted authority cannot be passed to submitted jobs.
- Adopted authority cannot be used to run the CHGJOB (Change Job) command.

Security Recommendation

If you use programs that adopt authority, make sure they perform only one task (e.g., change a password, update prices, retrieve personnel information) and that they don't call other programs (which might propagate authority unnecessarily) or that they call only programs that have USEADPAUT(*NO) specified. Be sure to document the programs thoroughly and make them easy to identify.

Create user profiles specifically to own the programs that adopt authority. Implement these profiles with PASSWORD(*NONE) and with only enough authority for the programs to perform their tasks. Don't use QSECOFR to own programs that adopt authority.

Authorities and Save/Restore Functions

One subtlety in the issue of public and private authorities concerns how the authorities are saved and restored. IBM i saves and restores public authorities with the object, as part of the object header. In contrast, it maintains private authorities as part of the user profile; private authorities aren't saved with the object. When you restore an object and it already exists on the system (i.e., when the object you restore replaces an existing object), IBM i maintains the existing object's public authorities, which may or may not be the same as the authorities of the object on the media as well as the object's private authorities. When you restore a new object, IBM i restores the public authority from the media along with the object, but if the object had private authorities when it was saved, they are not saved in releases prior to V6R1. In V6R1, if you have *SAVSYS or *ALLOBJ special authority, you can specify that the private authorities associated with the object be saved. Then, if you have *ALLOBJ, you can restore the private authorities when restoring the object.

What happens when you restore a user profile? When you restore one or more user profiles, you must also run the RSTAUT command to restore the user profiles' private authorities. This explains why, if you restore your complete system, including all user profiles and objects, but don't run the RSTAUT command, only the public and primary group authorities work. No private authorities exist until you run RSTAUT.

Another interesting point concerning the private and public authorities' save/restore functions relates to users who have *SAVSYS special authority. Let's assume you've given your system operator *SAVSYS authority. Let's also assume that you have a library called PAYROLL that holds all payroll files and that you've used library authority to secure those files. In other words, the library's public authority is *EXCLUDE, and you've authorized only a few users (not including your system operator) to the library. However, you've specified *CHANGE public authority to the files in library PAYROLL. In effect, you've secured the library from all users except those with private authorities. But if an unauthorized user can get inside the library, he or she can have public access to the files.

This security scenario is a "normal" and somewhat typical practice for securing some libraries. However, you need to be aware of a security exposure in this scenario related to saving and restoring the files that reside in the library. If the system operator (who has *SAVSYS authority) were to restore to a nonsecure library files saved from the PAYROLL library, he or she would then have complete public access to the files in the unsecured library. Because public authorities are saved with the object, the files' public *CHANGE authority would be restored with the files to the unsecured library, and the system operator (or any other role or individual with *SAVSYS) could retrieve, modify, and delete records in that library. The system operator could then simply save the modified version of the file and restore that version to the PAYROLL library.

You will probably want to give *SAVSYS special authority to system operators so they can perform system backups, so what's the solution to this exposure? First, you must have some level of trust in the system operator. Second, you should audit the restore operations performed on your system and monitor for any unusual restoration of files into libraries — unscheduled restores or restores into nonsecure libraries, for example. Last, using object-level security for objects that contain private or confidential data (e.g., files containing salary information or cardholder data) will eliminate this exposure.

Object Ownership

Every object on the system has an owner. When an object is created, the object's owner is the user profile that created the object. You can use the CHGOBJOWN (Change Object Owner) command to change the ownership of an object. But what exactly does it mean to "own" an object?

The user profile that owns an object is automatically granted *ALL authority to the object when the object is created or when ownership is transferred to a new owner. Having *ALL authority gives the owner all object and data rights. The owner or another user who has sufficient authority (e.g., *ALLOBJ special authority) can revoke one or more of the owner's authorities to the object. For example, the owner can revoke his or her own data rights to the object, thus protecting the data but retaining the ability to control the object. If for some reason the owner lacks sufficient authority to the library in which the object resides, the owner can't access or manage the object regardless of which object and data authorities the owner has to the object itself.

Object ownership isn't an authority per se, but because ownership is so powerful, you need to control and audit it as you would authority settings. In general, the best way to do so is to create a few user profiles to own specific applications.

Let's say your system has the following libraries, applications, and files:

- FINOLIB — All financial applications
- FINFLIB — All financial database files
- MFGOLIB — All manufacturing applications
- MFGFLIB — All manufacturing database files
- DEVLIB — All development files and applications

In this scenario, I recommend three owners: one for all financial objects, one for all manufacturing objects, and one for the development objects. I would implement this strategy in two steps:

1. Create three user profiles with PASSWORD(*NONE), SPCAUT(*NONE):
    - FINOWNER (financial owner)
    - MFGOWNER (manufacturing owner)
    - DEVOWNER (development owner)
2. Change the ownership of the libraries and corresponding objects as follows:
    - FINOWNER owns libraries FINOLIB and FINFLIB.
    - MFGOWNER owns libraries MFGOLIB and MFGFLIB.
    - DEVOWNER owns library DEVLIB.

The names of the user profiles used as owners should indicate their use. The sole purpose of these profiles is to own the applications and associated objects. The profiles should not be group profiles, and users should not use them to sign on. To ensure that users can't use the profiles to sign on, create the profiles with PASSWORD(*NONE). If you use purchased software, you'll notice that most third-party applications use one user profile to own all the application's objects.

In a development environment, depending on how your change control software works, you may need to assign ownership of development objects to a group profile. This approach lets programmers easily work with objects created by other members of the team without requiring you to implement an elaborate authority scheme. When each member of the development team is a member of the group profile, the group profile does all the work. Some organizations create their developers' profiles with OWNER(*GRPPRF) so that when a developer creates an object, the object is owned by the developer's first group profile. This method works fairly well if all the users create objects in the QSYS file system. However, IBM i ignores the OWNER parameter for objects in other file systems, and the user retains ownership. Given today's environment, in which programmers might be writing Java programs and creating stream files, this method doesn't work well. In this case, you can use the profile-swapping APIs (which I discuss in Chapter 12) or create an automated nightly process that sweeps application directories and sets the appropriate ownership.

Many organizations use QSECOFR as the owner of production objects. This strategy is unwise for two reasons. First, it encourages the use of the QSECOFR profile to implement and modify production objects. QSECOFR should be used only to install or upgrade the operating system or in the case of a system emergency. Second, when QSECOFR owns production objects, there's a temptation to use that profile to create programs that adopt authority. Allowing programs to adopt the authority of QSECOFR, the most powerful profile on the system, is very dangerous. Programs that adopt authority should be owned by a profile with limited authority created specifically for that task.

Security Recommendation

Don't have QSECOFR or other IBM-supplied profiles own production objects.

This chapter has described the specific authorities defined by IBM i along with the features available for implementing these authorities. What I haven't discussed yet is how to implement these features in a comprehensive security scheme that meets your organization's requirements for securing the data that resides on the system That discussion begins in Chapter 12, after an explanation in Chapters 7 through 11 of the tools available for securing the IFS, securing output queues and spooled files, implementing encryption, and establishing network and Internet security.

Limit User Function

When you're designing the security aspects of a new application, you need to determine the ownership and *PUBLIC authority of the application's objects and decide how to control users' abilities to perform tasks within the application. A user's ability to perform a task, or function, within an application often depends on whether the user has authority to an object, such as a command or a file. However, sometimes you need to control whether a user can perform a function within an application when there is no associated object to which you can check the user's authority. The User Function Registration APIs let you define and register a function. You use the APIs in the application to determine whether the user has authority to perform the registered function.

Function definition isn't limited to user-written applications. For example, System i Access defines several functions that let you control which of its features are displayed on a user's desktop. IBM i defines functions to control access to some service tools, such as some DASD management functions and Backup, Recovery and Media Services (BRMS). You should consider functions defined to control what appears on the desktop (e.g., data transfer) to be a modern-day implementation of menu security. In other words, in this case, the support is more a usability enhancement than a security mechanism, and you shouldn't count on those functions to secure your system. However, functions that control access in IBM i (e.g., the DASD management, Service trace, and BRMS functions) are relevant to security and can be considered part of the implementation of your security scheme.

To administer the IBM i functions, you use IBM Navigator for i. Right-click the name of the system you're connected to in Navigator, and select Application Administration. The resulting dialog box, shown in Figure 6.7, lists all registered functions, categorizing them under three tabs. The first tab (labeled "iSeries Navigator" in the figure) lists functions defined by Navigator and its plug-ins. The second tab, Client Applications, lists all other client functions (sometimes referred to as applications), including System i Access. The third tab, Host Applications, lists applications that reside on IBM i, not on the client. Business partners and other application providers can define and register their own functions.

Figure 6.7
Navigator for i Application Administration Window

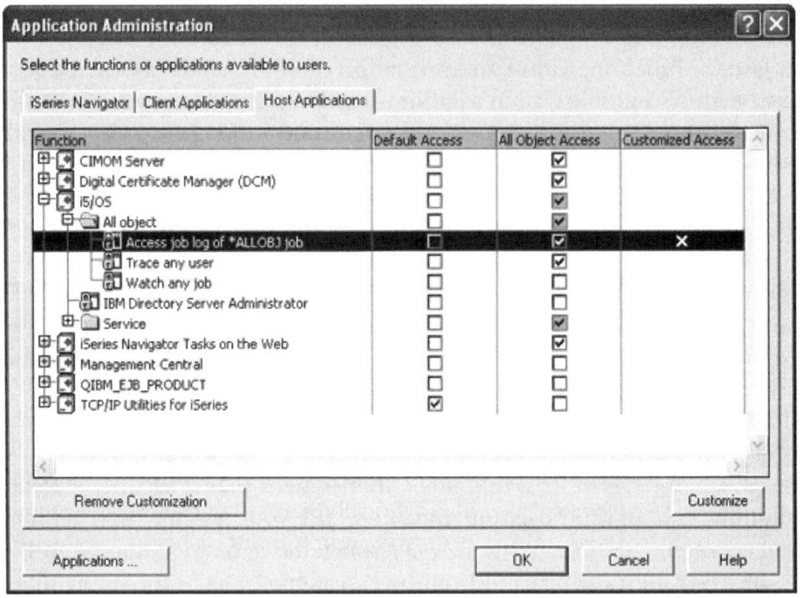

From Navigator for i, you can see which functions a particular user is allowed to perform. To do so, select the system you're connected to, expand Users and Groups, select All Users, right-click the user you're interested in, select Capabilities, and click the Applications tab. Figure 6.8 shows a sample window.

IBM also provides three CL commands for managing application administration: DSPFCNUSG (Display Function Usage), CHGFCNUSG (Change Function Usage), and WRKFCNUSG (Work with Function Usage). You may wonder why the command names use the words "Function Usage" rather than "Application Administration." Navigator for i uses the Function Usage APIs to implement Application Administration. But because "AppAdmin" is easier to say, pronounce, and remember than "FuncUsage," and because AppAdmin was the only way to manage this information until i5/OS, AppAdmin became the way the function was generally referred to. Regardless of the name confusion, these commands let you manage the same information you can manage using Application Administration in Navigator for i.

Figure 6.8

Viewing a User's Capabilities in Navigator for i

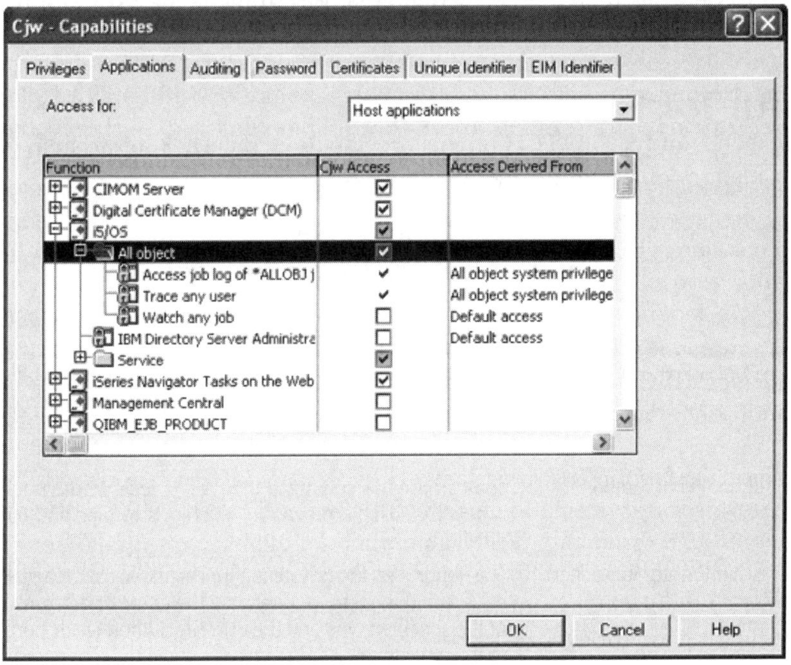

Administrators and auditors who want to determine who is allowed to access which AppAdmin functions (e.g., trace, cluster administration, disk units, service dump) will find the DSPFCNUSG command useful. The CHGFCNUSG command lets an administrator change who is allowed to access which AppAdmin functions and other attributes, such as whether *ALLOBJ authority should be honored when a check is performed. The WRKFCNUSG command lists the AppAdmin functions for which current settings can be displayed or changed.

Here's an example of the CHGFCNUSG command, using my favorite function in Application Administration:

```
CHGFCNUSG  FCNID(QIBM_ACCESS_ALLOBJ_JOBLOG) +
           DEFAULT(*DENIED)                 +
           USER(ALLISONB)                   +
           USAGE(*ALLOWED)
```

This command denies access to the QIBM_ACCESS_ALLOBJ_JOBLOG function for the default user. (Think of the default user as the "*PUBLIC" authority for the function.) The command also allows the ALLISONB user profile to use the function. Why is this my favorite function? Because it gives users who have *JOBCTL special authority the ability to see the job log of an *ALLOBJ user. This feature has allowed many of my clients to remove *ALLOBJ authority from their on-call programmer profiles. Without this feature, if you want to view the job log of a failed job that is running as an *ALLOBJ user, you must have *ALLOBJ authority yourself.

Helpful Tools

The SECTOOLS and SECBATCH menus provide three security tools to help you effectively manage object authorities on your system (Table 6.3).

Table 6.3
Tools for Managing Object Authorities

Security tool	Description
PRTADPOBJ (Print Adopting Objects)	Use this command to print a list of objects that adopt the authority of a specified user profile. You can specify one profile or specify a set of profiles generically (e.g., specify PGMR* to see all user profiles that begin with PGMR).
PRTPUBAUT (Print Public Authorities)	You can use this command to list all objects whose public authority is not *EXCLUDE. You can select objects by type and library or directory.
PRTPVTAUT (Print Private Authorities)	This report lists the private authorities that exist for a specified type of object in a library or directory. The command can list all objects of the specified type and their private authorities, list only changes to the objects since the last time you ran the report, or even list the private authorities that have been deleted since the last time you ran the report.

Navigator for i

Navigator for i provides an interface for managing object authorities. To manage object authorities, which Navigator refers to as a user's *permissions* to an object, navigate to the object, right-click it, and select Permissions. Navigator will display a window similar to the one shown in Figure 6.9.

Figure 6.9
Navigator for i Permissions Window

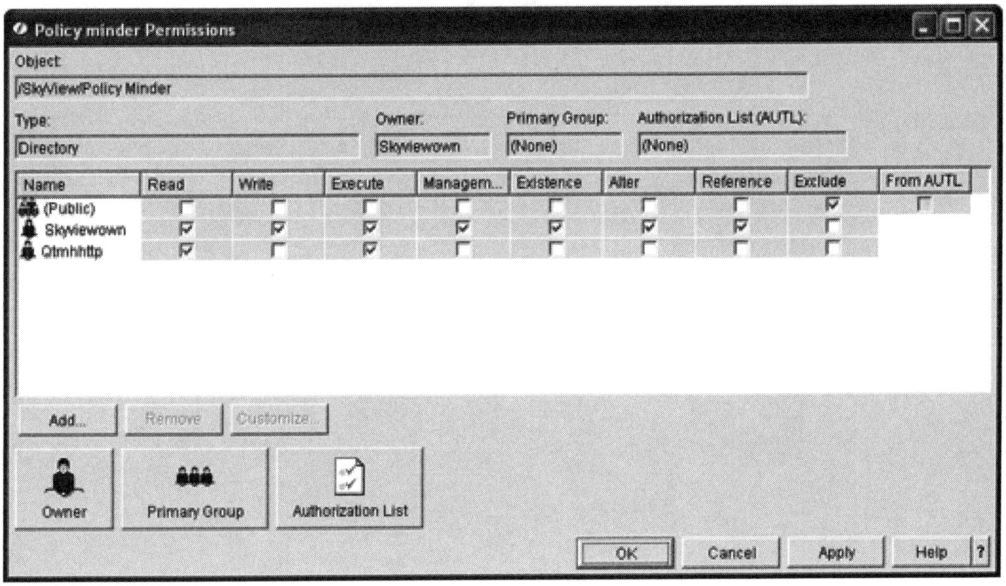

Chapter 7

Security Considerations for the IFS

The *integrated file system*, or IFS, is really a set of file systems that are available on IBM i. The primary use of the IFS is to store objects such as stream files, images, Hypertext Markup Language (HTML), and Portable Document Format (PDF) content that cannot be stored in traditional IBM i libraries.

Table 7.1 lists the file systems encompassed by the IFS. The IFS always includes the root ('/') and QSYS.LIB file systems (QSYS.LIB uses the traditional *library_name/object_name *objtype* syntax to refer to objects) and may include the Network File System (NFS), the QNTC file system, and other file systems.

Table 7.1
File Systems in the IFS

File system	Description
/ — "Root" file system	Serves as an "umbrella" over the other file systems
QOpenSys — Open systems file system	Contains Unix-based files and directories
UDFS — User-defined file system	Contains user-defined and -managed file system(s)
QSYS.LIB — Library file system	Contains IBM i libraries and other types of IBM i objects
Independent ASP QSYS.LIB — Independent ASP library file system	Supports the server library structure in any independent auxiliary storage pools (ASPs) you create and define
QDLS — Document library services file system	Contains document library objects (DLOs) and folders
QOPT — Optical file system	Contains stream files stored on optical media
QNTC — Windows NT Server file system	Provides access to data and objects stored on a server running Windows NT (4.0 or higher), including Integrated System x Servers or System x servers attached via the Integrated System x Adapter
QFileSvr.400 — i5/OS File Server file system	Provides access to other file systems that reside on remote System i systems
NFS — Network file system	Provides access to data and objects stored on a remote Network File System (NFS) server

Most people use the term "IFS" to refer to the parts of the IFS that include the root, QOpenSys, and user-defined file systems. These file systems use a different authority scheme than QSYS.LIB uses — one based on Unix. Other file systems use their own authorization method. This chapter describes the security considerations for root and the other file systems that follow the Unix authority scheme. When I use the term "IFS" here, I'll be referring to the file systems that follow the Unix approach to security.

Security for the IFS is an issue that should concern all IBM i security administrators because of the way the IFS comes secured — or, to be more accurate, *not* secured — from IBM. By default, the IFS is shipped wide open — with the equivalent of *PUBLIC(*ALL). Unfortunately, when it comes to managing access and appropriate authorities, the IFS is one of the most ignored areas of the system. Although it's rarely necessary to rework the security scheme of the *entire* IFS, most systems usually have one or two application directories, directories containing WebSphere or other Web application configuration files, or directories with stream files containing payroll, tax, credit card, or other sensitive information that need to be secured. This chapter should help you determine how best to secure these directories and their objects.

IFS Authorities

The Unix security scheme used by the IFS file systems defines the following authorities:

- **R* (read authority) permits a user to list the contents of a directory or read an object (e.g., a stream file).
- **W* (write authority) lets a user add a new object (e.g., a stream file, a subdirectory) to a directory.
- **X* (execute authority) permits a user to traverse a directory.

You can use these three basic authorities in combination to provide different levels of access to IFS objects and data. **RWX* authority, for example, is equivalent to the IBM i authority of *CHANGE. **RX* authority equates to *USE. Table 7.2 shows how the IFS's Unix authorities map to the IBM i authorities you're used to.

Table 7.2
IFS Authorities and IBM i Authorities

Authority	*RWX	*RW	*RX	*R	*WX	*W	*X
Object							
*OBJMGT	—	—	—	—	—	—	—
*OBJEXIST	—	—	—	—	—	—	—
*OBJALTER	—	—	—	—	—	—	—
*OBJREF	—	—	—	—	—	—	—
*AUTLMGT	—	—	—	—	—	—	—

continued...

continued...

Authority	*RWX	*RW	*RX	*R	*WX	*W	*X
Data							
*OBJOPR	X	X	X	X	X	X	X
*READ	X	X	X	X	—	—	—
*ADD	X	X	—	—	X	X	—
*UPD	X	X	—	—	X	X	—
*DLT	X	X	—	—	—	—	—
*EXECUTE	X	—	X	—	X	—	X

When managing access to objects in the IFS (remember, I'm talking about objects in the file systems that use the Unix security scheme), you must deal with two sets of authorities: object authorities and data authorities. As you can see in the table, the Unix authorities *R, *W, and *X (and their combinations) map to data authorities. (Note that the table lists *OBJOPR authority, which is an "object authority" in IBM terms, with the data authorities that make up the implementation of the Unix authorities on IBM i.) However, even though you may be working with a Unix application and its objects, "under the covers" the objects are still IBM i objects, so you can't ignore the fact that all IBM i authorities must be accounted for — thus, the presence of the object authorities in the table.

Managing Authorities to IFS Objects

To manage the authorities to objects in the IFS, you can use either green-screen CL commands or the Navigator for i interface.

The WRKAUT (Work with Authority), CHGAUT (Change Authority), and DSPAUT (Display Authority) commands provide the green-screen interface. When issuing these commands, you identify the object by specifying a path name rather than using the traditional *library_name/object_name *objtype* naming convention you're accustomed to. For example, to work with the authorities to the object proj1.txt in the directory /home/cjw/projects, you'd use this command:

```
WRKAUT  '/home/cjw/projects/proj1.txt'
```

If you're uncomfortable typing a path name with the CL commands, you can opt to use the WRKLNK (Work with Object Links) command to navigate through the directory structure to the desired directory or object. When you reach the directory or object for which you want to view or change authorities, enter option 9 (Work with authority) beside it. The WRKAUT command screen will be displayed with the path name filled in for you.

Your other choice for managing IFS objects is to use Navigator for i. To view the authorities for the root directory using this method, go to My Connections | *System_name*|File Systems|Integrated File System. Right-click on Root, and choose Permissions. Navigator will display a window equivalent to the WRKAUT command display.

Figure 7.1 shows the WRKAUT command's display of the default access for the root directory as IBM ships it. Note that *PUBLIC has data authorities *RWX and all object authorities. In "traditional" IBM i security terms, this authority is equivalent to *PUBLIC(*ALL). Figure 7.2 shows Navigator's view of the same information — the authorities in force for root.

Figure 7.1

WRKAUT Command Display: Default Authority to Root ('/')

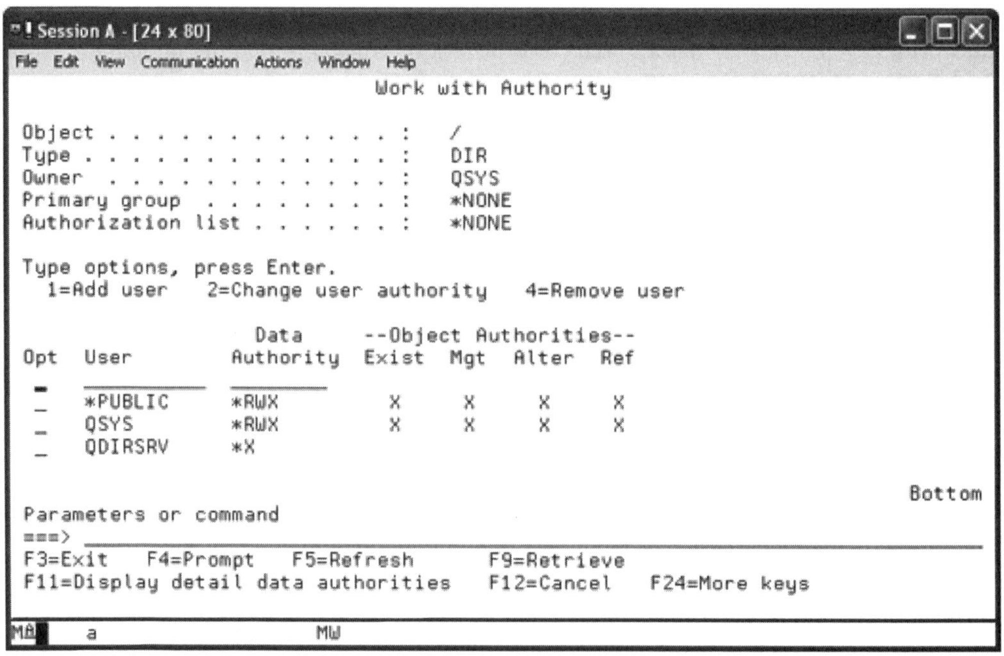

Figure 7.2

Navigator for i Permissions Window: Default Access for Root ('/')

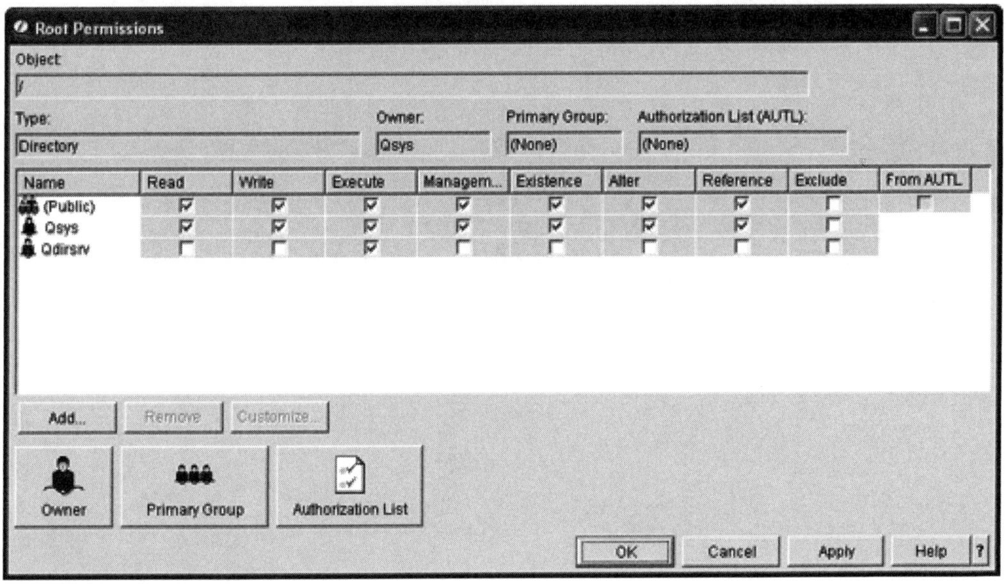

You can use the green-screen or Navigator interface to manage the authority of any object, including objects in the QSYS.LIB file system. For example, the screen shown in Figure 7.3 lets you manage authority to the APPDEVSRC/AR_SOURCE file by using the CHGAUT command. In this example, the IBM i object is identified using a path name.

Figure 7.3

Managing Authorities to an IBM i Object with the CHGAUT Command

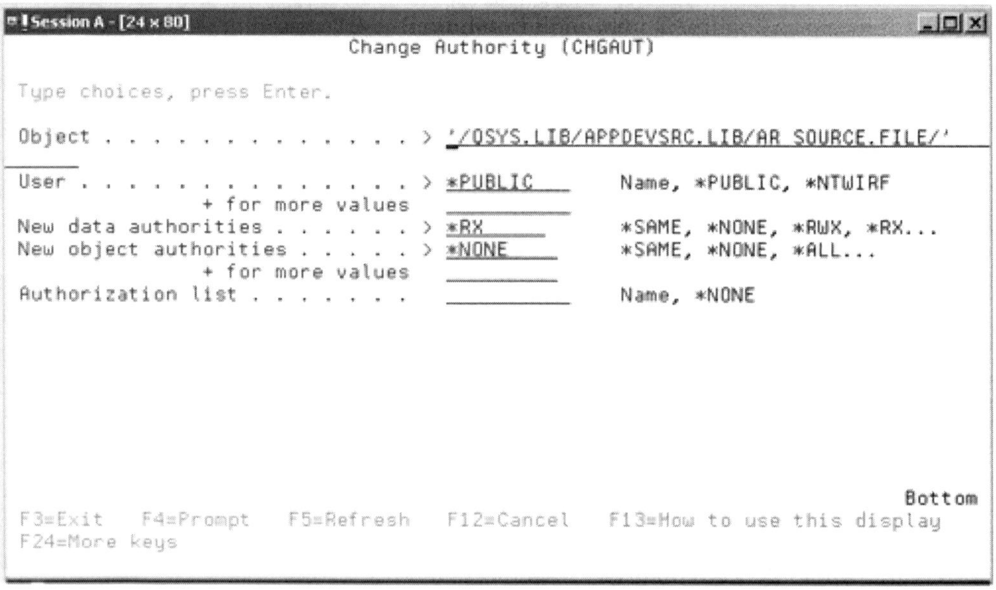

File Attributes

Objects in the IFS possess some PC-like properties called *file attributes* that contribute to the way their access is determined and how the objects are managed. File attributes are properties of a file that work in addition to (not as a replacement for) the IBM i authority settings for the file. There are six file attributes:

- *Read-only* — The file or folder can only be read. It cannot be updated, changed, or deleted when this attribute is on.
- *Hidden* — The file or folder is hidden and can't be worked with unless you know its name.
- *System* — The file is a system file. (A folder cannot have this attribute set.)
- *Archive* — This attribute indicates whether the file or folder should be archived. Some programs use this attribute to determine which objects are backed up.
- *Archive* (i5/OS attribute) — This attribute has the same meaning as the Archive attribute. It is always selected and cannot be changed.
- *Allow save* — Determines whether the file will be saved.

To set the file attributes for a file in the IFS, navigate through the IFS to the desired file or folder. Right-click on the object, and select Properties. A window similar to the one shown in Figure 7.4 is displayed. You use the check boxes at the bottom of the window to set the file attributes of the object.

Figure 7.4
Setting File Attributes for an IFS File

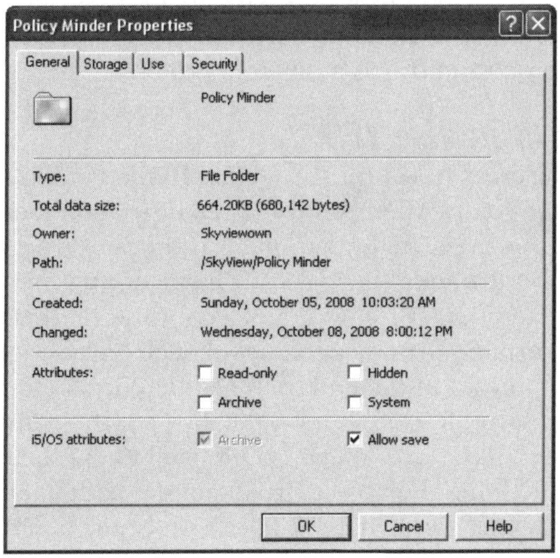

You can set the file attributes for objects that reside on IBM i by using PC interfaces or by running the CHGATR (Change Attribute) CL command.

Adopted Authority and the IFS

If you've written a traditional application that runs with adopted authority, you know that as long as the owner of the program that's running either owns or has authority to the object (e.g., a file) being accessed, the person running the program can access the file. However, this scheme doesn't work for accessing objects in the IFS because the IFS ignores adopted authority.

To access an object in the IFS, the user running the process must have authority to the object. For such applications, you'll need to either explicitly authorize your users (e.g., give them a private authority) to the required objects or change the process or thread so that it runs as a different user who has sufficient authority to the IFS objects being accessed. To run as a different user, you have several choices: profile swap, profile token, or the UID and GID family of APIs. Chapter 12 discusses these approaches in detail.

Security Recommendation

Unless you need to allow directory access through an application (such as a user selecting a menu option to access a file in the IFS), I usually take the approach of granting authority to the users who need access to the directory. In most cases, only a handful of users need access to a directory that is being locked down, and they typically need access for a File Transfer Protocol (FTP) or Open Database Connectivity (ODBC) connection rather than through an IBM i application. When the access is a direct access rather than application access, I find it easier to grant the users authority rather than writing the code to switch the process to run as a different profile.

Auditing Objects in the IFS

Just as you can audit objects in IBM i (or the QSYS.LIB file system), you can audit objects in the IFS. The challenge comes when you try to read the audit journal entries for these objects. The objects' names are actually path names and can vary quite a bit in length; therefore, they don't fit in the audit journal's 10-character object and library entry fields.

You'll recognize an audit entry for an IFS object by the *N that appears in place of the object name. To determine the object name for such entries, refer to the Absolute Path Name field, which appears at the end of the audit journal entry. If the path name is longer than 5,002 bytes, you'll see the first 5,002 bytes of the path name in this field. If you want to see the rest of the path name, you'll need to use the file ID (FID), a number that uniquely identifies the object. To obtain the path name from an FID, you must use the API Qp0lGetPathFromFileID() (Get Path Name of Object from Its File ID). Fortunately, not too many path names exceed 5,002 bytes!

File Shares: Accessing Objects in the IFS

File shares are what make the IFS, or part of the IFS, available for viewing or manipulation via the network. You use file shares when you want all users on the network to have access to something. For example, you may want to use the QIBM file share (which is shipped with System i Access) so that users can easily find and download the latest version of System i Access — unless letting users download updates is totally against your policy. If that is the case, you probably want to remove that share.

It's important to note that although an individual file or the entire root directory may be shared on the network, users who attempt access still need the appropriate authority to a directory or file to perform the intended request. Just because a file is shared doesn't mean everyone can read it. That ability is still determined by the object authority settings on the directory or library being shared. (Yes, you can share a library!)

When you view directories or libraries under "Integrated File System" in Navigator, a hand beneath a directory or library name indicates a share. In Figure 7.5, the hand under the QIBM directory in the left pane indicates that a share has been defined for QIBM (as shown in the pane on the right). (You may find the hand a little hard to recognize in the figure.)

Figure 7.5
Share Defined for Directory QIBM

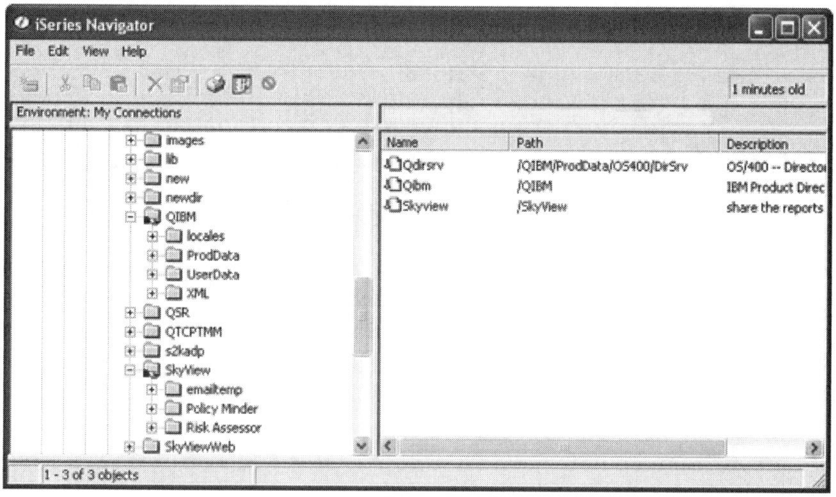

A file share can be defined as either read-only or read/write. The IBM-shipped file shares Qdirsrv and Qibm are read/write shares. That means that if you map a drive to the IBM i system, users can write to these directories (assuming the users have sufficient IBM i authority to do so). The Qca400 file share is read-only; users can read from this directory, but they can't write to it. Even if a user has *W (write) authority that lets the user write to the directory through IBM i interfaces, defining the share as a read-only share permits that user to read from the directory but not to write to it.

You can choose to remove any of the IBM-defined file shares and/or create file shares of your own. To create a new file share, you use Navigator. Go to My Connections|*system_name*|File Systems|Integrated File System, right-click the directory or library you want to share, and select Sharing. You can choose to make the share read-only or read/write. Obviously, the more secure setting is read-only because it lets users only read the data and not update it. (Again, even if you do create a read/write share and a user uses it, the user has to have IBM i object authority to the directory as well as to any object for which access is attempted.) You can also choose how many users can use the share at one time.

To see the existing file shares and the type of access defined for them, go to My Connections|*system_name*|File Systems|File Shares. Right-click a share, and choose Properties to see whether the share is defined as read-only or read/write. Figure 7.6 shows the share properties defined for the root directory. As you can see, in this example root is defined with read/write access, and an unlimited number of users can use this share.

Figure 7.6

Share Properties Defined for the Root ('/') Directory

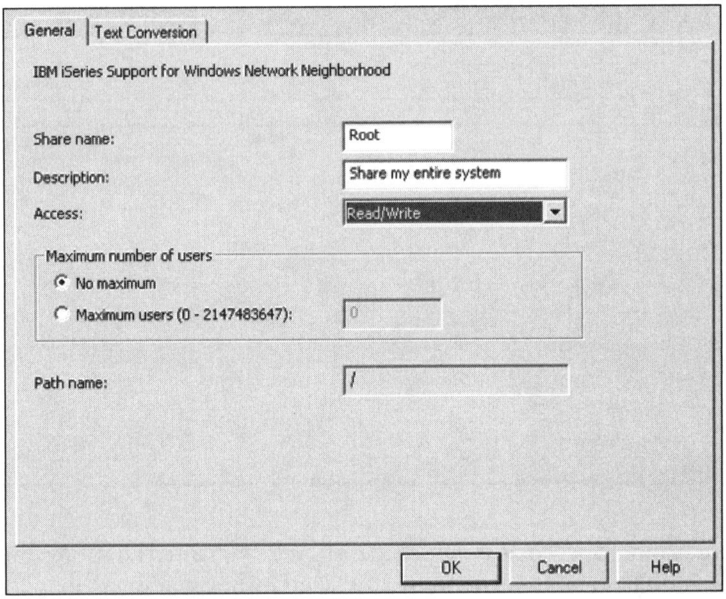

Creating a file share to the root directory can cause significant risk to your system. Sharing root ('/') exposes the directory structure of your entire IBM i system onto your network. If you don't have object-level authority set properly, any user who can map a drive can access any file on the system whose *PUBLIC authority is set to *R (read) authority or *USE authority. On many systems, this would mean that most, if not all, files were accessible. Unlike on Windows, you can't define particular shares for particular users. Once you've created a share, it is available for all network users who possess an IBM i user profile and password.

Gotchas and Helpful Hints

This next section is meant to help you avoid some of the common "gotchas" associated with the integrated file system.

General Cautions

- It is very easy to over-secure the IFS. Be sure to document the authorization steps you take in case you need to back out of them.
- As I've said, adopted authority isn't honored within the IFS. If a user lacks authority to an object, the application must accommodate that user through a version of a profile swap or (my preference) through a private authority to the user's or the group's profile.

- The IFS does not honor the attribute of a user profile that automatically assigns ownership of newly created objects to the user's group (the OWNER parameter). The user — not the group — always owns a newly created object in the IFS.
- Displaying the contents of a directory object requires *RX data authority to every directory in the object's path. Changing the current directory requires only *R data authority to the directory. Therefore, it's possible for users to change their working directory to a directory for which they can't display the contents!
- Don't remove the private authorities that IBM i grants for some IBM-supplied profiles unless you know why they've been added.

Creating New Objects

- In general, a new object inherits the authorities of its directory. (The exception is when you create IFS objects through APIs such as mkdir(), open(), or creat(). With these APIs, you can specify the data authorities for the owner, primary group, and public authorities.) Inherited authorities include the primary group authority, authorization list, and private authorities. The only difference would be the owner if a user other than the directory owner creates the object.
- Removing the owner's authority to a directory can produce undesirable results when users create objects into that directory. When an object is created into a directory, the owner of the new object receives the same authority as the owner of the directory. Therefore, if a user creates a file into a directory whose owner has had his or her authority removed, the user will have no authority to the newly created object.
- Given that most IFS APIs let you specify the data authority settings for the owner, primary group, and *PUBLIC on the API call, you may need to educate your developers to ensure that they specify appropriate values for these parameters. Note that because most of these APIs are Unix APIs, there is no concept of object authorities; therefore, the object authorities are inherited from the parent directory.
- You have the option of specifying to have the primary group and its authorities set based on the user creating the object.

Copying Objects

- The CL COPY command creates an exact duplicate of an object, including its authorities (except for the owner, of course). The CPY command does the same thing. COPY is actually an alias of the CPY command, and you can use these two commands interchangeably.
- Because the copy commands duplicate authorities, to retrieve the authority information about the original object a user must be the owner or have *OBJMGT authority to the object being copied.

- Both the COPY and CPY commands let you make the owner of a copied object the same as the owner of the object being copied. However, the default is to make the user who is making the copy the owner. Depending on how these commands are used in your environment, it may be helpful to educate developers about this fact and retain the ownership of the object being copied (or you could change the command default).

Appendix D of IBM's *System i Security Reference* (available as a PDF file under the InfoCenter's Security topic) lists other authority requirements for objects, under the heading "Integrated File System Commands." This documentation can be very useful in determining what authority is required.

CPYTOSTMF and CPYTOIMPF

The biggest IFS-related issue that I see in most organizations occurs when they are running the CPYTOSTMF (Copy to Stream File) and CPYTOIMPF (Copy to Import File) commands. That's because, by design, these commands default the owner's authority to *RWX and set the primary group and *PUBLIC authorities to *EXCLUDE. If the directory into which the files are being created is secured with an authorization list or has private authorities, neither of these authorities is copied. Subsequent attempts to access the objects will work only if the user is the one who created the objects — which usually is not the case. So, what do you do?

If your organization has written the code, you'll want to immediately follow the creation of these files with a CHGOWN (to change the owner) or CHGAUT (to change the *PUBLIC authority) command. If the files in question are created via a vendor package, your choices are to implement the IFS exit points and write a program that fixes the authorities after the file is created, write your own routine that runs periodically to fix the ownership/authority, or buy a vendor package that resets the authorities.

New in V6R1

Relief comes in V6R1 when the CPYTOSTMF and CPYTOIMPF commands provide the ability to specify that authorities should be inherited from the directory.

Virus Scanning

Although IBM i itself isn't affected by PC- or Unix-based viruses and worms, the objects stored in the IFS can be. In addition, the IFS has proven many times that it can very efficiently and effectively store and propagate viruses. Today, most viruses are caught by firewalls, routers, or e-mail scanners and never reach the IFS. But some organizations still require that the IFS be scanned for viruses. You can accomplish this by purchasing a vendor package or by pointing your Linux or Unix antivirus solution toward the IFS or porting it to run natively in the Portable Application Solutions Environment (PASE).

Two security-related system values pertain directly to the IFS: QSCANFS (Scan File Systems) and QSCANFSCTL (Scan File Systems Control). These system values work in conjunction with some file system exit points and are described in more detail in Chapter 4. While the motivation for IBM to provide this capability was to give vendors the ability to perform realtime virus scanning, use of QSCANFS and QSCANFSCTL is not restricted to that function. With a little imagination and some programming, you could write a program to fix the authorities and ownership of objects, solving some of the pre-V6R1 trials involved with creating stream and import files.

Security Recommendations

I conclude this chapter with some advice for securing the IFS appropriately. For more information about the IFS itself, see the "Files and file systems|Integrated file system" topic in IBM's Information Center.

*PUBLIC Authority for Application and User Directories

The approach you take to setting authorities on directories should be no different from the approach taken with libraries and applications whose objects reside in libraries. Just as you can set the *PUBLIC authority of an application library to *EXCLUDE and grant authority only to the group whose users have a business need to run the application, you can do the same for directories. This translates into setting directories that contain private or confidential data to *PUBLIC authority OBJAUT(*NONE) DTAAUT(*EXCLUDE) and granting authority to only those users (or, my preference, their group) who have a specific business need to access objects in the directories. A good example is a directory containing a stream file or text file of credit-card transactions to be sent to the bank for processing, a file containing payroll information about to be sent to an outsourced payroll processor, or an uploaded file of users' tax information. Obviously, such a directory should be available only to the group of users who have a direct need to access or process this information. To ensure the appropriate users can access the information, grant their group DTAAUT(*RWX) OBJAUT(*ALL), or have the group own the directory and the files in it.

*PUBLIC Authority for IBM-Supplied Directories

Setting authorities on IBM-supplied application directories is much like setting authorities on system-supplied libraries. Just as many operations would fail if you were to set *PUBLIC(*EXCLUDE) on QSYS, many applications — especially Web and Lotus Domino applications — will fail if you exclude *PUBLIC from the root directory.

IBM's recommended *PUBLIC authority setting for root is OBJAUT(*NONE) DTAAUT(*RX). (You can also apply this recommendation to the /QOpenSys directory.) It's possible to get by with DTAAUT(*X), but some applications need to read the contents of the root directory and therefore need *R authority. Setting DTAAUT to *RX will prevent users and applications from creating subdirectories into the root directory.

With the exception of root and QOpenSys, most IBM-supplied directories have their *PUBLIC authorities set appropriately — that is, *PUBLIC OBJAUT(*NONE) DTAAUT(*RX). However, you'll want to review each of these directories to make sure. None of them should default to *PUBLIC OBJAUT(*ALL) DTAAUT(*RWX).

In addition, most IBM application directories have given the necessary private authorities to the appropriate system-supplied user profiles. For example, the QTMHHTTP profile has been given private authority to the directories and objects it needs to run the Apache Web server. You do not want to remove or otherwise alter these private authorities unless you've thoroughly researched the ramifications of doing so.

Determining Appropriate Authority

To determine the appropriate authority setting for a directory, you first need to determine the directory's purpose. If a directory is associated with an application, set *PUBLIC authority as noted above. Users who need to access the application require the following settings:

- OBJAUT(*NONE) and DTAAUT(*X) to all directories in the path to the application objects
- OBJAUT(*NONE) and DTAAUT(*X) for the directories containing the application objects (users may need DTAAUT(*RX) if the application needs to read the contents of the directory)
- OBJAUT(*NONE) and DTAAUT(*RWX) for the directory if the application creates objects into it
- OBJAUT(*NONE) and DTAAUT(*WX) for the directory if the application renames or deletes objects
- OBJAUT(*OBJMGT) at the object level for objects being copied or renamed
- OBJAUT(*OBJEXIST) at the object level for objects being deleted

Home Directory

You may want to create a home directory for each of your developers (e.g., /home/cjw) and let them use that directory as their "playground." Doing so is analogous to giving developers their own IBM i library to use. Do you let your developers create whatever libraries they want to create? Usually not — especially not on a production machine. Therefore, you'll probably want to establish the following structure:

- Set *PUBLIC authority to /home to OBJAUT(*NONE) and DTAAUT(*X).
- *Set PUBLIC authority to /home/developer_name to OBJAUT(*NONE) and DTAAUT(*EXCLUDE). If other users need to be able to access objects in the user's directory, you can give them (or their group) DTAAUT(*RX) to /home */developer_name.*
- Have the developer either own or grant OBJAUT(*ALL) DTAAUT(*RWX) to /home*/developer_name.*

Web Applications

The QTMHHTTP user profile needs authority to the directories containing the Web server configuration files. The profile under which a Web application runs needs authority — at least DTAAUT(*X) if not DTAAUT (*RX) — to the directories containing the HTML, graphics, and so on. This profile is typically QTMHHTTP; however, an application profile can be specified, or the application can run under the IBM i user profile of the user requesting access to the application (i.e., the URL). The user profile under which the Web server accesses resources for an application is specified as part of the Web server configuration directives in the Web server configuration file. (QTMHHTTP has already been given the appropriate private authorities.)

QPWFSERVER Authorization List

If you want to prevent users from accessing the QSYS.LIB file system through System i Access file serving (i.e., via Navigator for i), NetServer file serving, and file serving between IBM i systems, you can do so by removing the users' authority to the QPWFSERVER authorization list, which IBM ships with IBM i. The *PUBLIC authority to QPWFSERVER defaults to *USE. Setting *PUBLIC to *EXCLUDE prevents users from accessing QSYS.LIB through interfaces such as Navigator.

Another way to think of this restriction is as preventing access through directory structure interfaces such as Windows Explorer. Interfaces such as File Transfer Protocol (FTP), Open Database Connectivity (ODBC), and Distributed Data Management (DDM) do not check the authority to the QPWFSERVER authorization list.

I recommend setting the *PUBLIC authority to the QPWFSERVER authorization list to *EXCLUDE and giving only those users who have a business need *USE authority to the authorization list.

Review (and Remove) File Shares

Many systems — especially development systems — have an overabundance of file shares. Most of the shares have been added without any realization as to what is being made available to the entire network. You should review the file shares that have been defined on your systems and remove those that are inappropriate.

You'll also want to add a section to your security policy regarding shares:

- Determine the policy for who can create shares.
- Determine the policy for what should be shared across the network.
- Communicate these policies to developers, network administration, tech support, and security officers.
- Set appropriate authorities on all directories and their objects, especially those that are shared across the network.

If at all possible, do not share the root directory. If you must share it, make sure you have an appropriate object-level security scheme in place. In addition, place a dollar sign ($) at the end of the share name (e.g., root$) to prevent its name from being broadcast throughout the network via Windows Network Neighborhood.

To prevent users from being able to create file shares, change the QZLSADFS (Add File Share) and QZLSCHRS (Change File Share) APIs from their default setting of *USE to *PUBLIC *EXCLUDE.

Final Advice

As I said in the beginning of this chapter, the IFS is the most ignored part of IBM i — especially when it comes to security. Don't propagate the problem! Just because you may be unsure how to clean up what's already been placed into the IFS, don't let it get worse. Go ahead and change the settings on the directories, set the standards, and then communicate them to the appropriate groups (e.g., development, tech support, security team) within your organization. As you communicate and enforce these standards, some of the "fallout" will help you determine what is already in the IFS and how you need to clean it up. At this point, the worst action is no action.

Helpful Tools

The IBM i SECTOOLS menu provides two tools to help you manage authorities to IFS objects. Both the PRTPUBAUT (Print Public Authority) command and the PRTPVTAUT (Print Private Authority) command let you specify a path name for the object name. When specifying a directory, you also have the option to search the subdirectories and include those objects in the report. Warning: If you run the report for root ('/'), the printed report for either command can be huge.

Two other commands can be helpful in determining ownership and other object attributes for objects in a directory. The RTVDIRINF (Retrieve Directory Information) command produces an intermediate file of information, which you can then print using the PRTDIRINF (Print Directory Information) command. Unfortunately, the file produced by running RTVDIRINF is in a format that renders the command unusable for running queries or using SQL statements to produce customized reports.

Chapter 8

Securing Your Printed Output

Many organizations fail to consider the fact that they may have confidential or private information that gets printed. If your security policy requires you to secure all confidential or private information, you need to think about whether or not this type of data is part of your printed output. If it is, you'll need to assess the security settings on the output queues that contain the spooled files (printed output).

In many organizations, printers are everywhere — in public areas as well as on employees' desks — and they produce countless documents and reports containing significant customer and corporate information. For some companies, the proliferation of printers and printouts could constitute a vulnerability and a compliance issue. It's not at all unusual for competitors to try to bribe programmers and operators to obtain printed reports containing strategic information. Almost everyone in an organization can print a plethora of reports and walk right out of the building with the information. And hackers have been known to "dumpster dive" — to literally shift through the garbage in a dumpster looking for reports that include private information (e.g., credit-card numbers, social security numbers, social insurance numbers).

To reduce the chances of printed output getting into the wrong hands, you may need to secure output queues, reorganize menus, and even move printers to more secure locations. Then, after the information has been printed, the printouts should be destroyed appropriately — most likely through shredding. As part of your employee awareness program and training about your organization's security policy, users need to be educated not to just toss these reports into their wastebaskets.

Output queues (outqs) are a functional necessity in IBM i, but that doesn't mean all users need to have access to data in all output queues. Yet that's exactly the way many organizations' outqs are configured. To appreciate the importance of controlling access to output queues, remember that you can use output queues not only to print spooled files but also to display them. Output queue security is one of the least obvious forms of security in IBM i. To achieve the desired results, you must carefully plan and design your output queue strategy.

Security-Related Output Queue Attributes

You create output queues using the CRTOUTQ (Create Output Queue) command. The values you specify for a few security-related CRTOUTQ command parameters determine how you control user access to output queues and spooled data. You can change these output queue attributes using the CHGOUTQ (Change Output Queue) and WRKOUTQD (Work with Output Queue Description) commands. (Before you can change the attributes for an output queue, you must first stop any writers attached to it.)

Four CRTOUTQ command parameters — DSPDTA (Display Data), OPRCTL (Operator control), AUTCHK (Authority check), and AUT (Authority) — work with an output queue's existing public and private authorities to determine who can and cannot view and manipulate the output queue's spooled files. Before you start to design your output queue security strategy, you need to fully comprehend the role of each of these attributes. You'll also want to understand the authority required to start and stop the writers.

Don't feel badly if you don't grasp all the nuances of securing outqs the first time you read this chapter. The attributes interact with each other differently depending on the value of each attribute. I often have to go back and look at the charts before I can achieve the desired results.

DSPDTA (Display Data)

The CRTOUTQ command's DSPDTA parameter specifies the access that users with at least *READ authority have to the outq. A value of *YES means that users who have *READ access can display, copy, and send the data from any spooled file on the queue — using the SNDNETSPLF (Send Network Spooled File) or SNDTCPSPLF (Send TCP/IP Spooled File) command, for example. A value of *NO specifies that users who have *READ authority to the output queue can display, copy, and send the output data only from their own spooled files, unless

- They have *JOBCTL special authority and the output queue's OPRCTL attribute value is *YES, or
- They have *CHANGE authority to the output queue and the output queue's AUTCHK attribute value is *DTAAUT, or
- They own the output queue and the AUTCHK value is *OWNER.

If the DSPDTA value is *OWNER, only the owner of a spooled file can display, copy, send, or move the spooled file. *OWNER is the appropriate DSPDTA value for operators who manage output queue entries but shouldn't be able to view the contents of spooled files. When you specify OPRCTL(*YES) and DSPDTA(*OWNER) for an output queue, a user profile with *JOBCTL special authority can hold, change, delete, and release spooled files on that output queue but cannot display, copy, send, or move the spooled files.

OPRCTL (Operator Control)

The CRTOUTQ command's OPRCTL parameter specifies whether a user who has *JOBCTL special authority can manage or control the files on an output queue. When you specify OPRCTL(*YES), a user profile with *JOBCTL special authority can control the output queue (e.g., hold, release, start a writer to the queue) and manage the spooled file entries (e.g., hold, release, change, delete). As described above, when the output queue's DSPDTA attribute value is *NO, user profiles with *JOBCTL special authority can still display, copy, send, and move all users' spooled files.

Before assigning *JOBCTL special authority to users' profiles to give them the capability of starting their own print writer, you should consider providing them with a program that adopts authority and starts the writer for them. Programs that adopt authority also adopt the owner's special authorities, such as *JOBCTL. Assigning users *JOBCTL gives them significantly more capabilities than just starting and stopping writers. *JOBCTL lets users hold, change the priority of, or even end any job on the system. Providing a program that performs the starting and stopping of writers permits users to take these actions but doesn't give them the full capabilities associated with assigning *JOBCTL to their user profiles.

AUTCHK (Authority Check)

The CRTOUTQ command's AUTCHK parameter specifies whether the commands that check a requesting user's authority to an output queue should check whether the user owns the output queue (*OWNER) or just has data authority (*DTAAUT) to the output queue. When the AUTCHK value is *OWNER, only the user who owns the output queue can change or delete spooled files owned by other user profiles. When the value is *DTAAUT, any user with *READ, *ADD, and *DLT authority to the output queue can change or delete other users' spooled files on that output queue.

AUT (Authority)

Last, the CRTOUTQ command's AUT parameter specifies the *PUBLIC authority of the outq. You can use the EDTOBJAUT (Edit Object Authority), GRTOBJAUT (Grant Object Authority), or RVKOBJAUT (Revoke Object Authority) command to modify this authority.

*SPLCTL Special Authority

The last, and perhaps most important, point to remember regarding output queue security is that a user who has *SPLCTL special authority has complete access to all output queues and spooled files, regardless of which outq attributes and authorities are specified. Think of *SPLCTL as the "*ALLOBJ" of spooled files. This means that if any user on your system has *SPLCTL special authority, you truly can't secure any output — that person *will* have authority to view everything in every outq. Some organizations may not have any confidential or private data in their spooled files. However, if you have even one outq that contains confidential reports, you need to take action (i.e., alter the attributes of the outq) and remove the user's *SPLCTL authority.

Table 8.1 reproduces a chart from IBM's *System i Security Reference* that describes the printing functions and the parameter values required to run them. I sometimes get confused reading this table, so here's a hint: The "Printing function" column lists the commands available for working with outqs, spooled files, and writers. The other columns list the values that must be set if you are to run one of the commands. You can read the

"Printing function" column to determine the commands that can be run, but to make sense of what values must be set, read the rest of the columns straight across, one line at a time — like you're reading a book. For example, to run the DSPSPLF (Display Spooled File), CPYSPLF (Copy Spooled File), SNDNETSPLF, or SNDTCPSPLF command, the DSPDTA parameter must be set to *YES and the user must have at least *READ authority to the outq (and it doesn't matter what the values of the AUTCHK and OPRCTL parameters are, nor does the user require any special authorities) *or* the DSPDTA parameter can be *NO, but in that case the AUTCHK parameter must be *DTAAUT and the user must have *READ, *ADD, and *DLT authority to the outq, and so forth.

Table 8.1
Authority Required to Perform Printing Functions

Printing function	Output queue parameters			Output queue authority	Special authority
	DSPDTA	AUTCHK	OPRCTL		
Add spooled files to queue	Any Any	Any Any	Any *YES	*READ Any	None *JOBCTL
View list of spooled files (WRKOUTQ command[2])	Any Any	Any Any	Any *YES	*READ Any	None *JOBCTL
Display, copy, or send spooled files (DSPSPLF, CPYSPLF, SNDNETSPLF, SNDTCPSPLF[2])	*YES *NO *NO *YES *NO *OWNER	Any *DTAAUT *OWNER Any Any Any	Any Any Any *YES *YES Any	*READ *READ, *ADD, *DLT Owner[3] Any Any Any	None None None *JOBCTL *JOBCTL Any
Change, delete, hold, and release spooled file (CHGSPLFA, DLTSPLF, HLDSPLF, RLSSPLF[2])	Any Any Any	*DTAAUT *OWNER Any	Any Any *YES	*READ, *ADD, *DLT Owner[3] Any	None None *JOBCTL
Change, clear, hold, and release output queue (CHGOUTQ, CLROUTQ, HLDOUTQ, RLSOUTQ[2])	Any Any Any	*DTAAUT *OWNER Any	Any Any *YES	*READ, *ADD, *DLT Owner[3] Any	NONE None *JOBCTL
Start a writer for the queue (STRPRTWTR, STRRMTWTR[2])	Any Any	*DTAAUT Any	Any *YES	*CHANGE Any	None *JOBCTAL

[1]This authority is required to direct output to an output queue.
[2]Using these commands or equivalent options from a display.
[3]You must own the output queue.
Source: System i Security Reference, *V6R1, Copyright 2008, International Business Machines Corporation.*

Output Queue Ownership

The issue of output queue security also raises the question of who should own output queues. This may seem like a simple question, but it's important for two reasons: First, the owner of the output queue can modify the outq's attributes as well as grant and revoke authorities to the outq, which means the owner controls who can view and work with spooled files on that queue. Second, the AUTCHK parameter checks the ownership of the output queue as part of the authorization test when the queue is accessed.

The system operator should be responsible for creating and controlling output queues that hold data considered public or unsecured. You can use this ownership with the authority parameters on the CRTOUTQ command to create an environment that lets users control their own print files and print on various printers in their work areas.

For secured data (e.g., payroll, human resources, financial statements), the department supervisor profile or the supervisor's role is a good candidate for owning the outq. This person (or role) is the one that determines which users should be allowed to see the reports contained in the outq.

Sample Output Queue Security Implementation

For example purposes, consider an organization that has outqs in the IT department, the financial area (e.g., accounts payable, accounts receivable, general ledger), human resources, the order/sales departments, and the warehouse. Let's assume you want to secure the financial spooled files — an output queue named FINOUTQ — but you want the operator to manage the queues. Let's also say you want to secure the human resources output queue — output queue HROUTQ — and spooled files. You want no one except HR personnel to manage this output queue and its spooled files. The remaining outqs should have minimum security and allow complete access to any spooled file and to the outq itself. How do you accomplish this?

First, create all the outqs except FINOUTQ and HROUTQ, and make the system operator profile the owner of the queues. You'd enter the following CRTOUTQ command for each output queue you create:

```
CRTOUTQ  OUTQ(outq_lib/outq_name) +
         DSPDTA(*YES)             +
         OPRCTL(*YES)             +
         AUTCHK(*OWNER)           +
         AUT(*USE)
```

The combination of DSPDTA(*YES), AUTCHK(*OWNER), and AUT(*USE) for the public authority means that any user can display, copy, or send any spooled files on the output queue and that only spooled file owners can change, hold, release, or delete their own spooled files. (The owner of the output queue can change, hold, release, or delete any spooled file in the queue, regardless of who owns it.) The one exception to the limitation of AUTCHK(*OWNER) is that a user profile that has *JOBCTL special authority and OPRCTL(*YES) can change, hold, release, or delete any spooled file. This implementation lets you give system operators *JOBCTL authority to manage these outqs while limiting other users to managing their own spooled files. However, all users with at least *READ data authority can display, copy, and send any spooled file on the queue.

Next, create FINOUTQ, which will hold confidential financial reports. You want all profiles to have access only to their own reports, and you want the operator to manage the queue without being able to view the spooled files. You'd create FINOUTQ as follows, with FINUSER as the owner:

```
CRTOUTQ  OUTQ(outq_lib/FINOUTQ) +
         DSPDTA(*OWNER)         +
         OPRCTL(*YES)           +
         AUTCHK(*OWNER)         +
         AUT(*USE)
```

Although everyone, including the system operator, can access this output queue, each user profile can display, copy, or send only that user's own spooled files. The system operator, by virtue of *JOBCTL special authority and the OPRCTL(*YES) attribute on the output queue, can manage the queue but can't look at any reports other than the ones the system operator places on the queue. The DSPDTA(*OWNER) attribute ensures that only the owner of a spooled file can display, copy, or send that spooled file.

The last output queue is HROUTQ, which is extremely confidential. Only those in the HRUSER group should be able to access this output queue or these spooled files. To create this output queue, you'd use the following CRTOUTQ command:

```
CRTOUTQ  OUTQ(outq_lib/HROUTQ) +
         DSPDTA(*YES)          +
         OPRCTL(*NO)           +
         AUTCHK(*DTAAUT)       +
         AUT(*EXCLUDE)
```

Make the HRUSER group profile the owner, and revoke all other authorities from the queue.

Profiles with *JOBCTL authority can't access this queue because of the queue's OPRCTL(*NO) attribute. Only the owner has any object or data authority to the queue. Only the owner can start or stop a writer to the outq. All users in the HRUSER group can display, copy, or send any spooled file on the outq. As a result of the AUTCHK(*DTAAUT) attribute and the fact that the HRUSER group profile owns the output queue, all users in the group can also change, hold, move, or delete any spooled file.

Helpful Tools

The PRTQAUT (Print Queue Authority Report) security tool, available on the SECTOOLS and SECBATCH menus, prints the security settings for both output queues and job queues on your system. The full report lists all the queues that meet the selection criteria. The command also provides a report that lists any changes since the last time you ran the report and any new output or job queues that have been added in the interim.

Navigator for i

One use of Navigator for i is to manage spooled files. If you need to e-mail a job log to IBM or your vendor for debugging, it's a snap. Simply click on the name of your system in Navigator and choose Basic Operations|Printer Output. Your spooled files will be displayed in the panel on the right, as shown in Figure 8.1. From there, you can select the desired spooled file and drag it to your desktop.

Figure 8.1
Managing Spooled Files in Navigator for i

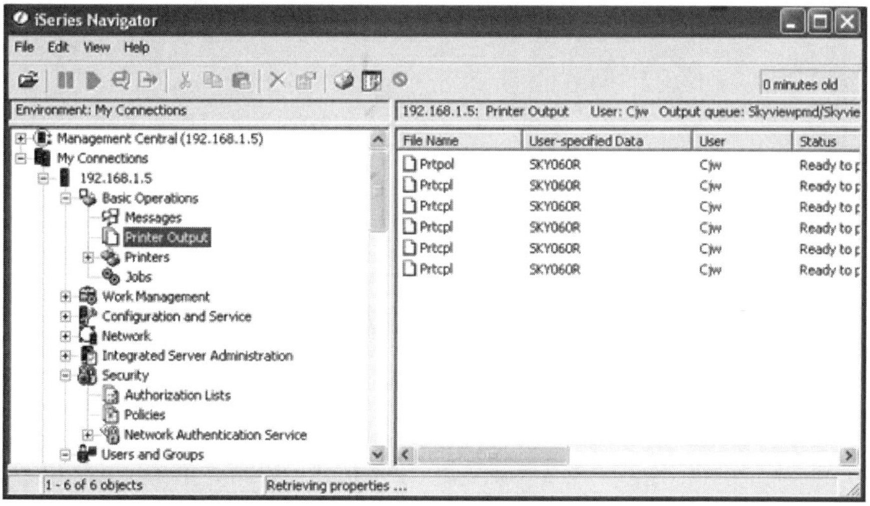

Chapter 9

Encryption

Like many other aspects of security, the need to implement encryption is often driven by the need to be in compliance with some law or regulation. This chapter discusses the issues you'll need consider to ensure a sound encryption implementation.

I typically see encryption implemented in four areas:

- Transmission of data
- Data in a file (either all or selected fields in the file)
- Backups
- Disk

I'll discuss each of these areas in detail. First, though, let's go over some basic encryption concepts.

Encryption Basics

Encryption consists of three components: an encryption algorithm, a key, and the data to be encrypted. The algorithm you use is typically dictated by the situation. For example, some algorithms are more efficient (perform better) for encrypting large amounts of data than others. I discuss the common algorithms and their characteristics a bit later.

The encryption key is (pun intended) key to the whole encryption process. Without it, you're missing part of the encryption equation, and data cannot be encrypted or decrypted. Should the key fall into the wrong hands or be discovered, the person possessing the key can decrypt your encrypted data. In other words, it's not the algorithm that must be kept secret when you encrypt data; it's the encryption key. Therefore, you should limit who has knowledge of or access to the key to those who have a business need — in other words, to only a very few people. In addition, the "strength" of an encryption key determines how quickly someone without knowledge of the key may be able to determine the key and therefore be able to decrypt the data.

Before we take a look at the types of encryption algorithms available, let me stress that you should never — ever — attempt to create your own encryption algorithm if you ever hope to have encryption add any measure of security to your data, much less be in compliance with any law or regulation. Cryptologists spend years developing their algorithms, which other cryptologists then spend years trying to break. Instead of creating your own, use a well-known algorithm. As I said earlier, it's not the algorithm that needs to be kept a secret — it's the encryption key.

Now, let's take a look at the types of encryption algorithms in use, along with the situations in which they are typically employed. In Chapter 10, you'll learn about the various ways you can use encryption to provide secure communications to and from your IBM i system.

Public Key Infrastructure

Public key infrastructure (PKI) uses encryption to let you send data securely and privately. When you use an encryption algorithm, you feed the algorithm the encryption key, which is used to encrypt the data. Figure 9.1 illustrates the differences between symmetric keys, asymmetric keys, and hash algorithms.

Figure 9.1
Types of Encryption

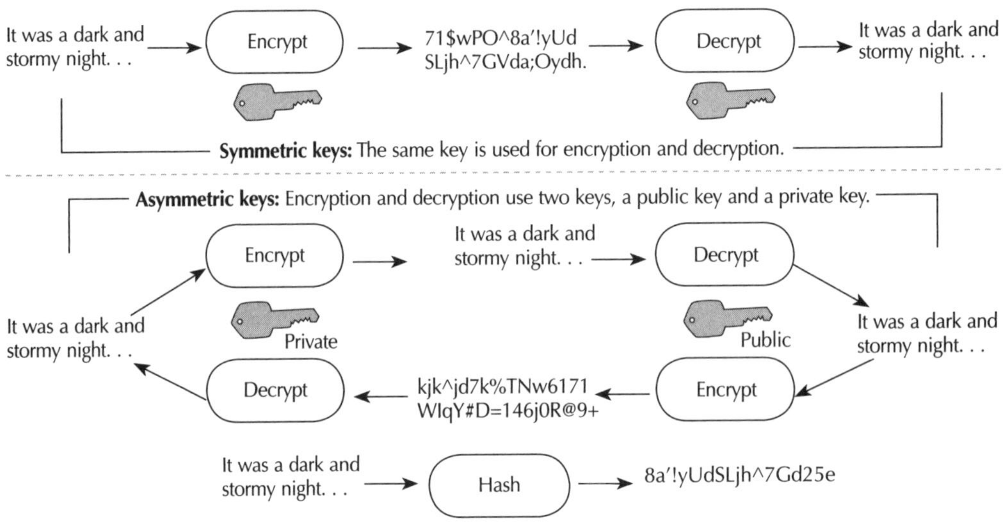

Symmetric key algorithms use the same key to both encrypt and decrypt data. These algorithms are generally faster than asymmetric key algorithms and are often used to encrypt large amounts of information. Some symmetric key algorithms you'll hear mentioned include TripleDES, RC5, and AES (the standard for the U.S. government).

Although symmetric keys are simple and fast, they have drawbacks. Say you encrypt an e-mail message with a symmetric key. How do you securely give that key to the e-mail recipient? Send it in another e-mail? Not very secure. You might instead choose to use an algorithm that requires an asymmetric key.

Asymmetric key algorithms use one key to perform one function and another key to perform the opposite function. In other words, the same key cannot be used to both encrypt and decrypt the data. The two keys, one a *private key* and the other a *public key*, are together known as a *public/private key-pair*. Asymmetric key algorithms are typically used for authentication. For example, they are used during a Secure Sockets Layer (SSL) "handshake" to verify that the client knows the server to which it is trying to connect. RSA and Elliptic Curve are two common asymmetric key algorithms.

Hash algorithms produce a result that is not decryptable. They are typically used to compare two values to ensure that the values are the same. For example, the verification of a digital signature uses a hash function to ensure that the signed document hasn't been altered. Because a hashed value cannot return the original value, this method provides the most secure form of data storage and is the most appropriate method when the data's original form (i.e., clear-text data) is not required. MD5 and SHA-1 are examples of hash algorithms.

Transmission of Data

Whether it's making the connection from home to your office, collecting or sending confidential or private data via a Web application, or transmitting a file to a trading partner, bank, or government entity, the most familiar form of encryption is that in which the session is encrypted so that any data flowing over the connection is encrypted and not sent in clear text.

Encrypting the transmission of data is something most of you have already implemented — probably via an SSL-enabled server or some type of virtual private network (VPN) connection. In the next chapter, I'll get into a more detailed description about how these types of connections work, but suffice it to say for now that encrypted sessions typically use two types of encryption algorithms: one for authentication and another to encrypt the data. For example, when an SSL connection is being established, the client making the request uses an asymmetric algorithm to authenticate the server to which the connection is being made. Once the client knows it can trust the server to which the connection has been made, the two ends of the session exchange "session keys" and switch to a symmetric algorithm because such algorithms are significantly faster for encrypting and decrypting data. In fact, the slowest part of an SSL session is the initial authentication phase, or handshake.

Encrypting Data in Files

IBM provides several native options for encrypting data in a database field. The most rudimentary one is coding to the CIPHER machine instruction. If you are running V5R3, APIs integrated into i5/OS provide software encryption functionality. (A subset of this function is available via PTFs in V5R2.) The ENCRYPT_RC2 keyword is also available, but use of this function (along with any other "roll-your-own" solution) requires careful attention to the key management implementation.

For the most secure method, as well as the most complex, you can install a Federal Information Processing Standard (FIPS) compliant cryptographic coprocessor card and write to the APIs provided with the card. Before V5R4, the crypto card is the only IBM solution that provides encryption key management features. Although the crypto card is the most complex option, in V6R1 IBM has added some user-friendly commands and APIs that make using and managing the crypto card significantly easier.

Last, in V5R4, IBM added integrated encryption key management to i5/OS along with the APIs required for implementation. These APIs give those of you implementing your own solution a secure way to create, store, and manage the encryption keys.

Identify the Scope of Your Project

As with most encryption projects, one of the biggest challenges when you encrypt data in files is ensuring that the solution is architected correctly. Another challenge is the management of the encryption key. One of the first steps I encourage you to take if you're considering encrypting file data is to identify the scope of your project. This may seem like an obvious step, but I've found that many organizations don't actually know where all the data that should be encrypted resides. To understand the project's true scope and ensure that all the data that should be encrypted actually is, you need to know where all the data is.

This typically is not a question one person can answer. It normally requires input from many individuals with a variety of roles. Why? Because no single person is aware of all the regulations that determine what constitutes private data, nor of the laws and regulations that govern the protection of private data, the ramifications of compromising or not protecting the data, all the places where the data is stored, and how the data is used. Only after gathering this input from all your sources can you know the true scope of your encryption project and decide how to proceed. Let's look at the steps you need to take to determine project scope.

Step 1: Determine what data needs to be encrypted. The laws and regulations with which your organization must comply might dictate what data needs to be encrypted. For example, if your organization stores cardholder data, the Payment Card Industry's Data Security Standard (PCI DSS) dictates what data must be encrypted. In other cases, the data owners may determine that they want their data to be encrypted.

Step 2: Determine where the data is stored. Figuring out which data to encrypt is the easy step. The more difficult task is determining all — and I do mean *all* — of the places where the data resides. Without fail, private or confidential data is located in more places than first seems obvious. To make sure your list is complete, you'll probably need to assemble a group of programmers, system administrators, data owners, database administrators, system analysts, and even representatives from your third-party software vendors. In making your list of all data occurrences, don't forget private information that resides in spooled files, outfiles, query files, text files, stream files, and printed reports. Also make note of databases that are replicated to high availability (HA) machines, databases residing on development machines (including copies of the production database made into developers' libraries), logical files that include the fields requiring encryption, and data residing on network servers and in desktop applications (e.g., Microsoft Excel spreadsheets).

After listing all the places where the private data resides, examine each instance and determine whether the data is actually required for that specific instance or can be eliminated. In many cases, you can significantly reduce the scope of your project (as well as the exposure to your organization's private data) if you stop storing the private data on servers, stop letting it be included in queries, create a logical file that does not include the fields containing private data, and eliminate the private data from printed reports and spreadsheets. Often, the data is included in these places simply because it's part of the information in the original file structure, not because it's actually necessary. In some cases, the data has only recently come to be considered susceptible to abuse. At one time, information such as Social Security numbers was not considered sensitive. How times have changed!

For example, do you really need to include credit-card numbers on a printed report of all returns to a store if the person viewing the report is interested only in what's being returned and the reasons for the returns? No, you don't. Viewing credit-card information isn't part of this person's job function, and therefore these details should be removed from the report. And, without question, removing the credit-card numbers from the report is much easier than adding logic to the application to permit this individual to see a masked or partial credit-card number. It also eliminates the need to put appropriate access controls on the file, add routines to encrypt the credit-card numbers, and deal with the proper disposal (shredding) of the printed reports.

Step 3: Determine how the information is being used. This step, while easier than Step 2, still usually takes an assembly of individuals to complete. You want to understand how the data is being used so you can determine what type of encryption is needed. For example, if the data is a Canadian social information number or a driver's license number used only for verification purposes, you can likely store it in a one-way hash. A hash cannot return the original data; it is one-way only. For this reason, hashing private data is much easier to implement and, even more important, more secure than implementing more complex encryption algorithms that allow decryption of data.

To ensure that encryption provides an added layer of protection on top of the access controls you have on the files containing your data, you'll want to limit the scenarios for which the data must be decrypted. For example, if the data is being sent to a bank for transaction processing or to a clearinghouse for verification, it is likely that the clear-text data is required. For this scenario, you'll need to determine how and when to decrypt the data so that the data is in clear text for as short a time as possible. Also, after transmitting the clear-text data over an encrypted session, you should remove or clear it as soon as possible.

Architecting Your Application to Use Encryption

Whatever method you choose for encrypting data in files, some amount of application changes may be necessary. The ideal solution is to re-architect your application and move the private data into its own database, using a separate database for each type of private data (e.g., one for cardholder data, another for human resources data). This approach makes it much easier to ensure that the encryption/decryption functions are called only when necessary, helping to keep the data secure. This technique also helps minimize encryption's effect on system performance because the data is decrypted only when necessary, not on every file open. In addition, it puts you in a much better position to respond quickly should encryption requirements change in the future.

However, this type of re-architecture is usually not an option for most organizations. If that is your case, you may need to declare a new field to hold the encrypted value because an encrypted value is often larger than the original value. Unfortunately, this means you'll have to recompile all programs that use the file in question. In addition, once you've populated the encrypted field, you'll need to take steps to neutralize or otherwise "wipe out" the clear-text version of the private data that currently resides in the file.

Several vendors offer solutions for database encryption, and some provide options that do not require adding a field to the file. Purchasing a solution may give your organization more flexibility, especially if re-architecting isn't an option. When investigating these solutions, make sure you understand their key management capabilities to ensure they meet your organization's compliance requirements. Many vendor solutions offer robust key management, often providing the ability to change keys without having to decrypt and re-encrypt data all at once. For most shops, it will be easier to buy an encryption solution than to try to implement one using the features provided in i5/OS and IBM i.

The Key Is Key Management

When you encrypt data in files — or data "at rest," as the laws and regulations often refer to such data — good encryption key management is essential. In fact, the key is so important to a secure encryption implementation that the PCI DSS includes specific requirements for encryption key management. I expect this issue to only grow in importance. In other words, it won't go away, so it's best to understand what constitutes best practices for key management and to implement those practices if at all possible. I cannot emphasize enough that without good key management, your encryption project will likely be a disaster and will provide no security benefit to your organization's data. I'll stop my lecture for now and start the discussion of what constitutes best practices for encryption key management.

Key Generation

Use a random number generator to generate the key. Using your child's name, the name of your favorite sports team, or even a seemingly random string of numbers and letters does *not* constitute a random number. Truly random encryption keys are significantly stronger than keys based on a phrase or what seems to you to be a random string.

Separation of Duties

The same person should not create the keys, use the keys (i.e., write programs to use the keys for encryption or decryption), and be the custodian of the keys. For large shops, this requirement shouldn't be an issue, but for small shops it may present a challenge. Either way, be prepared to explain your separation of key management duties to any auditor who asks.

Storing the Keys

Appropriate placement and access control of stored keys is vital. Storing keys in clear text in the program source or a data area or as the first record in the file is not acceptable. You should use the integrated key management APIs and create your own key management system, use a hardware crypto card that provides key management, install a key management appliance (a separate piece of hardware that securely stores and serves the keys), or use a vendor product that includes key management. The critical components of a key management system are that the keys are separated from the data they are encrypting and that the keys are stored securely.

Controlling Access to Keys and Encryption/Decryption Routines

Without good access controls on the key storage facility itself (this means you've implemented object-level security if you're using a software implementation), the mechanism that retrieves the key, and the encryption routines (especially the decryption routine), you gain no additional security benefit by encrypting data. If everyone on the system has access to the encryption key, what's the point? In addition, if you have a software-based key management system, it's likely that everyone with *ALLOBJ can retrieve the encryption key. That's because users with *ALLOBJ special authority can access any object on the system. (That's one reason it's good to use hashes wherever you don't need the clear-text data.)

About the best you can do to thwart retrieval by *ALLOBJ users is to throw up some roadblock in an attempt to make retrieving the encryption key and accessing the decryption routines more difficult. One such roadblock is to check the process to see whether it's running with *ALLOBJ; if it is, reject the retrieval or decryption request. Another is to check the program that's calling the decryption routine and decrypt only if the calling program is named in an approved list. Understand, though, that if your organization is required to restrict access to keys by *ALLOBJ users, you'll probably need to implement some type of hardware key management system, such as an encryption appliance or a hardware crypto card.

Backing Up the Keys

Careful planning must go into your strategy for backing up the encryption keys. The keys should not be on the same media as the data. However, you must develop a scheme so that you can associate which key was used on which date, or you may not be able to retrieve your data if you need to bring it back from long-term storage or perform a disaster recovery.

Changing the Keys

I've saved discussion of this aspect of key management for last because it's one of the least-settled when it comes to what constitutes best practices. In addition, depending on your implementation, it can be one of the more difficult to implement as an administrator. I'm talking about the process of changing the encryption key that's been used to encrypt data in files. One advantage of moving the encrypted data to another file or adding another field to your files is that you can add a field that keeps track of the encryption key used. Note that this field does *not* contain the encryption key itself. Rather, it contains a key name or "token" – that is, some other information used to identify the key used to encrypt the data. If your implementation includes a way to identify the key, you can change the encryption key, the application can start to use it, and then you can run a background job to decrypt and re-encrypt the records using the new key. If you don't have a way to keep track of the key, you'll have to decrypt all files and then re-encrypt them with the new key. This task can be tricky because the update must take place when no application processes are using the file. If any are, or if an application starts to use the file before the key change is complete, the application may not be able to access some information because the application will be using the new key but not all of the data has been decrypted and re-encrypted.

When encryption first became popular, this aspect of key management wasn't such a huge effort. At that time, best practices said you should change the key every year or two. The exception to this rule was if you felt that the chance of compromising the data by decrypting and re-encrypting it was greater than leaving it encrypted with an old key. However, best practices are evolving, and it's becoming more common to see auditors require keys to be changed annually. Look for more guidance from PCI on this issue as well as the regulations requiring keys to be changed more frequently.

There is one issue, though, about which there is no debate: If a breach occurs or you become aware that the encryption key has been compromised, you must change it immediately. Whether you change keys on a regular basis or not, you need to establish and document a policy reflecting your key management change practices.

If you have to change the key, the question arises, "Do I need to go get my back-up media and decrypt and then re-encrypt that data?" The answer is that no law or regulation requires this — at least not now.

Encrypting Backup Media

Two basic methods exist for encrypting backup media. One is to first save the data to a save file (*SAVF), then encrypt the save file, and then save the encrypted file to the media. This approach requires that sufficient storage be available for the save file. Depending on how much data you're saving, this method can use a significant amount of storage, which may present problems on some systems.

The other method for encrypting backup media is a hardware solution in which a device (often called an appliance) performs the encryption before the data is written to media. This alternative removes the requirement for the intermediate save file.

A third method that some consider "encryption" is the use of compression software. Compression software reduces the space required for the backup data and produces what I call "mangled" data. Although mangled data is not easily read, it is *not* the same as encrypted data, and it probably won't meet the requirements of the laws or regulations with which you need to comply. Be sure to take this point into consideration before choosing this option.

Whichever method you select, a significant factor in your planning must, again, be key management. The best practices described in the preceding section must be applied here — even the requirement to periodically change the keys. But because you're encrypting backups, you must take special care to keep track of which encryption key you used for which backup. In other words, you need a process in place that ensures you can retrieve the key should you have to bring back the media from your storage facility or perform a disaster recovery.

Another consideration to key retrieval is where your key management system is run. You cannot run a key management system on the same system and have it be backed up with the rest of your backup. If you do, you won't be able to recover your system because you'll never be able to access the key management system — because it's encrypted and the key to decrypt it is in the encrypted key management facility!

Last, although it's not a security issue, when planning you'll want to consider the performance impact of encrypting backups and the subsequent impact on your system operators' schedules.

What IBM Provides

In V5R4, IBM introduced a very limited encrypted backup solution. It requires the use of the TS1120 tape drive inside a TS3500 tape library (i.e., a Backup Recovery Media Services, or BRMS, managed tape library). In V6R1, the BRMS support is expanded to include software encryption and no longer depends on any specific hardware being in place. IBM (and vendor) offerings in this area are likely to expand as compliance requirements increase.

Encrypted Auxiliary Storage Pools

In V6R1, IBM introduced the concept of *encrypted disk* — specifically, the ability to encrypt auxiliary storage pools (ASPS) and independent storage pools (IASPs). The benefit of encrypting ASPs is an additional layer of security, which is especially valuable in a Storage Area Network (SAN) environment, in situations where a disk drive might be stolen, or in the event that a drive fails and you're unable to wipe the drive of your organization's data.

Disaster Recovery Considerations

Regardless of the type of encryption project, don't forget to take into account its impact on your disaster recovery plans. If the media is encrypted, you'll need to have keys available at your alternate location so that the media can be read and data restored. If you used a hardware encryption method (either for encrypted backups or for encrypting data in database files), the same hardware must also be available at your recovery site. If you have encrypted ASPs, your key management system will need to be available. Last, if you encrypt data while it's being transmitted, the appropriate digital certificates must be installed in your servers or VPN configurations.

Success Depends on Planning

Don't undertake an encryption project lightly. Significant planning and a well-architected implementation are necessary to ensure success. Determining the regulatory issues with which you must comply as well as the scope of your project are critical to the project's success. This information will then help you decide whether to implement one of the options from IBM or to evaluate vendor solutions.

Helpful Resources

Documentation for all the hardware and software encryption APIs provided by IBM can be found in IBM's Information Center. An excellent Redbook written by a stellar set of encryption experts is *IBM System i Security: Protecting i5/OS Data with Encryption.*

Chapter 10

Connecting to the System

To understand security problems and solutions in this networked world, you must comprehend a variety of security issues, including those associated with physical network security, network configuration security, network attributes, the nuances of TCP/IP and some TCP/IP applications, System i Access for Windows and for Web, secure communications, and wireless technology. This discussion covers quite a few topics, so grab some caffeine, sit tight, and we'll get through this rather lengthy chapter together!

Physical Security

Physical network security helps prevent inadvertent damage, as well as intentional abuse and misuse. If you don't physically secure your network, inevitably a janitor, an employee's child, or someone else will trip over a wire and possibly sustain personal injury as well as damage your equipment. Some parts of a network can't be physically secured (e.g., telephone lines, ISP-controlled routers), but by taking some practical, common-sense steps, you can achieve an acceptable level of physical security.

To begin, you should put all network equipment in a secure location. The principle of allowing access only to users who have a business requirement applies to physical components as well as to database files. If you operate a host computer that supports remote access, place all communications equipment (e.g., firewalls, routers, cables) in your locked computer room or at least in a locked closet.

Similar precautions are also necessary — and often even more important — at remote sites. Many remote sites that have access to the host have a small remote controller, router, or system but no IT personnel on site. In such cases, you must rely on the people at the site to keep the network equipment secure. With a little planning and by moving equipment to a safe and secure location, you can prevent many accidents.

System Values

Two system values control whether network devices are auto-created and thus have implications for security. System value QAUTOCFG (Automatic Device Configuration) determines whether devices such as printers, tape drives, and optical drives are configured automatically. While I'd never propose that you configure these devices by hand, in most organizations it's rare for new devices to be added frequently. So I recommend leaving this system value off (0) unless you are specifically installing devices.

The more interesting system value in today's environment is QAUTOVRT (Automatic Configuration of Virtual Devices), which controls whether virtual devices are auto-configured (as well as how many can be configured). If QAUTOVRT has been set to *NOMAX for a while, you probably have all the virtual devices you need, and you should be able to change the value to a number or to 0 (off) without any adverse effects. IBM advises setting QAUTOVRT to a value that is double the current number of virtual devices on your system. However, only you can determine how many virtual devices your organization needs. If you feel you have enough devices, set QAUTOVRT to 0. The system won't delete any virtual devices that have already been configured; it just will prevent the configuration of additional devices. By turning off QAUTOVRT, you can prevent someone from overloading your system with thousands of unwanted virtual devices.

*IOSYSCFG Special Authority

*IOSYSCFG special authority is required to configure all lines, controllers, and devices on the system, as well as to configure and manage all TCP/IP servers and configurations. *IOSYSCFG is quite powerful, and you'll want to carefully manage and monitor who has this special authority.

Network Security Attributes

Some of the network configuration settings from the original days of OS/400 are defined in the system's network attributes. Three network attributes relate specifically to security:

- JOBACN (Job action)
- PCSACC (PC Support access)
- DDMACC (DDM access)

You can use the DSPNETA (Display Network Attributes) command to view the values for these attributes. If you want to change the network attributes, you must have *ALLOBJ and *SECADM authority to run the CHGNETA (Change Network Attributes) command. Let's look at each of the security-related network attributes.

JOBACN

JOBACN is an attribute used with Systems Network Architecture (SNA) communications that determines how the OS/400 operating system responds to incoming job streams from other systems in your network. In other words, it probably is not very applicable to today's environment. This network attribute can have one of three values: *REJECT, *FILE, or *SEARCH.

*REJECT causes the target system to reject the input job stream. When a job is rejected, the system sends a message to both the originating user profile and the intended recipient stating that the input stream was rejected.

*FILE causes the system to file the input job stream in the recipient's network file queue. The intended recipient (a user profile) can display the job stream, cancel it, receive it into a database file, or submit it to a job queue for processing. Again, the system sends a message to the sender and receiver notifying them that the incoming stream was filed.

*SEARCH instructs the system to search the network job action table for information about how to control the incoming job stream. The network job action table holds information about how the system is to handle specific requests. Each record in the table consists of the sender's ID, the action the local system is to take for that sender's requests, the user profile to use when submitting the job if the desired action is *SUBMIT, and the job queue to use for submitting the job.

Security Recommendation

Specify a value of *REJECT for the JOBACN network attribute unless the remote system uses Systems Network Architecture Distribution Services (SNADS) to submit jobs to your system. If SNADS is used, specify a value of *SEARCH to explicitly control the job actions.

PCSACC

The PCSACC attribute was created to control actions from PC Support and the original Client Access clients. PCSACC can have one of four values: *REJECT, *REGFAC, *OBJAUT, or the name of an exit program. The only value that has any meaning in today's environment is *REGFAC. This value tells System i Access clients to look in the registration facility for exit programs that may be defined for the exit points associated with System i Access. The system ignores all other settings for PCSACC.

DDMACC

The DDMACC attribute determines how your system processes Distributed Data Management (DDM) requests from remote systems. The local system receives a DDM request when a remote system tries to access data using a file, data area, or data queue whose type is *DDM. The three possible values for the DDMACC attribute are *REJECT, *OBJAUT, or the name of an exit program.

Security Recommendation

If you don't use DDM or Distributed Relational Database Architecture (DRDA) or you run either of them over TCP/IP (rather than SNA), set the DDMACC network attribute to *REJECT.

Security Considerations for TCP/IP

Because TCP/IP (which stands for Transmission Control Protocol/Internet Protocol) is an integral part of today's networks, there are some fundamental configuration steps you'll want to take to protect your system. After discussing these steps, I'll go over a few features of TCP/IP as well as some of the more well-known TCP/IP applications and their associated security issues. Chapter 11 discusses security considerations for the Web server.

Starting TCP/IP Servers

The basic security rule for TCP/IP servers is this: Until you need the function, don't start the server. To follow this rule, you need to understand and control which application servers are started when you start TCP/IP.

Two CL commands control which TCP/IP servers are started: STRTCP (Start TCP/IP) and STRTCPSVR (Start TCP/IP Servers). When you run the STRTCP command, the system examines each server's AUTOSTART (Autostart server) attribute and starts the servers whose AUTOSTART value is *YES. (You can set the AUTOSTART value using the appropriate change command, such as CHGFTPA, or Change FTP Attributes.) Several of the TCP/IP servers — including DDM, File Transfer Protocol (FTP), Line Printer Daemon (LPD), Simple Mail Transfer Protocol (SMTP), Simple Network Management Protocol (SNMP), and Telnet — default to auto-start. If you don't want them to start automatically, you can change their AUTOSTART value to *NO.

The STRTCPSVR command, however, overrides the auto-start setting of all servers and starts all servers. To ensure that you don't start servers you don't need, you can use the CHGCMDDFT (Change Command Default) command to change the STRTCPSVR command's SERVER (Server) parameter to specify which server you want to start. (You can specify only one server to start on the CHGCMDDFT command.) For example, the command

```
CHGCMDDFT  CMD(STRTCPSVR) NEWDFT('SERVER(*FTP)')
```

starts only the FTP server. Another option is to set the STRTCPSVR command's SERVER parameter to *AUTOSTART to have only those servers with AUTOSTART(*YES) started.

IBM ships the STRTCP and STRTCPSVR commands as *PUBLIC(*EXCLUDE), and the IBM-supplied profiles QPGMR, QSRV, QSRVBAS, and QSYSOPR are granted *USE private authority to these commands. You'll want to monitor who can run STRTCP and STRTCPSVR. (This is one example of the type of task programmers can run when you make them a member of QPGMR. You must ask yourself whether starting and stopping TCP/IP is really a function the programmer role should perform. Sorry — I digress!)

Session Time-out Value

While considering the auto-start attribute of the TCP/IP servers, you should also think about the session time-out value. The Telnet server takes its value from the QINACTITV (Inactivity Time-out Interval) system value, but servers such as FTP define their own time-out value. You'll need to evaluate the appropriateness of this setting by considering possible performance impacts and how you use these servers. (By the way, not all servers have a time-out value.)

Securing Ports

The Internet Assigned Numbers Authority (IANA) assigns specific ports to well-known TCP/IP applications. To determine the ports defined on IBM i, run the CFGTCP (Configure TCP/IP) command and select option 21 (Configure related tables) followed by option 1 (Work with service table entries).

All users can use programs that listen on ports. To limit the damage someone could do by eavesdropping on a port, you can restrict any ports you don't use. You can restrict a port by configuring it to run under a profile that has been deleted. To do so, first be sure that you don't want to use the port's application, because restricting the port causes the application not to start. Then, create a profile:

```
CRTUSRPRF   TCP_USER PASSWORD(*NONE)
```

After creating the profile, follow these steps to restrict the port:

1. Go to the TCP/IP Configuration menu (GO CFGTCP).
2. Select option 4 (Work with TCP/IP port restrictions).
3. Select option 1 (Add).
4. On the Add TCP/IP Port Entry display, enter the port number associated with the application you're restricting as the lower port range value.
5. For the upper value, specify *ONLY.
6. For the protocol, specify *TCP.
7. Specify the user profile you just created to use this port.
8. Repeat these steps, specifying *UDP for the protocol.

After following these steps for each port you want to secure, delete the user profile:

```
DLTUSRPRF   TCP_USER
```

The port restriction will take effect the next time TCP/IP is started. If an application running as QSECOFR issues a bind() operation to that port, the bind will fail.

Internal Addresses

One way network administrator's help protect their internal network is to use internal or "private" addresses. Private addresses are outside the range of addresses that are valid for use on the Internet. (In other words, they are in reserved ranges, such as 10.*x.x.x* or 192.168.*x.x*.) Several security mechanisms let you control processing based on IP address. Using private addresses within your organization helps you separate "inside" requests from "outside" requests.

IP Packet Filtering

IP packet filtering, a function that is integrated into IBM i, lets you prevent unwanted packets from coming onto or leaving your system. To filter packets, you set up filter rules that specify the types of packets you want to allow. The information you filter is contained in the packet's header and can include source IP address, protocol (e.g., TCP; User Datagram Protocol, or UDP), destination IP address, source port, destination port, IP datagram direction (i.e., inbound, outbound, or both), forwarded or local, packet fragments, and TCP SYN (synchronization) bit. For example, you might want to allow inbound FTP requests only when they come from a particular TCP/IP address in your network.

After you define filters, any request that doesn't match a filter is expressly denied. Therefore, if someone requests a Telnet connection to your system and your filter rules don't allow Telnet, the request is denied.

Although I don't recommend using packet filtering as the sole replacement for a standalone firewall product, native IP packet filtering provides another layer of defense in your network security configuration.

NAT

Using Network Address Translation (NAT), you can hide or map TCP/IP addresses. NAT lets you join networks that implement different addressing schemes and hide internal TCP/IP addresses from the outside world by mapping them to public addresses. It's important to keep your internal addresses private. If you publish your internal addresses externally, you give intruders a roadmap to your network and make it easier for them to launch an attack. I recommend you use private TCP/IP addresses internally and map them to a public address when going outside your internal network.

Packet filtering and NAT are discussed under the same topic in IBM's Information Center. If you are considering using this integrated function, be sure to review the planning and configuration information under the "Networking|TCP/IP applications, protocols, and services|IP filtering and network address translation" topic. Specific planning and implementation guides are described there in detail.

FTP

File Transfer Protocol lets users download files from and upload files to your system. FTP also provides facilities to rename, add, and delete files. Before manipulating a file, the user must provide a valid IBM i user ID and password (this information is sent in clear text). Remember, however, that because the user doesn't sign on via a Sign On display, the user's initial program and initial menu are never run. This is another case in which menu-based security is insufficient.

FTP ignores the setting of the QMAXSIGN (Maximum Number of Sign-on Attempts) system value. It permits five attempts to provide a valid user ID and password before dropping the connection (this limit is hard-coded, and you can't change it). However, FTP does honor the QMAXSGNACN (Maximum Sign-on Action) system value when you set it to 2 or 3 (the values that disable a user profile). In other words, if a user hits FTP's limit of five unsuccessful attempts to provide a user ID and password, the system can disable the user ID.

FTP uses the INACTTIMO (Inactivity timeout) parameter in its configuration to set the length of time an FTP session can be inactive before the system closes the connection. It ignores the QINACTITV system value.

FTP lets users submit remote commands through its RCMD (Remote Command Facility) subcommand. However, unlike with the System i Access for Windows remote command server, limited-capability users can't run FTP remote commands.

You can use exit points to implement *anonymous FTP*, which lets a system define a well-known user ID and password for guests to use to sign on and download data. You can use these exit points to enable a tightly controlled FTP environment that lets users download only a predefined set of files. For example, a software company might enable anonymous FTP to let customers download beta code for its newest product.

You can also use exit programs to disable downloading of specific files and to prevent execution of RCMD while still permitting file transfer. Or you can use the TCP/IP Utilities, one of the Host Applications defined in Navigator's Application Administration function, to control access. This feature has the basic characteristics of the vendor software packages that let you control FTP. Figure 10.1 shows Application Administration's Host Applications tab, which lists the host applications and the FTP server settings you can use. To reach Application Administration, right-click the name of the system you're connected to in Navigator, select Application Administration from the shortcut menu, and go to the Host Applications tab.

Figure 10.1
Host Applications: TCP/IP Utilities

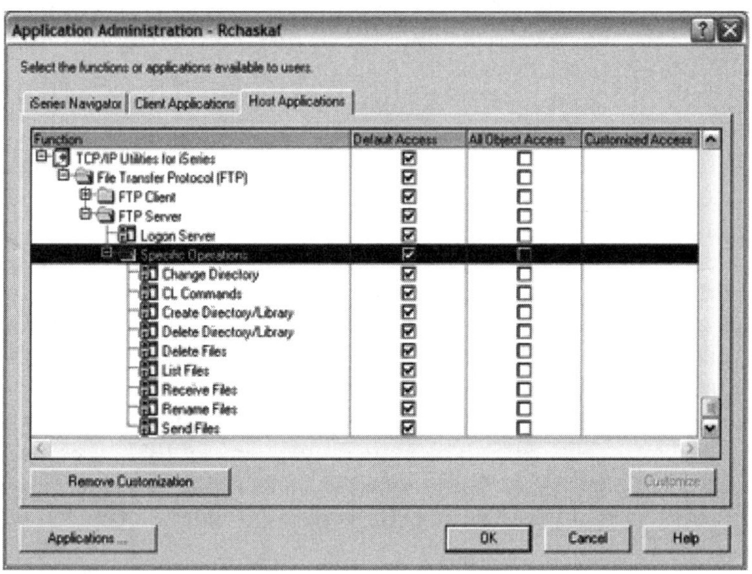

FTP is an example of why it is so critical to implement object-level security appropriately. Any user with *USE authority to a file can download it. Any user who has *CHANGE authority can download or upload the file. So, a file that contains inventory data and has *PUBLIC(*CHANGE) authority can be downloaded, changed using a PC spreadsheet, and then uploaded, compromising the integrity of your production data.

One tactic intruders like to use is to store stolen data (e.g., credit-card numbers, pornographic photographs) on someone else's system. This practice is called *parking*. To protect your system against parking, you might want to write your exit program to make sure a user can't execute both PUT and GET subcommands to the same library or directory (in other words, if you allow PUTs to a directory, disallow GETs, and vice versa). This precaution removes some of the allure of using your system because intruders can't retrieve any information they store on your system.

SMTP

Simple Mail Transfer Protocol lets you send and receive e-mail. Flooding a system with misdirected e-mail (i.e., e-mail addressed to an invalid recipient) is a popular denial-of-service attack. To protect your system from being a router for misdirected e-mail, don't use the *ANY *ANY entry in the system distribution directory. Removing this entry causes IBM i to reject misdirected e-mail when it isn't addressed to a user in your network.

You can also have a problem if your system redirects misdirected e-mail to another system on the Internet. If your system has been found to do this, other systems on the Internet may "blackball" your system and refuse to accept e-mail from it. If the blackballed system is your Web server, which is sending out order confirmations via e-mail, you have a problem. On the other hand, you want to protect your own system from systems that are known to route misdirected e-mail by configuring your system to control SMTP's RELAY and CONNECTIONS functions.

Another hacker trick is to *mail-bomb* a system, flooding it with e-mail from many locations. If you configure SMTP to automatically register users in the system distribution directory, such an attack will also flood the system distribution directory. To avoid this exposure, specify *NO for the AUTOADD (Automatic registration) parameter in the SMTP configuration.

Viruses attached to e-mail also can be a real nightmare. You can configure the SMTP server through either Navigator or the CHGSMTPA (Change SMTP Attributes) command to scan e-mail attachments and the subject line of all e-mail that comes in or goes out through the SMTP (or POP) server. This support lets you define a filter; the system then scans all incoming and outgoing mail. When a match is found, you can either delete the attachment or send it to a specified location for possible repair.

You can also use an alias table to hide the profile name of IBM i users who send and receive e-mail. If hackers can find user profile names, they're halfway to signing on to your system. Use the WRKNAMSMTP (Work with Names for SMTP) command to define aliases.

POP

The name of this TCP/IP application, Post Office Protocol, pretty well describes its function. POP is an electronic mailbox that receives e-mail and stores it until the user retrieves it. To retrieve e-mail, a user must have an entry in the system distribution directory and an IBM i user profile and password. The password is passed to the operating system as clear text. You can't control whether clients (e.g., Web browsers) store the IBM i user profile and password. If a user has a reason to access your system other than to access e-mail, you might want to create two profiles: one for accessing the user's personal mailbox and another for the user to use to sign on and access IBM i applications. If you use this approach, create the mailbox profile with the attributes shown in the following table so that the user can't use the mailbox profile to sign on to and exploit your system.

Attribute	Value
Initial program	INLPGM(*NONE)
Initial menu	INLMNU(*SIGNOFF)
Limited capabilities	LMTCPB(*YES)
Special authorities	SPCAUT(*NONE)
Group profiles	GRPPRF(*NONE)

Unfortunately, a user who can't sign on also can't change passwords. Therefore, you need to set the password expiration interval to *NOMAX (so it never expires) or write a program that changes the password and runs at a regular interval (the latter approach is preferred).

To prevent your system from being mail-bombed, be sure to set the size of your ASPs to an appropriate value. The value should be large enough to store and serve your users' e-mail but small enough so that the limit is reached without wasting too much space on unwanted e-mail in the event of a mail bomb.

POP stores e-mail in the integrated file system (IFS). To prevent users from snooping through other users' e-mail, you'll need to monitor the authority of the IFS subdirectory /QTCPTMM/SMTPBOX. E-mail sent to the system through SMTP is stored here before being delivered to a user's mailbox. The files that store the e-mail use the naming convention QTCPTMM/MAIL/user_profile_name. Anyone who has *RX authority can see the contents of these files, and anyone who has *RW authority can update or change the files. Navigate to the subdirectory using the WRKLNK (Work with Object Links) command, and take option 9 to modify or see who has authority to these files.

DNS

TCP/IP's Domain Name Service (DNS) provides name lookups: Given a host name, DNS finds the corresponding IP address, and vice versa. IBM recommends that the DNS server integrated into IBM i be used within your intranet only. That's because it may not be able to be updated quickly enough when vulnerabilities in this protocol are discovered for use outside your network.

You shouldn't use the same DNS for both intranet and Internet (or public) systems. Think of DNS as a telephone directory. You typically don't publish your employees' internal phone extensions in a public directory, but you publish your company's toll-free telephone number widely. The DNS on IBM i is your "private" directory; your "public" directory resides on your firewall or is provided by an Internet Service Provider (ISP). To an intruder, a DNS server provides a topology of your network, complete with IP addresses.

Some networks are partitioned into subnets. You might not want users in one subnet to know the topology of the other subnets. To define which IP addresses or network interfaces can view your topology, you can define protection directives for your DNS.

To protect your DNS configuration, monitor who has authority to the \QIBM\UserData\OS400\DNS and \QIBM\UserData\OS400\DNS\TMP directories and the files they contain. Only the user who created the DNS files or users who must maintain those files should have authority to them. Public authority should be *EXCLUDE.

REXEC

TCP/IP's Remote Execution (REXEC) facility provides the same type of function as FTP's RCMD subcommand. Just like FTP, REXEC requires a user ID and password and checks the value of the user profile's LMTCPB (Limit capabilities) parameter before running any commands. If the user profile is a limited-capability user (LMTCPB is set to *YES), REXEC rejects the request. You can also use exit points to validate incoming requests.

If you don't normally let users access an IBM i command line, don't start the REXEC server. When you start this server, you in effect provide a command line.

SNMP

Simple Network Management Protocol lets IBM i act as an agent to manage gateways, routers, and hosts in the network. An SNMP agent gathers information about the system and performs functions requested by remote SNMP network managers. Some of the information gathered could be information you've hidden by using aliases or your DNS server. An intruder could use SNMP to gain information about the topology of your network and possibly to alter the network.

Access to SNMP requires a *community name.* A community name is somewhat analogous to a password, but it isn't encrypted. To limit the possibility of someone masquerading as a legitimate SNMP network manager, change the value of the

INTNETADR (Internet address) parameter on the ADDCOMSNMP (Add Community for SNMP) command from *ANY to the IP address of your SNMP network manager.

SNTP

Simple Network Time Protocol (SNTP) lets IBM i act as a client or server for synchronizing device clocks. The PCI DSS standard requires that clocks in your network be synchronized. This is so that audit and log entries can be coordinated across systems. If clocks aren't synchronized, re-creating the sequence of events when investigating an incident becomes very difficult. In addition, if you're going to use Kerberos for authentication, you'll need to synchronize clocks to minimize problems with Kerberos tickets timing out.

DRDA and DDM

Distributed Relational Database Architecture defines the protocols for communications between an application program and a remote relational database. DRDA on IBM i uses Distributed Data Management to communicate with the remote system that contains the relational database.

The security aspects of DDM and DRDA running over TCP/IP differ significantly from running over SNA. That's because SNA had some inherent security features of which DDM and DRDA were able to take advantage. TCP/IP doesn't have these security features, though, so you don't have as many security configuration options when running either DDM or DRDA over TCP/IP.

DDM

When you configure DDM, you must configure security on both the source system and the remote system. On the source system, the user making the DDM request must have a Server Authentication Entry, defined using the ADDSVRAUTE (Add Server Authentication Entry) command. For example:

```
ADDSVRAUTE  USRPRF(application_requester_job_profile) +
            SERVER('QDDMSERVER')                      +
            USERID(server_job_profile)                +
            PASSWORD(password)
```

The USRPRF (User profile) parameter specifies the user on the source system. The USERID (User ID) parameter names the user under which the request will run on the remote system. This user needn't be the same as the user on the source system. In fact, you may not want users on the source system to have a profile on the remote system. In this case, name a different profile under which to run the remote request. Depending on the configuration on the remote system, a password may or may not be required.

To configure security on the remote system, you modify the remote client authentication properties of the DDM server using either Navigator or the CHGDDMTCPA (Change DDM TCP/IP Attributes) command. You have four options for specifying what is required on the connection:

- No password is required on the connection.
- A password is required (this is the default).
- An encrypted password is required.
- A password is not required, but if one is passed it must be valid.

You must use the "no password" option if the profile on the remote system doesn't have a password — that is, if it is defined as PASSWORD(*NONE). However, I don't recommend this setting. Consider this: The ADDSVRAUTE command authority is *PUBLIC(*USE). That means that any user on the system has access to the command. No authority checking is performed if you leave the USRPRF parameter at the default value of *CURRENT. No authority checking is ever done on the USERID parameter. Therefore, with the remote system client authentication set to "no password required," anyone can run the ADDSVRAUTE command and specify that the user's request run as QSECOFR on the remote system. Now, when the user submits a DDM remote command, it will run on the remote system as QSECOFR.

With the "no password required" setting, if a password is passed, the remote system ignores it. The last option — that no password is required, but if one is passed it must be valid — isn't much better than the "no password required" setting; therefore, I don't recommend its use either.

Requiring a password (the default setting) is a much better choice. Unfortunately, this setting introduces a synchronization problem. The user's password on the remote system must match the password specified in the server authentication entry. So either you define a password that never changes (not something I recommend or that would pass any type of formal audit or comply with the PCI DSS or other regulations), or you have the challenge of changing the password on the remote system and having to remember to also change it in the server authentication entry on the source system.

Requiring an encrypted password is one of the better options, but it still presents the challenge of keeping passwords in synch. The most secure option is to configure the source and remote systems to use a Kerberos ticket for authentication. (For more information about using Enterprise Identity Mapping, or EIM, and using Kerberos for authentication, see the Security topic under the appropriate release of the operating system in the IBM Information Center.)

Technical Note

If you configure the remote client authentication to require a password, you must set system value QRETSVRSEC (Retain Server Security) to 1 (allow decryptable passwords) so that the clear-text password can be obtained from the server authentication entry.

DRDA

When you run DRDA over TCP/IP, it uses the connections provided by DDM. The remote client authentication setting on the remote system then determines the authentication requirements. As I noted, the default is to require a user name and password to be sent to the system on which the relational database resides. The user can provide the user ID and password on the SQL CONNECT statement when connecting to the relational database. Or, you can use the ADDSVRAUTE command to define a server authorization entry as described above.

If you choose to use the encrypted password option, you must use a client that supports something called *password substitution*, a technique that lets the server and client pass the password in a secure fashion. The Java Toolbox and System i Access OLE DB Provider are clients that support password substitution. DRDA also supports Kerberos as an authentication mechanism; just as it was with DDM, this is the most secure method of authentication.

Security Considerations for PCs

PCs and the technology they enable are one of the biggest challenges facing IBM i security administrators — especially given the portability of laptops and other mobile computing devices. Nothing underscores the inadequacies of menu-based security more than using mobile-device–based software to access IBM i databases. If you still rely solely on menu security to control your sensitive and confidential information (in other words, if you don't use object-level security), your security configuration has holes. Let's look at System i Access for Windows and some of the security issues you'll have to deal with there.

System i Access for Windows

System i Access for Windows provides several applications, including Open Database Connectivity (ODBC), data transfer, Navigator for i, and 5250 emulation. These applications make it convenient to manipulate IBM i data using desktop applications, such as Microsoft Excel. When you install System i Access for Windows on a PC on which Excel has been installed, an icon is added to Excel to enable the downloading of IBM i data into a spreadsheet.

System i Access for Windows does present some security challenges, but it does not create security exposures in and of itself. What it is excellent at doing, however, is exposing a system that does not have a sound object security implementation. Users cannot access any object through System i Access that they don't have authority to. Depending on how you or your vendor software has implemented object-level authorities, your data might now be accessible through client applications because the default access (i.e., *PUBLIC authority) has been left wide open.

You must couple the use of System i Access for Windows with good resource security to ensure users can access only the data you want them to access. I discuss System i Access security in this section, but I also encourage you to monitor the System i Access Web site at *http://www-03.ibm.com/systems/i/software/access* for information about the latest features and issues.

Data Transfer and Remote Command Issues

System i Access for Windows provides the ability to transfer data and run remote commands. If you let users take advantage of these capabilities (i.e., if this functionality is installed on their PCs and the servers are started), you need to be aware of the consequences for the confidentiality of your data. Remember, all files with *PUBLIC(*USE) authority can be downloaded. Files with *PUBLIC(*CHANGE) authority can be downloaded, changed, and uploaded. You cannot guarantee the integrity of the data in files left at *PUBLIC(*CHANGE).

Letting users run remote commands is like letting users submit batch commands: IBM i doesn't check the user profile's LMTCPB parameter. If you're tempted to use an exit program to totally disallow use of the remote command server or to change its autostart feature and never start it, you'll effectively disable most of Navigator for i as well because IBM uses the remote command server to implement Navigator. That's another reason why I prefer to use object-level security to protect data.

Technical Note

See Informational APAR II12227 for a list of System i Access for Windows functions and the servers required to perform each function. This APAR can help you identify which functions will no longer work if you choose not to start a server.

Host Servers

By default, running the STRTCP command also starts all the host servers. (These are the servers used by System i Access for Windows to process requests, such as the license management server, sign-on server, database server, and so on.) You can use Navigator to customize which servers are started when you run STRTCP, or you can edit the QATOCSTART file in QUSRSYS to customize which servers you want to start. Note that if TCP/IP fails or the ENDTCP (End TCP/IP) command is run, you must use the ENDHOSTSVR (End Host Server) command manually to end the host servers.

Limiting What Users See from the Desktop

System i Access for Windows provides several ways, including Microsoft Group policy objects (GPOs), Application Administration, and selective install, to limit what users can do from their desktops. You should think of these alternatives as modern-day forms of menu access control. In other words, none of these methods is a substitute for object-level security. Nevertheless, limiting what appears on a user's desktop is a good practice because it reduces complexity and confusion for the end user.

Microsoft Policies

Support for Microsoft GPOs is inherent to the Windows operating system. Just as an administrator can control whether new software can be loaded onto the PC, an administrator can set a group policy for which System i Access functions are allowed.

Application Administration

The Application Administration interface, which is included in Navigator, lets you administer parts of System i Access for Windows. The effect is the same as enforcing Microsoft policies, except that an IBM i system administrator rather than a Windows administrator usually configures Application Administration.

Several CL commands help you manage application administration: DSPFCNUSG (Display Function Usage), CHGFCNUSG (Change Function Usage), and WRKFCNUSG (Work with Function Usage). These commands let you manage the same information that you can manage using Application Administration in Navigator. However, I find it much simpler to use the Application Administration interface rather than these commands.

Selective Install

System administrators can selectively configure which components of System i Access for Windows they want installed on users' PCs. This is a good place to begin to control what function is available from users' desktops, but this configuration isn't foolproof. If Joan has more capabilities than Dennis, Dennis might simply go to Joan's PC and copy the additional System i Access for Windows components onto his own PC (unless there's a Microsoft group policy that prevents him from doing so). Or, if you've left in place the QIBM file share that's shipped with IBM i, a user could map a drive using this share and download the components. If you choose to use selective setup, remove this share and create a new share for your customized install image.

ODBC Security Considerations

Open Database Connectivity is a set of standard interfaces developed to provide easy access to databases. Many applications use ODBC or the Java-compatible Java Database Connectivity (JDBC). Again, any database file that has public authority of at least *USE can be read through ODBC.

Some organizations attempt to use an exit program to help control unauthorized use of ODBC. The problem is that depending on the function being requested, somewhere between three and five host servers are used. There is no server named "ODBC" for which you can write one exit program and gain control. In addition, vendors using ODBC may use the DDM or DRDA server, or they may write their own server. In other words, it's difficult to administer and securely control ODBC requests. At the risk of sounding like a broken record, I must remind you that this is a reason to make sure you have a sound object security implementation.

System i Access for Web

System i Access for Web appears to be much less of a security challenge for administrators when all you're trying to do is gives users 5250 access to IBM i. It provides much more administrative control than System i Access for Windows over which users can perform extended functions such as running remote commands or transferring data. This control isn't "smoke and mirrors," like Application Administration. If the user doesn't see a function in his or her browser interface, that function is not available to the user. That's because the checks for the function occur within IBM i and not on the user's PC, as they do when the system checks Application Administration to see whether functions should appear on a user's desktop.

System i Access for Web also eliminates the need for some exit programs because the function runs through the Web server and there are no exit points associated with the Web server. But System i Access for Web definitely needs some of your attention. As shipped, it provides access to all its functions. Before deployment, you'll want to remove access from the *PUBLIC group for all functions you don't want to provide to your end users. Figure 10.2 shows the window you use to set security policies in System i Access for Web.

Figure 10.2
System i Access for Web

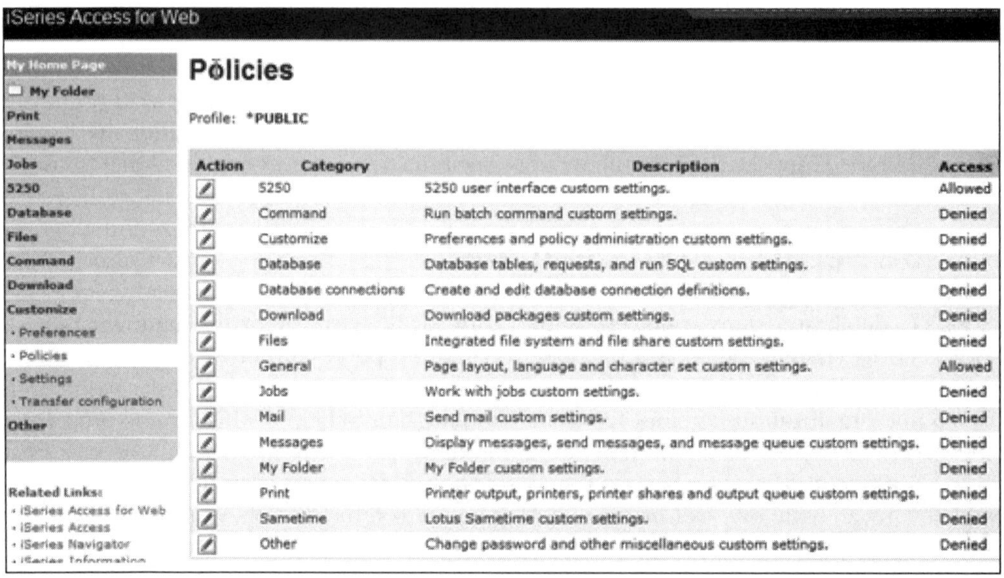

The configuration files for System i Access for Web reside in the IFS. Depending on how you've implemented this function, you may need to look down a different path to monitor authority to these configuration files.

If you're using the Integrated Web Application Server *ADMIN instance with the IWADFT Apache HTTP server, look at

```
/QIBM/UserData/Access/Web2/intappsvr/Admin/...
```

If you're using an Integrated Web Application Server instance created via the Web Admin pages, substitute the name of the Integrated Web Application Server instance in place of <instance> in the following path:

```
/QIBM/UserData/Access/Web2/intappsvr/<instance>/...
```

Last, if you're using WebSphere, substitute the appropriate values for <appsvrtype>, <wasprf>, and <appsvr> in the following path based on the values used for those parameters on the CFGACCWEB2 (Configure Access for Web) command:

```
/QIBM/UserData/Access/Web2/<appsvrtype>/<wasprf>/<appsvr/...
```

You'll want to monitor the permissions to these files to ensure they remain DTAAUT(*RX) and OBJAUT(*NONE).

Using Exit Points

I've mentioned exit points and exit programs several times thus far. Exit points are places throughout IBM i where the operating system "exits" its processing and calls a user-written program. This user-written program is given control for a moment, and then control is returned to the operating system. The concept of an exit program is unique to IBM i. Exit points are defined for servers, such as FTP and Telnet, as well as for functions, such as creating a user profile or changing a password. You register most exit programs in the Registration Facility (using the WRKREGINF, or Work with Registration Information, command), although some legacy exits are defined in network attributes (CHGNETA command) and system values (WRKSYSVAL, Work with System Values, command).

Use of exit programs have been made popular by security vendors that claim to be able to secure your system solely by using exit points. Or, they claim you must have exit programs in place, or else your system isn't secure. Contrary to these claims, native IBM i object security is the key to a sound security implementation. There is *no* substitute for object-level security — no matter what any security vendor tells you. That said, there are times when you might choose to implement an exit program. If you want to add an extra layer of security beyond object-level security, you might think of implementing an exit program.

If I'm going to use an exit program, it typically is not to add an extra layer of security; it is to take advantage of the additional logging of network traffic, beyond what is available through IBM i auditing, that exit programs provide. IBM i doesn't log transactions that occur through these network interfaces. The information provided by exit programs is very useful when you're reworking an application security scheme, helping you understand which processes and users are accessing the application database files from outside the application itself. (For more information about using this information, see Chapter 16.)

As I've said, exit programs are *not* a substitute for object authority, but they do have their place in some security configurations. If you're going to use exit programs, I don't recommend writing your own. Exit point formats — especially the ones dealing with database access and SQL — are complex and vary from exit point to exit point. Numerous vendors provide this type of function, and there is likely a solution out there that meets your requirements as well as your budget. Consider such solutions to be "commodity" software.

For those adventurous types out there who want to write their own exit programs, you'll find the exit point documentation and examples scattered throughout IBM's Information Center. In the V6R1 InfoCenter, the TCP/IP exit points are discussed under the individual servers, as in "Networking|TCP/IP applications, protocols, and services|FTP|Securing FTP on i5/OS|Using SSL to secure the FTP server on i5/OS| Managing access using FTP exit programs." Telnet isn't quit so buried; you'll find the Telnet exit point documentation under "Networking|TCP/IP applications, protocols, and services|Telnet|Managing the Telnet server|Using Telnet exit point programs." To find the exit point formats for the host servers for System i Access for Windows, follow these InfoCenter links: "Connecting to System i|System i Access|System i Access for Windows|Administration|Host server administration|Use server exit programs." The other exit point formats are documented under their respective topics in "Programming|application programming interfaces." For example, if you want to look up the security-related exit points, go to "Programming|Application programming interfaces|APIs by category"; then click on "Security APIs" to see the list.

Management Central

Management Central has several features that let you manage users across multiple installations of IBM i. You can create a user on one system and schedule a task within Management Central to have the user profile propagated to specific systems in your network — including the user's password. Or, you can take an existing profile and propagate it to another system. This method is much easier than taking the results of your last SAVSECDTA (Save Security Data) command execution and restoring that to another system.

Despite its usefulness, Management Central has several aspects that you should keep your eye on. One feature of creating profiles on systems around your network is that once the task is complete, you can have a "post" command run. This command could call a program or run a command. You'll want to monitor what is run during this post command so that no one alters it to run malicious commands. Monitor who has authority to file QAYPSJDFN in QMGTC, and monitor the file's contents as well. The file should be set to *PUBLIC(*EXCLUDE) with no private authorities.

Secure Communications

Various laws and regulations have raised the bar for secure communications. Many require that private data (e.g., credit-card numbers) be transmitted over public networks in encrypted form. The easiest way to implement this requirement is to encrypt the entire transmission. Chapter 9 discussed encryption basics. In this section, I focus on how you can use encryption for communicating to or from your IBM i system.

Digital Certificates

A digital certificate has many similarities to the passport that you use to get into other countries — but in electronic form. Like a passport, a digital certificate must be issued by a particular agency — in this case, a Certificate Authority (CA) — and is valid for a particular time period. You must present some form of identification to have a CA issue you a digital certificate. Once issued, the certificate can be used to identify a user, server, company, or other entity. And just as some countries don't allow entry by people holding passports from certain other countries, servers can choose to disallow digital certificates issued by particular CAs.

Part of the process of creating a digital certificate includes the generation of a public/private key-pair. Users typically store private keys in their browser on their PC. A public key is included as part of the digital certificate. Let's look at how a very common Internet technology makes use of digital certificates and the public/private key-pairs associated with them.

Secure Sockets Layer

Secure Sockets Layer (SSL) is a protocol that encrypts a communications flow. FTP, Telnet, and HTTP (Hypertext Transfer Protocol) are examples of applications that use SSL to encrypt the data that flows between a client and server.

When an application requests SSL services from a server, the client always verifies the server. To do this, the client sends some random data to the server, along with a challenge that in effect says, "Tell me who you are." When the server receives the challenge from the client, the server uses its private key to encrypt the random data and then returns the encrypted data and the server's digital certificate to the client. The client then takes the public key from the server's digital certificate, decrypts the random data, and compares it with the data originally sent.

Also, as part of verification, the client determines what CA signed or issued the server's digital certificate and checks whether the CA is included in the client's list of trusted signers. In other words, the client makes sure it trusts the CA that issued the server's digital certificate. If the random data wasn't decrypted properly or if the server's digital certificate isn't in the client's list of trusted CAs, the connection isn't established.

When the decrypted random data is the same as the original data and the server trusts the CA that issued the digital certificate, the client and server negotiate a session key. The client and server then encrypt the rest of the transmission using a symmetric key.

These steps are always followed when a client requests an SSL session. Optionally, after server authentication, you can have SSL authenticate the client. For client authentication, the steps described above are reversed; in other words, the server sends the client random data and a "Tell me who you are" request, and authentication proceeds as it did in the other direction.

Most of IBM's servers are enabled for SSL — Apache, Telnet, FTP, DDM, DRDA, Lightweight Directory Access Protocol (LDAP), Management Central, Java Toolbox, ODBC, and all the System i Access for Windows servers (e.g., data transfer, data queues, remote queues, print servers). To configure these servers to use SSL, you must use IBM i's digital certificate facility, called Digital Certificate Manager (DCM), which I'll discuss in a moment.

For encrypted Web sessions, you must go into the *ADMIN instance of the Web server and enable each server instance to use SSL. Then, in your HTML, you code the link to the page that requires an SSL connection to be HTTPS (HTTP secure) rather than HTTP. You can also use a protection directive to ensure certain pages aren't served if the connection isn't SSL.

For the System i Access for Windows servers, you must assign a digital certificate for the various servers to use for authentication. Then, you need to configure each client to make SSL requests for the appropriate connections (e.g., ODBC).

Some of the IBM i servers allow client-side authentication. Client-side SSL authentication requires users who make a request of one of these servers to have a digital certificate so that the server can validate the users. Client-side SSL authentication is a powerful feature that lets the system administrator very tightly control who can access the system. As system administrator, you can configure the server to accept only digital certificates issued by a particular CA. You could, for example, choose to set up the IBM i system as your CA and then issue the certificates to only the individuals you want to connect to your system.

This approach provides one method for making secure Telnet sessions from outside of your network. For example, say you have four system administrators who rotate on-call duty. By issuing each administrator a digital certificate for his or her home PC, you can let those administrators sign on from home to manage the system and perhaps save them trips to the office in the middle of the night. Because those administrators are the only ones you've issued certificates for, they will be the only users able to establish a session to the Telnet server. (Note that this approach is necessary only if you don't have a virtual private network or other encrypted connection that the administrators use to connect to your organization's network.)

For instructions on configuring your HTTP server instance to be SSL-enabled, follow the V6R1 InfoCenter links to "Networking|HTTP Server|Scenarios|Adding Secure Sockets Layer (SSL) protection." For instructions on SSL-enabling your System i Access host servers, go to "Connecting to System i|System i Access|System i Access for Windows|Administration|Secure Sockets Layer (SSL) administration."

Digital Certificate Manager

The Digital Certificate Manager lets you set up your IBM i system as a CA, create digital certificates, and receive digital certificates from well-known CAs, such as VeriSign and Thawte. You use DCM to associate a digital certificate with SSL-enabled servers, such as Telnet, IBM HTTP Server, Java Toolbox, and various System i Access for Windows servers, so that your system knows which digital certificate to use when a client makes an SSL request. For instructions on configuring your system as a CA, see the "Security|Digital Certificate Manager" topic in the Information Center.

Using your IBM i system as a CA is a good option when first implementing SSL connections for System i Access for Windows or System i Access for Web. That's because you can try the function and understand the configuration requirements before buying a digital certificate from a well-known CA. But because an IBM i CA is not in the clients' list of trusted signers, rolling out an implementation across your entire organization using your system as the CA will likely be more work than you want. In this case, it will be more cost-effective to buy a certificate from a well-known CA. If you are rolling out the implementation only to your administrators, using your system as a CA may be a viable solution because there aren't many administrators and this type of user is more comfortable changing the configuration of client software.

Virtual Private Networks

Virtual Private Networks (VPNs) are another method of ensuring that data is encrypted during transmission. VPNs let you establish a secure, private connection over a network. While IBM i has integrated VPN capabilities, most people establish a VPN connection elsewhere in their network.

The difference between a VPN and SSL is that VPNs encrypt in the TCP/IP stack so that once the session is established, everything that flows is encrypted. With SSL, every application using a connection must be enabled for encryption. VPNs are often used for employees who work remotely or for administrators who are on-call and must connect to the organization's network from home.

Secure Shell

An alternative to SSL for some applications is Secure Shell, or SSH. This function is available for IBM i through the IBM Portable Utilities for i5/OS (product 5733-SC1). OpenSSH is the open-source implementation of the SSH protocol suite. OpenSSH is widely available for use on many other platforms, including AIX, Linux, and z/OS.

The OpenSSH portion of 5733-SC1 that you might consider includes the following utilities:

- ssh — A secure Telnet replacement that lets an IBM i user connect as a client to a server running the sshd daemon. You can use this tool to connect to the Hardware Management Console (HMC) on the eServer 5*xx* iSeries models.
- sftp — A secure ftp replacement.
- scp — A secure file copy program; basically an alternative to sftp for copying a single file in the IFS.

For more information about SSH on IBM i, go to *http://www-03.ibm.com/servers/enable/site/porting/tools/openssh.html*.

Wireless Considerations

Technology for securing wireless communications is evolving rapidly. Because of that, I'm not going to talk about specific technologies here. Rather, I'll point out some of the general issues with which to concern yourself.

First, familiarize yourself with the technology and what is currently available to secure wireless connections. Use all techniques that make sense and are cost-effective for your organization. Don't limit your research to just wireless LAN connections. For example, personal digital assistants (PDAs) and mobile telephones are also wireless devices and can transmit and receive data and images that could consist of vital corporate data. In fact, this is one of the biggest areas of data loss today.

Second, understand the threat. If you or your employees are using wireless devices, the security threat is very real. *Sniffers* are available for wireless networks, and tools are available for detecting access points. There are documented cases of people driving around neighborhoods (a practice known as "war driving") and hopping on ("LAN-jacking") any network they detect. Imagine that one of your employees has a wireless network in her home that provides Internet access from any room in her house. From the comfort of one of her deck chairs, she can log into your corporate network and catch up on her e-mail. Unfortunately, you have no requirement to use an encrypted session, so the teenager next door also has access to your corporate network. Don't believe it? My business partner often has a better connection to his neighbor's wireless network than to his own. The threat is real.

Third, understand your compliance requirements. The PCI DSS has specific requirements for wireless networks and tests that must be performed quarterly. In addition to being a compliance requirement, these tests are simply good security practices. Some of the worst cases of ID theft have originated with the thieves gaining access to corporate networks through wireless devices.

Last, make sure your security policy accounts for wireless technology. Whether you permit its use at all is the first consideration. If you do, you'll want to define its appropriate use and the appropriate placement of access points as well as address the following questions:

- Whether employees who use wireless networks at home have any additional security requirements (e.g., mandatory use of a VPN when accessing corporate networks from home over a wireless network)
- Whether accessing the corporate network from public "hotspots," such as those found in coffee shops and airports, is allowed
- Whether additional security considerations are required (e.g., file sharing is not allowed, confidential data stored on laptops must be encrypted)

Helpful Tools

The WRKREGINF (Work with Registration Information) command is a tool that can help you monitor aspects of network security. This command prints a list of all registered exit points and the programs associated with them. You can use it to track exit programs in the registration facility.

Navigator for i

Navigator for i can simplify the management of TCP/IP servers. In fact, you can administer some functions, such as IP packet filtering and VPN configuration, only through Navigator. Figure 10.3 shows the view of TCP/IP servers presented by Navigator.

To manage TCP/IP through Navigator, expand the icon for the system to which you're connected, and then expand Network. Expand IP Policies to configure IP packet filtering as well as VPN connections. Expand Servers and then select either TCP/IP or Client Access to manage the respective servers. To change the auto-start or inactivity time-out value, right-click one of the servers and choose Properties.

Figure 10.3
Viewing TCP/IP Servers

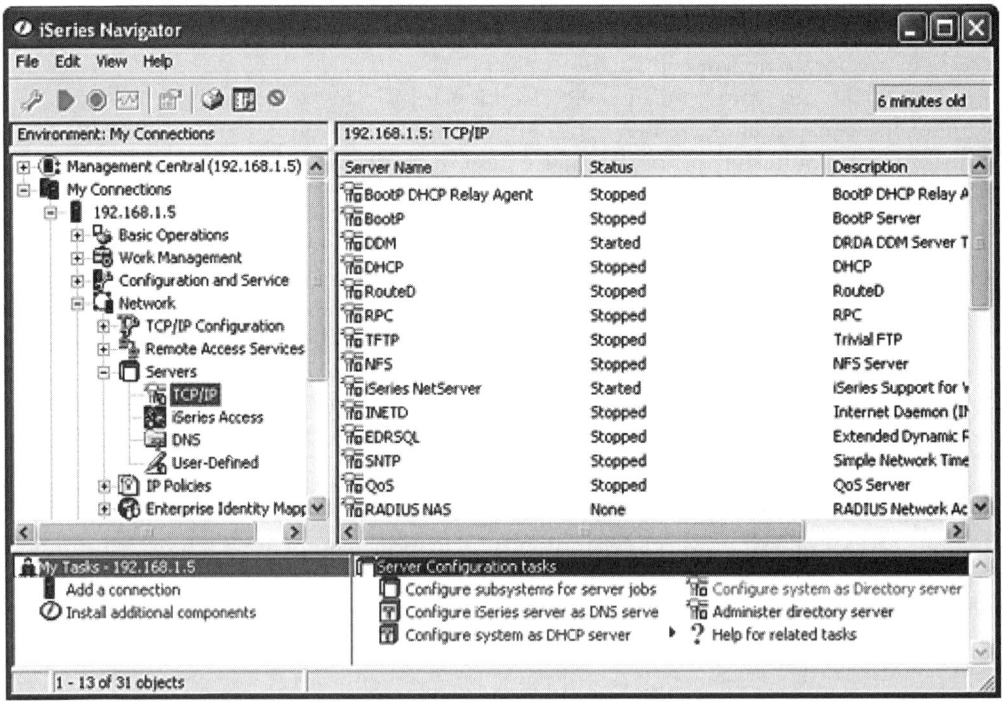

Chapter 11

Internet Security

When I first started writing about the topic of Internet security, it was rare to find an organization that allowed an Internet connection directly into its business. Today, it's rare when there *isn't* an Internet connection. With Internet access now a given, the implications for the security of your network and, more important, your data cannot be ignored. Let's talk through some of the issues to make sure they're being addressed appropriately in your organization.

Determine Your Risk

Your first task in assessing your Internet security needs is determining what the risk to your organization is should your data be compromised. Obviously, this assessment needs to occur even without consideration of the Internet. You need to determine the data's value to your organization, identify the vulnerabilities — both from external hackers and from internal threats (e.g., a disgruntled employee) — and then, based on those assessments, determine the lengths to which you're going to go to protect the data.

This is one of those times when you must weigh the risks against the cost of securing systems and data. That cost might be the additional funds that are required to buy more tools or hardware devices to secure the data. It might be the extra personnel needed to implement and maintain a more robust security scheme. Or it might be the cost of lost opportunity if the proposed interface is rejected because of the expense of securing it or the risk it poses to the organization's data. Few absolutes exist in the world of security. Developing and implementing a security policy consists of a series of tradeoffs.

In the past, Internet access was one of those decision points or trade-offs — added function versus a significant risk of exposure to data. For most organizations today, the benefits of the new functions and business opportunities provided by the Internet are worth the risk of potential data exposure. To mitigate that risk, organizations have implemented numerous layers of security. So, what you need to do now is determine whether your organization's current security scheme provides a sufficient number of layers to reduce the risk of data loss to a tolerable level. The rest of this chapter discusses the considerations you'll want to make during this risk-determination process.

The Process

So, where do you start? The most important step is the first one: determining your level of risk tolerance. In other words, how resolved are you, and to what lengths are you willing to go, to prevent your system from being broken into? To answer this question, imagine that someone breaks into your system. What damage could they do? What laws might be broken? Who would you need to notify about the breach — the board of directors, the government, the people whose information was compromised?

What data stored on your system is accessible via the Internet? If the data is sensitive (e.g., credit-card numbers), highly confidential (e.g., business data that your competitors would like to get their hands on), or private (e.g., patient medical records), or if laws or regulations have the potential to be violated, your risk tolerance will be lower than if your organization doesn't deal with credit cards and it outsources payroll processing.

What is the cost to your business if your system or network is breached — both from a reputation point of view as well as the cost of recovering from the breach (the forensics that must be performed, immediate reactions required to recover from a breach, legal counsel, and so on)?

Your level of risk tolerance will determine how many layers of defense you need to implement to protect your network, systems, and data. If you were dressing for a cold January day in Bozeman, Montana, you'd need many layers of clothing — long underwear, thermal socks, turtleneck, wool sweater, flannel-lined pants, down coat, scarf, hat, mittens, and boots — to stay warm. Likewise, the lower your security risk tolerance, the more layers of defense you need to dress your system and network in to be as safe as possible from outside attack. Your risk tolerance level feeds directly into your security policy requirements.

Corporate Security Policy

Chapter 2 touched on the fact that you need to update your corporate security policy to take into account your use of Internet technology. The lower your risk tolerance, the more restrictive your security policy will be, and the more layers of protection your systems will require.

To formulate your policy, you need to decide how you're going to deploy Internet technology throughout your network and identify what security technology will be included. You also need to form a policy about employees' appropriate use of the Internet, as I discussed briefly in Chapter 2. Put the policy, along with the consequences of noncompliance, in writing and communicate it throughout your enterprise. Here are some of the topics you may want to address:

- Where are you going to deploy firewall technology?
- What TCP/IP applications are required?
- Are you going to allow downloads from or uploads to your Web site or other Internet connection?
- Does your Web site require realtime data from systems in your internal network (as is the case with many WebSphere applications that use IBM i as their back-end data store)? If so, how are you going to implement the connection to your internal network? Must it be an encrypted connection?

- Will you let users connect to your system over the Internet? How are you going to protect that connection? With a virtual private network (VPN)? With Secure Sockets Layer (SSL)?
- Are you going to let employees access the Internet? Some organizations by default permit no access but grant access to users in roles that require it — for example, to let programmers search for problem resolutions or to let administrators obtain information about new technologies.
- Are you going to monitor the Web sites your users visit? Do you want to use a filter to prohibit access to Web sites that contain specific content?
- What does the company want to do if objectionable material is found on users' workstations or if users repeatedly visit inappropriate Web sites?
- How should employees handle suspected "social engineering" probes? That is, do employees understand what to do with e-mail or telephone calls that solicit personal or confidential corporate information?

After you've determined your level of risk tolerance and updated your security policy, it's time to decide how you're going to implement the policy. The rest of this chapter describes some of the layers of defense you can use to protect your network. You'll need to examine each potential layer and determine whether it makes sense for your organization to deploy it. Understand that there isn't one right answer. Each business must determine for itself what approach will serve its business needs. Not every company will find it practical or necessary to implement every defense layer. You might also discover and want to deploy some layers of defense I don't discuss here. The important thing is that, given your business needs, you and your management feel comfortable with the layers of protection you put in place and are confident that enough layers are in place to mitigate the business risk to your organization's risk-tolerance level.

Internet Service Provider

Your first layer of defense lies with your Internet Service Provider (ISP). When choosing an ISP, one of the first features you'll want to understand is the security benefits of the ISP's service. For example, what will the ISP do for you if your organization comes under a denial-of-service attack or if you suspect the organization has been hacked? A cooperative ISP is a must when you're trying to track down a breach or discover where a denial-of-service attack originated.

Another question to ask prospective ISPs is what their business continuity plans are. And I encourage you to validate that they are following them. The ISP that my business used to use claimed to back up its clients' Web sites, but at some point after signing the contract, the ISP stopped doing that and didn't bother to inform the clients! The ISP subsequently suffered a major hardware failure. Fortunately, we had a backup of our Web site, but I'm certain that other, less prepared clients lost their entire sites.

Firewalls

Firewall technology has evolved significantly over the years. The functions of very basic firewalls include using Network Address Translation (NAT) to hide your system's TCP/IP address from the rest of the Internet world (thereby protecting the system's identity) and filtering unwanted incoming and outgoing requests or IP packets. Firewalls can also monitor for attacks and take action if an attack is detected. In addition, some firewalls include VPN capabilities and virus scanners to scan incoming e-mail. Where you implement the technology in your network depends on the functions you want, how much network traffic you anticipate, and how many layers of defense you want to employ.

To determine where to implement firewall technology, you must take your network configuration into consideration. Figure 11.1 depicts a network configuration that uses a dedicated IBM i system as a Web server for a company that performs e-commerce (perhaps an online retailer or a site that accepts online payments). To protect the production system in this scenario, you'd put the firewall between the Web server and the internal network. Note that these systems could be physically separate systems, or they could be separate logical partitions (LPARs) on the same physical piece of hardware. Depending on the number of connections coming back into the internal network, a firewall appliance might be sufficient. Usually, though, you'll need a more traditional hardware firewall. You might also want to use IP packet filtering in either your router or your dedicated IBM i system to block unwanted IP traffic.

Figure 11.1
Dedicated Web Server Running IBM i

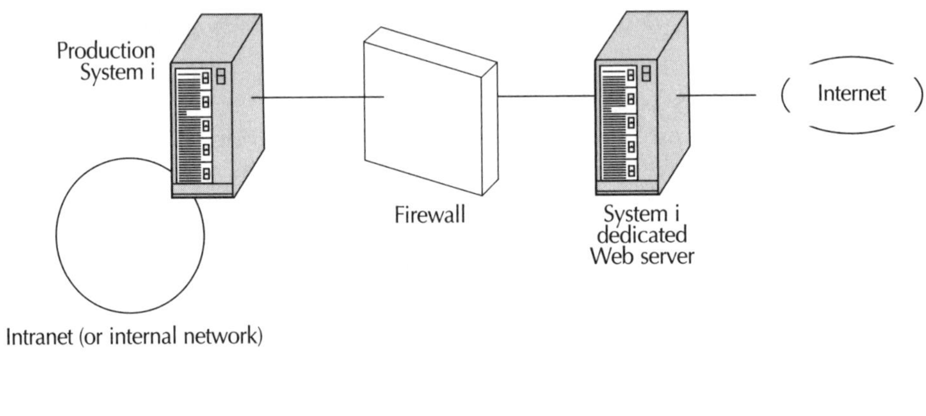

The configuration depicted in Figure 11.2 provides Internet access as well as production operations from a single IBM i system. I see this configuration becoming more and more prevalent. An enterprise that uses this type of network configuration might have a VPN connection from the production system to suppliers over the Internet.

Another configuration might have the production system doing double duty as an Internet Web server. Although doing Web serving and production work on the same server isn't as secure a configuration as having separate Web server and production systems, I've seen some companies that have made the business decision to use this configuration. In the case of a VPN connection, a router with IP packet filtering might meet your needs in this scenario. However, if your production system also acts as a Web server, you probably want the kind of services, such as proxy or SOCKS servers, that are found in traditional firewalls.

Figure 11.2
Production System Running IBM i

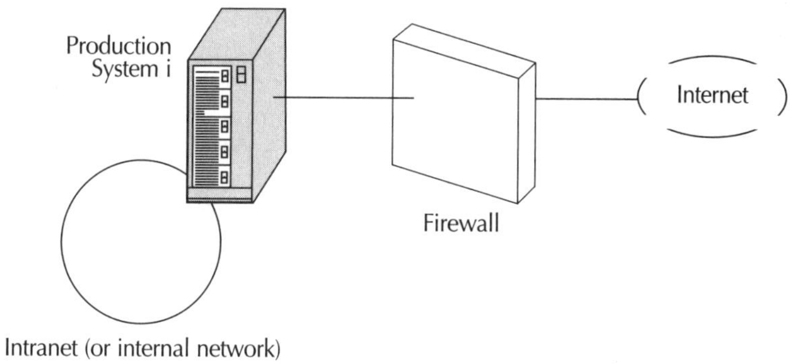

Before you buy a firewall, determine which features you really need and decide where to implement the technology. If you need a traditional firewall — a standalone system with firewall software — compare several vendors' firewalls that provide the features you need. Keep one thing in mind: The more functions and capabilities built into the firewall, the more complex it can be to configure. Most firewall breaches are due to configuration errors, not bugs in the firewall software. Because of this fact, if you don't have firewall expertise within your organization, consider hiring a reputable firm that specializes in firewall configuration and maintenance.

System Values

Although a firewall is your primary line of defense when connecting your system to the Internet, a firewall alone won't protect you. Errors in firewall configuration can leave your system vulnerable. And hackers are always looking for new ways to penetrate firewalls or fool them into thinking that an intruder's requests are valid ones. You need to assume that, at some point, your firewall won't keep out intruders. In other words, your first layer of defense fails. What then? What's your next line of defense?

Setting the IBM i system values to the recommended settings given in Chapter 3 is critical if you connect your system to the Internet. For an IBM i system connected to the Internet, you'll also want to consider using strong passwords. The more strict your password composition requirements are, the less likely that a hacker will be able to guess a valid password. This is an excellent scenario for implementing passphrases — that is, setting the QPWDLVL (Password Level) system value to 3 to enable passwords that can be up to 128 characters long. Using passphrases makes guessing passwords extremely difficult. Just remember that when you make it hard for a hacker to guess a password, you also make it hard for legitimate users to compose passwords and to remember them once they do. This is an example of making a trade-off between a more secure implementation and convenience for your users.

In addition to setting the system values as recommended, you'll want to examine the INACTTIMO (Inactivity time-out) attribute for each TCP/IP application and make sure the setting makes sense for your configuration. To set the time-out value, you use the appropriate change command to change the application's attributes. For example, to set INACTTIMO for a File Transfer Protocol (FTP) application, you'd use the CHGFTPA (Change FTP Attributes) command. Note that not all TCP/IP applications support the INACTTIMO parameter. For some applications, a timeout value doesn't make sense; others, such as Telnet, rely on the system value QINACTITV (Inactivity Time-out Interval) setting.

After you've considered the settings of your system values, you need to examine your next line of defense: the users you've defined for the system.

User Profiles

As you learned in Chapter 4, you must have a user profile to sign on to IBM i. To reduce the possibility of someone maliciously reaching your system from the Internet, you want to restrict the number of user profiles on your Internet-connected system. In other words, don't perform a SAVSECDTA (Save Security Data) command on your production system and then do a RSTUSRPRF(*ALL) on your Web server system to restore all user profiles there.

You also want to ensure that the IBM-supplied profiles aren't usable. IBM profiles are well-documented in both IBM and hacker publications and Web sites. Allowing IBM-supplied profiles to sign on is a wide-open door for hackers to exploit. Run the ANZDFTPWD (Analyze Default Password) command, and make sure the passwords for QPGMR, QSRV, QSRVBAS, QSYSOPR, and QUSER are set to *NONE.

QSECOFR is another well-published profile. To make it unusable to hackers, create a profile and make QSECOFR its group profile. Then set QSECOFR to a status of *DISABLED. In the event of an emergency, you can always sign on as QSECOFR at the console even when QSECOFR's status is *DISABLED.

For the few remaining user profiles on your system, consider using a cryptic naming convention. Profile names such as ALAN, MARY, MASTER, OPERATOR, and SECOFR are too easily guessed. User profile names can always be guessed, but the more cryptic they are, the longer the guessing will take. Increase the odds that an intruder will run out of time before making a successful guess on your system.

For those of you whose system doubles as a Web server and a production system, make sure you are very timely in removing the profiles of users who leave the company. It's even more critical to quickly delete the profiles when users are dismissed or laid off. To protect all active profiles on your system, you can use the CHGACTSCDE (Change Activation Schedule Entry) command to set the status of these profiles to *DISABLED during hours when you don't expect them to be used.

Resource Security

A vital layer of defense — especially if you're using your production IBM i system for your Web server — is to implement object-level security. Although more organizations are addressing this issue, most organizations I talk to have not. In other words, the access control settings on most objects allow users too much access.

My recommended approach is that of "least privilege" access or "deny by default." In other words, objects — especially files containing private or confidential data — are set to *PUBLIC *EXCLUDE, and only those users with an explicit business need to access the data from outside an application are given direct access to the objects.

By locking down your system, you also prevent hackers from using it to store inappropriate data. Hackers don't always keep stolen data, pornography, illegal copies of software, or hacking tools on their own systems; they often find other systems — including yours, if they can get to it — and store such data there. As you learned in Chapter 10, the term for this illegal activity is *parking*.

Controlling What Goes On

Don't be lax about what happens on your system. If you don't want an intruder to be able to perform some function and you don't need the function for business purposes, don't allow it on the system. The more services and applications available on your system, the more likely it is that an unintentional avenue exists into it. If a tool or application doesn't exist on the system, a hacker can't use it to break in or cause damage.

In addition to the system value recommendations provided in Chapter 3, I advise taking the following steps:

- Start the TCP/IP servers for only the applications you're running. For example, if you don't use Line Printer Daemon (LPD), don't start it. Or, start the FTP server only when you need to use it; don't leave it on all the time.
- Follow the directions in Chapter 10 to ensure that you start only the TCP/IP servers you need, and secure the ports used by any TCP/IP applications you don't use.
- Remove or secure compilers on your system, and secure the program-creation (CRT*xxx*PGM) commands.

- Turn off IP packet forwarding by specifying N (No) for the "Allow IP datagram forwarding" option when you add or change a configuration profile from the "Work with Point-to-Point TCP/IP (WRKTCPPTP)" display.
- Be careful how much validity you place in the data you receive over the Internet. Some companies use FTP exit programs on IBM i to permit or deny the use of FTP based on the user's IP address. But validating a user based on an IP address can be dangerous because IP addresses can be spoofed. (*Spoofing* occurs when someone alters an IP address to look like another, legitimate one.)
- Delete any software that is not being used.
- Promptly remove inactive profiles.
- Remove unneeded libraries and directories.

Secure Web Applications

IBM's HTTP Server (Powered by Apache) gives you another layer of Internet defense by providing the following mechanisms:

- Directives within the configuration file that provide access control to Web server resources
- Separation of who can access HTML source and CGI-BIN programs
- Secure Sockets Layer

The Web server doesn't serve information unless you specifically tell it to. The only Web page you'll see when you first start the Apache Web server is a welcome page. To tell the Web server what information to serve and to whom, you use directives within a configuration file. The Web server by itself can't grab confidential information out of your database files and serve it to the world.

Just as I recommend that you have a separate library for each of your native applications, I recommend that you separate your HTML source and CGI-BIN or Java programs from other applications that might exist on your Web server. By default, IBM i serves all CGI-BIN programs out of the QTCP library. *Don't* propagate this design. Instead, create a library for each Web application, and put that application's CGI-BIN programs there. Set public access for these libraries to *USE or *EXCLUDE to control who can add programs to the libraries. You should also create a separate directory for all your HTML source. Set public access for this directory to *R (the equivalent of *READ).

Separating HTML source and Web programs from other applications is necessary because the directives in the configuration file allow wild cards or generic rules. For example, you can add a directive to let users execute any program in a particular library. When you do that, any program in that library is accessible over the Internet. Another reason not to mix applications is that if you have programs and files with the same name or that both fit a generic directive, you might not be able to predict which program or file the system will serve.

Other directives enable basic authentication. An application that enables basic authentication challenges the requester for a user ID and password when a request comes to the server. For example, when checking the balance of a bank account, customers must enter a name and password to obtain the information. The Apache Web server lets you enable basic authentication in two ways — one way with IBM i users and another way with Internet users.

You can configure your Web application to require requesters to provide a valid IBM i user profile and password before serving any information to them. Although browsers don't transmit passwords in clear text over the connection, the passwords aren't encrypted. I don't recommend requiring IBM i user profiles and passwords for authentication unless you're operating in an intranet environment or using SSL.

Internet users aren't "real" IBM i users because they don't have a corresponding *USRPRF object, so an IBM i job can't run as one of these users. As you learned in Chapter 4, you implement Internet users using an object type called a validation list (*VLDL). Validation lists associate an entity (typically a user) with authentication information (e.g., a password, account number, digital certificate). You can specify a validation list for your Web application to use to authenticate users. When your application requires basic authentication, the system checks the user name and password against the information in the specified validation list. The Web application doesn't have to store or know the actual authentication information because the browser and server handle authentication between themselves.

Another feature of the Web server is the ability to configure Web pages to be served using SSL. When a secure HTTP (HTTPS) request initiates an SSL session, the browser verifies that it is talking to the server you pointed to. After the browser validates the server, the connection is established, and all subsequent flows are encrypted. Encryption is critical for ensuring privacy and integrity of confidential data over the Internet. All data sent via a nonsecure (non-SSL) connection is passed in clear text and is subject to being altered or sniffed (i.e., read from the communications flow). Because the Internet is a public network, you can't determine how many or which systems your data is routed through. (To be realistic, although sniffing certainly occurs, it's not something the average user can do. If your data is sniffed, the perpetrator is most likely someone inside your own organization who was able to attach a line-tracing device in your network.) Using SSL protects your data from both external and internal hackers. However, you should use SSL on only your confidential data because making an SSL connection and subsequently encrypting data incurs overhead.

Another feature of SSL — and an HTTP server configuration option — is client-side authentication. When you configure client-side authentication, the browser establishes an SSL connection as described above, and then the server challenges the browser for a digital certificate with which to authenticate the user at the browser. Client-side authentication uses digital certificates. A digital certificate can be associated with either a real IBM i user profile or an Internet user.

Whether you're using basic authentication with IDs and passwords or SSL client-side authentication with digital certificates, if the user you're authenticating is a real IBM i user, you can configure your Web server to run as that user. This is a powerful feature for intranet applications because it lets you use traditional IBM i object security to secure your data.

WebSphere applications present their own challenges — especially when it comes to passwords. WebSphere configuration files often contain the user ID and password used to connect to the database that is serving as the back-end data store. WebSphere does provide a facility to encrypt this password, but it's my experience that this feature is rarely used. The configuration files are in the application configuration directories in the integrated file system (IFS). You'll want to make sure that your programmers use the facilities provided by WebSphere to encrypt these passwords and that you've secured the configuration files appropriately.

If you must be in compliance with PCI DSS, your Web application must comply with PCI's Web application security standards. Even if your Web application doesn't accept credit cards, I suggest you read these requirements and implement them as appropriate for a more secure application. (For more information about the PCI's Web application requirements, go to *https://www.pcisecuritystandards.org.*)

Exit Programs

You can use exit programs to provide another layer of defense. Exit programs are best not for controlling access to objects, but for logging the activity that occurs for IBM i servers such as FTP, Distributed Data Management (DDM), and so on. Some vendors may claim that a system can be secured solely through the use of exit points, but that is absolutely not true. If you use exit points to control access to objects, you should think of them as another layer of defense on top of, but never as a replacement for, IBM i object-level security.

Monitoring

It's one thing to set up your system for an Internet connection, but it's another thing entirely to maintain the security of that system. To ensure that your security mechanisms remain effective and adequate, you'll want to perform several types of monitoring.

Intrusion Detection

Intrusion detection isn't just one type of monitoring, but three. First, you should analyze your system proactively for known vulnerabilities. For IBM i, vulnerability analysis includes making sure you've applied all the integrity PTFs. Starting only the TCP/IP servers you need and securing the ports of those you don't also fall into this category.

The second type of intrusion-detection monitoring is active monitoring. In V5R4, IBM added the ability to detect when the system is under attack at the IP stack level. Turning on auditing for intrusion-detection events (by specifying *ATNEVT for the auditing level system value, QAUDLVL) and configuring a script file in the IFS to tell IBM i what to alert you to will enable you to understand when the system is experiencing a denial-of-service attack, for example. In V6R1, the configuration is even easier, and you can receive realtime alerts about the attacks. In addition, you can monitor IBM i message queues such as QSYSMSG for messages about invalid sign-on attempts. Then, if you use a message management software package, you can set thresholds and send an alert when, for example, 50 invalid sign-on attempts occur in one minute.

For active monitoring to be successful, it must include a plan of action that you'll follow if you do detect an intrusion. Identifying the fact that your system is being attacked won't do you much good if you don't know what to do about it.

The third type of intrusion-detection monitoring is after-the-fact detection that someone has accessed your system without your permission. Part of this analysis includes determining what was accessed, changed, or stolen. IBM i provides several ways to log system accesses. The auditing capabilities that are integrated into the operating system (which I cover in Chapter 15) provide good information related to this type of intrusion detection. The Apache Web server also includes access and error logs that are useful in determining who accessed your system.

Typically, the problem isn't an inability to monitor. Rather, many companies that don't audit extensively or log Web accesses think that they don't need this type of information and consider such monitoring a waste of system resources. Unfortunately, you can't go back and re-create the audit entries after a break-in has occurred. Which objects and actions you're going to audit are choices you must make based on the needs of your business and your security policy — *before* a break-in occurs.

Security Configuration

Organizations aren't static; they change. And their systems and security needs change with them. You should regularly monitor the status of your security configuration — your user profiles, system values, and object authorities to be sure they remain valid. You can use IBM i's integrated security tools or a third-party product (such as SkyView's Policy Minder) to detect changes to your security configuration and to keep it current. Regardless of the method you choose for maintaining your configuration, the point is that you do it. I've seen too many systems where the security scheme had "degraded" because no process was in place to ensure it was maintained.

Testing and Evaluation

Test, test, and retest! Obviously, you should test your network configuration for holes before you go live with your Internet connection. But testing and evaluation is even more important after the Web application has been in place for a while. As proof of this point, PCI requires regular vulnerability scans for Web application and wireless networks. PCI as well as the Health Insurance Portability and Accountability Act (HIPAA) and other laws and regs require an annual vulnerability assessment of your system. Why? Because things change. You acquire a company and systems are merged. You implement a new enterprise resource planning application. You upgrade to the latest version of the operating system. All these events can cause vulnerabilities to appear. Unless you test and evaluate, these vulnerabilities will most likely go undetected and could be exploited.

Business Contingency Plan

Last, but certainly not least, are the changes you need to make to your business contingency plan. Here are some of the areas to consider:

- Does your backup or hot site have Internet access and the hardware needed to run your Web application? Does it have the bandwidth you require to support your critical business needs? Does it have the necessary encryption hardware or certificates needed to provide encrypted sessions?
- Do you have an alternative route to your ISP or an alternative ISP if your current ISP's connection is severed?
- In the event that you suspect or are informed that you've suffered a breach, do you have an incident response plan in place? For an in-depth discussion of creating an incident response plan, see Chapter 18.

The Good News

The Internet is an attractive place to do business, and it's rare for companies today not to have an Internet presence. Risk is part of life, and it's part of doing business. Connecting your system to the networked world shouldn't paralyze you. IBM i is robust and provides many layers of security you can implement. The secret to doing business safely on the Internet is to analyze your risks and then mitigate them by putting in place as many layers of defense as your business requirements dictate.

Chapter 12

Evaluating Applications' Current Implementations and Designing New Ones

We're now to the point of being able to put together all the topics we've talked about in earlier chapters — system value settings, user profile attributes, and object authorities. Separately, these might be interesting topics, but they aren't very useful in the practical sense. That's because they all must work together and be in balance to achieve the appropriate level of security that your organization requires and that supports and complies with the organization's security policy. Also, you don't purchase an IBM i system to look at file permissions, system values, or user profile attributes. You purchase the system to store data, run applications, and work with the data. So, it's time we take a look at how your applications have incorporated security and whether their authorization scheme complies with the requirements of your organization's security policy.

It's important to understand each application's authorization scheme — that is, to know how users obtain authority to run the application and access its objects. It's also important to understand how securely the application has been implemented — in IBM i terms, what is the *PUBLIC authority setting on the libraries, directories, and files? Unfortunately, most vendors don't implement their applications securely. Many leave their application objects at the default *PUBLIC authority of *CHANGE. The result is often that confidential or private data is available to anyone who has a user profile.

It's easiest to describe the principles of an application's security methods in the context of implementing an application the "right way." Even if you're not actually writing a new application, this explanation should help you understand what you need to look for in your existing applications to determine whether there are issues with the implementation of their security scheme.

From the Beginning

The right way to design an application recognizes that the security of the data, as well as secure coding practices, is one of the critical architecture design points. With this approach, data security is considered and evaluated at implementation as well as throughout the various testing phases. Security should be addressed at every phase of application development by default. Although it is becoming more prevalent, this is not a widespread practice. Often, entire applications pass through design, implementation, and testing with no regard for security. Then, at some point, someone discovers huge holes in the application's security scheme that must be fixed.

Those of you who've found yourselves in this situation know that it's very difficult to retrofit security into an application. Successful design and implementation of security methods within an application require thinking about security throughout the application development cycle. Beginning with the requirements phase and proceeding through design, implementation, and test, you need to ask — and answer — certain questions to be sure you're covering all the bases. Such questions include the following:

- Who needs to use the application?
- Do you need to allow the capability to define users within the application?
- Do you need to provide the ability to define roles within the application?
- How will you authenticate those users?
- Do the users need access to the application data from outside the logic of the application?
- What is the classification of the application data?
- What exposures exist in the chosen application runtime environment?
- How will security be verified, both at initial deployment and over the application's lifetime?

Addressing these questions across the duration of an application project will keep security in its rightful place as an essential design element.

Design Considerations

Security needs to be considered at the very beginning of an application's design: the requirements phase. To do this, make sure that at least one point in your list of requirements talks about security. For example, say you're writing an application to take customer orders over the Web. Among its requirements, your application might have the following:

- Retention and processing of cardholder data will be in compliance with PCI DSS requirements.
- Only the appropriate information will be retained.
- Information will be retained only for the life of transaction processing.
- Cardholder data will be encrypted at all phases, including storage in the database.
- Back-end data store systems (systems on which orders and status information are actually stored) can't be accessed directly from the Internet.
- Only authorized users can place orders, and orders must record each user who works on them.
- User authentication is always required and should time out after a short period of inactivity.

It's critical that the type of data (e.g., private, confidential) be identified in the initial design of the application. That's so that the design can be sure to include all the requirements of the laws and regulations for which this type of data must comply. Next, make security part of your application design template. This is what IBM has done for years to make security so robust in IBM i and its predecessors. Every IBM i team leader must address a set of security design points each time an enhancement or change is made to the operating system code that the team leader owns.

Here are some of the questions you should include in your application design template:

- Who (what roles) will use the application?
- What authorization scheme will be used?
- Which user profile owns the application, and is adequate logging in place?
- Under what profile(s) does the application run?
- Does the application need to adopt or swap to a "powerful" profile, and, if so, for what functions?
- What kind of audit trail does the application require?

By forcing the application's designer to answer these questions, you can force a discussion of the security design at a design review and ensure that the design meets requirements. You should also provide documentation that shows the security choices and tradeoffs made.

The following sections expand on some of these design questions.

What Roles Will Use the Application?

It's surprising how often designers overlook this critical question. After all, the type of application user will influence everything from your authentication model to how security limits an end user's activities. For example, an application intended for use by an order entry clerk likely won't require complex authentication, such as a digital certificate, and probably will restrict the user from entering system commands directly. On the other hand, an application to be used by a security officer may well have multiple levels of authentication and typically gives the user the ability to run commands.

Different roles will have different security requirements, depending on the data classification. A user in an unsecured area, such as a reservations clerk in a hotel lobby, likely won't routinely gain access to proprietary enterprise financial data. However, a manager in the same organization may need such access occasionally. In this case, the application would need to be able to accommodate multiple roles to control which role could perform which functions. It would also need to accommodate appropriate audit trails.

So take time to carefully consider what roles will be using your application. Determine from the outset what their various access methods will be (e.g., FTP, ODBC, 5250 emulation), and develop appropriate security policies that the application can support. Perhaps the most important aspect of any security design is how your application will enforce your organization's security policy.

Common Authorization Schemes

After working with clients and talking with many of you at conferences, I've determined that there are three basic authorization schemes that user-written and vendor-written applications typically use. Some applications use more than one of these schemes, and I've seen some applications require all three! By authorization scheme, I mean the method used to ensure that users of the application have sufficient authority to perform the tasks within the application.

While some authorization schemes are quite secure, they are not what is implemented in most IBM i organizations. I'm going to first discuss the typical authorization schemes and why they are problematic; then I'll discuss the schemes that are secure.

Menu Access Control

Menu access control is the way most end users see the system. It's typically a "green-screen" application that, when the user first signs on to the system, launches the user directly into the initial menu of the application. This is the "front end" for most of the authorization schemes I'll discuss. Menu access control assumes that security is achieved simply by presenting to each user only menus that provide the functions that the user needs to perform his or her job. Users can access anything on their set of menus and can access the objects and data associated with those menu options.

To ensure that application users have sufficient authority to perform their job functions, one of the three authorization schemes described next must be implemented for menu access control. That is, most application writers require that the application objects' *PUBLIC authority is left wide open (typically either *CHANGE or *ALL), that the application users are members of the profile that owns the application, or that the users have *ALLOBJ special authority. Figure 12.1 depicts the menu access control model.

Figure 12.1
The Menu Access Control Model

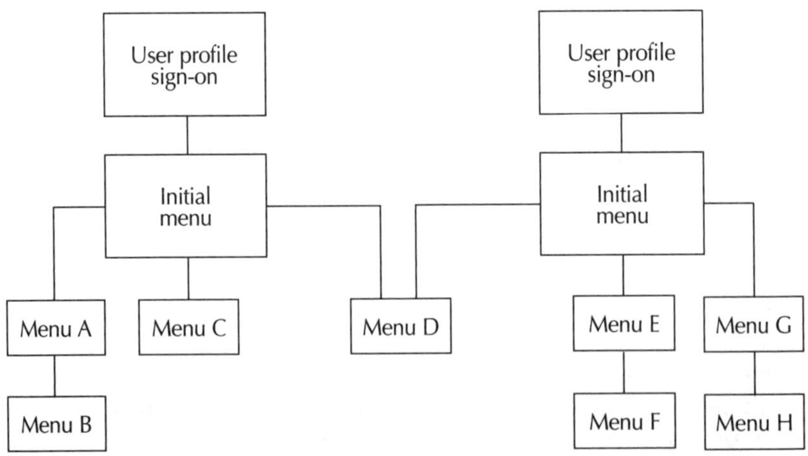

To implement menu access control, each user's initial menu is the "entry point" into the application. You design the initial menu for each user individually using the INLPGM (Initial program) or INLMNU (Initial menu) parameter of the user profile or, in the case of some packaged software applications, by means of initial user data entered and maintained by your organization's employees. In addition, application users are configured in their user profiles as limited-capability users (LMTCPB(*YES)) so that they can enter only a handful of commands if presented with a command line. The menu authorization model is the method of choice for many software vendors simply because it's the least complex for both them and their customers to implement.

I'm not trying to imply that you should get rid of your application's menu structure. Menus are a logical way to provide access to software applications. Because you don't want end users entering commands from a command line, you provide menus that logically group end-user functions. Whether the menu is a pull-down, pop-up, or plain full-screen menu, end users are much more productive when they can select tasks from a list than when they must memorize cryptic commands or program names. Also, by logically grouping the tasks users can perform, menus also secure the system by restricting users to the tasks on the menus.

Now, let's take a look at the most popular authorization schemes that implement the access controls for users' menus.

*PUBLIC Authority

In the *PUBLIC authority model, application objects' *PUBLIC authority is set so as to allow all application tasks. This translates into a setting of either *CHANGE or the one I see most, which is *ALL.

The problem with this scheme should be obvious. Literally any user on the system with access to a command line, such as operators, database administrators, programmers, and the handful of application users who might be allowed command-line access, can change application data from outside the protection of the application. Why do I say the "protection" of the application? Because most applications have logic built into them that prevents certain users from updating certain database fields. Or there may be application logic that lets fields be updated only within a certain range (e.g., a price field), and anything outside that range must be done by a supervisor. But in the case of wide-open *PUBLIC authority, the access is provided to literally every profile on the system, and every profile has the capability to at least update and possibly change any application object.

Group Authority

In the group authority model, one group profile owns the application. Users who want to use the application must be members of this group. Each group member also owns any object that the group profile owns.

The problem with this scheme is that ownership of objects is passed from the group to all users. In other words, if a group profile owns an object, every member of the group also owns the object. So when application users are required to be a member of the group that owns the application, every application user owns each application object — including all database files. That means that their access to these objects — including any sensitive or private data — is *ALL. Therefore, if users can get to the data from outside the application, they can replace programs with their own version, read the data, update the data, or perhaps even delete the entire file.

The argument I hear is that this scheme is not a problem because the users are limited-capability users and therefore can never access the application objects directly. Anyone who still actually believes this statement truly has their head in the sand. IBM i has become so open and supports so many protocols that there are now numerous ways for users to directly access application objects — including via Distributed Data Management (DDM), File Transfer Protocol (FTP), Java Database Connectivity (JDBC), Open Database Connectivity (ODBC), Web applications, sockets, and System i Access data transfer, to name a few.

The bottom line is this: Leaving application users as members of the group profile that owns the application poses a significant risk to the application data.

*ALLOBJ

In the *ALLOBJ authority scheme, application users are required to be assigned *ALLOBJ special authority in their user profiles.

The dangers of this scheme should be obvious. *ALLOBJ should be assigned only to the administrator role — never to end users. Users with *ALLOBJ can access anything on the system. And, again, it doesn't matter if the user is set to be a limited-capability user. As described for the group authority scheme, there are many ways to access application data on today's IBM i systems.

Authorization Schemes That Are Secure

Now that I've blown apart most of the IBM i application security schemes (!), let's take a look at a couple of schemes that provide for a secure implementation.

Application-only Access

One approach is to have a single profile own the application. This profile is authorized to access the data files associated with the application. Public access to the application objects is *EXCLUDE. Users are authorized to the initial menu or to run the command that invokes the application. Because the application uses adopted authority, end users can access the application data, but only through the application and not using other outside accesses such as DDM, FTP, ODBC, and so on. This model is very secure, but in today's world it is not very practical. I haven't seen a system in recent years that is not accessed using one of these outside access methods. Fundamentally, though, this is a good approach, so let's look at some variations that provide more flexibility.

Application-only Access with a Twist

This is the approach that SkyView Partners typically uses when we help our clients to rework their applications' security schemes. We use adopted authority so that the application users will have sufficient authority to run all parts of the application. But we accommodate the fact that there are users and processes that need to read, and sometimes update, some application database files. To accommodate this access, we secure all physical and logical files with an authorization list. Users and processes that need access are given authority to the authorization list. They are granted *USE authority if they need only to read the data and *CHANGE authority if the need is to update.

This is just the very high-level description of how this authorization scheme works and is implemented. I've developed a specific methodology that I use to take an application from using one of the three authorization schemes previously described to an authorization scheme in which the *PUBLIC access is set appropriately for the organization's security policy, approved users and processes can access the application data from outside the application, and application users have sufficient authority to run their application menu options. I'll describe this methodology in detail in later chapters.

Changing the Process

If you're accessing objects in the integrated file system (IFS), such as stream files or Hypertext Markup Language (HTML), adopted authority won't work because the IFS doesn't honor adopted authority. Your options in that case are to authorize the end user to the IFS objects or to change the process or thread so it runs as a different user — one with the authority to access the IFS object — when accessing these objects. To change the process to run as a different user, you have several choices: profile swap, profile token, and the UID and GID APIs. Let's take a brief look at each of these options.

Profile-swap APIs have been around for quite a while. These are the APIs that the IBM i servers (e.g., FTP, Telnet) use to get a job to run as the requesting user profile rather than the user profile under which the server was started. When a process's profile is swapped, all security attributes are swapped in, including the new user's special authorities, limited-capability attributes, group profile(s), and audit settings. The "swapped to" user owns any new objects created. Use of the profile handle that's generated before the profile is actually swapped is limited to the process that generated the handle.

Profile-token APIs have the same effect as the profile swap APIs, but they generate a token that can be passed between jobs. APIs can generate tokens as one-time use, multi-use, or multi-use regenerable types. These tokens time out and become invalid after a specified time.

The *UID and GID APIs* were added to aid in the porting of Unix applications running in the Portable Application Solutions Environment (PASE). The native IBM i versions — qsyseteuid(), qsysetegid(), and so on — give IBM i application developers a different option for an application's authorization model. Unlike what happens with the profile swap or profile token APIs, the UID and GID APIs swap in only a user profile (in the case of the UID APIs) or a group profile (in the case of the GID APIs). The UID is a numeric representation of the user rather than the user profile name. The GID is the numeric representation of a group profile.

How do these options help the application designer? Now, when a user accesses an application, part of the application startup can set the GID to the application owner. The application-owning profile is now the user's first group. The beauty of this solution is that the application owner becomes the end user's group for *that thread only*. The user's profile is never updated — just the thread. Before these APIs were introduced (in V4R5), the only way you could implement a similar model was to make a change to the user's group profile attribute in the user's profile (using the CHGUSRPRF, or Change User Profile, command, for example) and then use the swap APIs to swap to the same user to make the new group profile take effect in the current thread.

The problem with this old method is that if a user can start another process, the application profile is the user's first group profile in the new process, which lets the user access all the application objects with owner rights. This situation poses risks similar to those experienced when the user is directly defined as a member of the application owner's profile. You eliminate these risks when you use the GID APIs. Because the user's profile is never updated, starting a new job is done using the user's group profile list as defined in its profile, and it will not include the group profile specified in the GID API. Of all the profile-swap options, this method is my favorite because, like adopted authority, it provides temporary authority to application objects and, unlike adopted authority, it provides temporary authority to IFS objects — not just "traditional" IBM i objects.

Application Ownership

Getting back to the issues you need to address during the design phase, we turn now to the ownership of applications. Many IBM i administrators are livid when they discover third-party applications owned by IBM-supplied profiles, such as QPGMR and QSYSOPR. Especially irritating is when the application is owned by QSECOFR. If you're creating your own application, *do not* make one of these profiles the owner. Create a specific profile whose sole purpose is to act as the owner of the application. Create this profile with the following attributes:

- PASSWORD(*NONE) — You don't want people signing on using this profile.
- SPCAUT(*NONE) — Assign special authorities only if the application needs them to run its tasks.
- STATUS(*DISABLED)
- INLPGM(*NONE)
- INLMNU(*SIGNOFF)
- PWDEXPITV(*SYSVAL) — Some administrators set this attribute to *NOMAX, but that's not necessary.
- LMTCPB(*YES)

Which Profile Runs the Application and Is There Adequate Logging?

This may seem like a strange question when designing a traditional IBM i application because the user running the application always signs on to the system. But the issue becomes important when you consider Web and client/server applications.

If you're writing a Web application, you may want all the requests to run as an application profile rather than requiring the thousands of users logging on to your Web site to have valid IBM i user profiles. In this case, you can use validation lists as a secure way to store user identification and authentication information. Then the Web application's protection directives name an application profile under which the application will run.

In the case of client/server or WebSphere applications, most requests are, again, run as one profile. The problem with this model is that because all requests are run under the same IBM i user profile, the audit entries as well as all journal entries (if you're journaling the application files) are logged as this profile. Therefore, auditing these applications is difficult. That is, it's very difficult to determine which user actually made a specific transaction. In this case, make sure sufficient logging occurs on the client side of the application to ensure you have an adequate audit trail of who performed each transaction and when.

Does the Application Require a "Powerful" Profile?

The authorization methods I've described suggest that developers create an application profile with little or no special authorities. But what happens when an application needs to use an IBM i function that requires, for example, *IOSYSCFG or *ALLOBJ special authority? In these cases, you must swap to or adopt a powerful profile. If you're using adopted authority, the best way to do this is to create a program that is owned by a profile that has the required special authorities, performs only those actions that require the extra authority, and then immediately returns. If you're swapping profiles, obtain the profile handle, perform the swap, perform the required functions, swap back to the original user, and release the profile handles so they can't be used again. Also, make sure the swap back is part of the program's error-handling routine.

This requirement often arises when creating "utilities" for operators or help desk personnel. Operators may need to be able to specify a parameter on one of the restore commands that requires *ALLOBJ special authority. A short CL program that adopts an *ALLOBJ profile lets them perform this type of function without having to have *ALLOBJ assigned to their profile. Or help desk personnel may need to re-enable user profiles. In this case, you can create a program that is owned by a profile with *ALLOBJ and *SECADM special authorities and runs the CHGUSRPRF (Change User Profile) command with only the STATUS field available. When calling the program, the help desk personnel can reset the profile, but the capabilities have not been assigned to their profiles.

What Kind of Audit Trail Does the Application Require?

Audit trails are often an afterthought in application design. IBM i includes a broad set of auditing features. However, depending on the application you're writing, it might be advantageous to supplement the operating system auditing with audit entries of your own. You can configure IBM i to send an audit entry whenever a file is updated or changed, but the audit entry may not provide all the information you need, or it may not be specific enough for your particular application.

For example, you can configure IBM i to send an audit entry logging the activity if Linda downloads the entire human resources (HR) file. The audit entry will tell you the name of the file and the fact that Linda read it. It will also list the day and time when she read it. If Linda updates the field in the HR file containing social security numbers, the audit entry will give you the same information, noting that the file was changed (rather than read). But it won't tell you *what* changed.

Government regulations may require you to make a record of everyone who reads a particular field of a file or to maintain separate records of a field's original contents and its subsequently changed contents. For example, if John runs a query against a file of patient information, reading all fields except the one containing medical test results, no record is necessary. However, the moment he either changes or reads the field that contains the test results, a record of that fact must be logged. The system administrator running the application can enable this capability under IBM i by using a trigger program, but it's much easier for system administrators if this type of function is built into the application in the first place.

To meet the record-keeping requirements for most regulations, the record must identify the person causing the event and must include an accurate timestamp of the event. To have a record stand up to legal scrutiny, you must be able to show that the audit records couldn't be altered. The easiest way to do that under IBM i is to use the operating system's journaling features. If you send a journal entry with your application information, even a security officer can't alter that information. IBM i handles the timestamp and user identification. All you need to do is provide entry-specific data — that is, data that makes sense according to the application data being updated or read.

Implementation Details

Next, let's look at some specific IBM i implementation details that will help ensure a secure implementation of your application. These recommendations can also serve as a good checklist when you're evaluating vendor application implementations.

Set IBM i Authorities

What authority is required to run your application? Use Appendix D of the *System i Security Reference* manual as your guide. This appendix lists all IBM i commands and the authorities required to run each one. If your application resembles some IBM i function, check this appendix to find the authority requirement, and then require the same authority for your application.

It's easy to write your applications to demand too much authority. For example, as I've noted, some popular vendor applications require all users to have *ALLOBJ special authority. But unless your application is targeted exclusively at your system administrator, requiring *ALLOBJ is not appropriate. Similarly, some vendor applications require you to sign on with the QSECOFR profile; this is an especially popular requirement for product installation. But no application should require this profile's use unless the application performs a function for which IBM i requires QSECOFR as the process profile. I can think of only one function that makes this requirement: resetting the security officer password for the Dedicated Service Tools (DST).

The questionnaire on page 220 will help you assess the security impact of a third-party application. Have the vendor answer these questions before you make any new software selection.

Define *PUBLIC Authority

Application security used to be so easy. As long as you confined users to a menu and never let them access a command line, you had all the security you needed. But, as I discussed earlier, the fact that you can access application data from command lines and using other access methods (e.g., FTP, ODBC) means you must use an appropriate *PUBLIC authority setting for your application objects. The *PUBLIC authority setting is dictated by the classification of the data and your organization's security policy. Although you may think that *PUBLIC authority must always be set to *EXCLUDE, that's not always the case. Yes, it will be the case for human resources data, cardholder details, and health care information. But, depending on the organization's security policy, there are times when I've seen all other data on a system be set to *PUBLIC *USE. These organizations made sure that they had data integrity — that is, that the data could be changed only through the protection of the application logic — but all users of the system could view the data.

Secure Job Descriptions

If your application might be run on a system that is at security level 20 or 30 and you intend to create a job description that names a user profile, you should create it with *PUBLIC(EXCLUDE). At security levels 20 and 30, you don't have to specify authority for a user profile in a job description to use the job description. So, if you create a job description that names your security administrator's user profile, any user could use that job description and exploit your security administrator's user profile. By creating job descriptions with *PUBLIC(EXCLUDE), you can prevent such exploitation.

At security levels 40 and 50, you can't use a job description that names a user profile unless you have authority to the profile.

Manage Your Library List

If your application adds a library to a library list, make sure you remove the library from the list when the application ends or abends. Once a library is added to a library list, the

Security Questionnaire for Vendors

1. Is a special profile required to install the application? *[QSECOFR should not be required for installation.]*
2. What special authorities are required
   a. To install the application?
   b. To run the application?
3. Does the application modify or require modification of system values
   a. To install the application?
   b. To run the application?
4. Does the application create a profile (or profiles) at installation time? If so:
   a. What is the profile name?
   b. Is the profile created with special authorities? If so, please justify the need for each special authority.
   c. Does the profile need a password, or can the password parameter be set to *NONE? If a password is needed, are there special considerations when changing it? *[Some auditors require you to change application profile passwords on the same schedule as your user profiles. You may need this information to change the password from a default password so that you know how to change it regularly.]*
5. Is System Service Tools (SST) or Dedicated System Tools (DST) used or required to install, configure, or manage the application? If so, why? *[The answer should be No.]*
6. Does the application replace or change any internal objects, such as the System Entry Point Table or control blocks? *[The answer should be No.]*
7. Does the application contain code that replaces and runs instead of IBM i operating system code? *[The answer should be No.]*
8. Does the application register any exit programs? If yes, what do they do?
9. Does the application run at security levels 40 and 50? If not, why not?
10. How do you ensure that your application contains no Trojan horses, viruses, or backdoors? *[You're looking for an explanation of the vendor's development process. A sound process will involve code reviews, strict use of change control, virus scanning of distribution CDs, and so on.]*
11. Does your application have any code that runs in system state? *[Only IBM programs should be running in system state. If a vendor-written program runs in system state, the program has been altered.]*
12. What security scheme does your application follow? *[You're looking for a method of authorizing the user to run the application without requiring the user to be a member of the owning profile or have *ALLOBJ authority.]*
13. What is the *PUBLIC authority of the application's
   a. Programs? *[should be *USE]*
   b. Data files? *[should be *USE or *EXCLUDE if the files contain sensitive information]*
   c. Libraries? *[should be no more than *USE]*
   d. Directories? *[should be no more than *RX]*
   e. All other application objects? *[should be no more than *USE]*
14. Do you have references that can be called?

system doesn't check a user's authority to the library when the user (or a program the user is running) specifies *LIBL/*object_name* to access an object or call another program. This means that a user who runs the application retains authority to the library — even when the application isn't running.

Make Library-Qualified Calls

To ensure that your application calls the correct program or accesses the correct object, use library-qualified calls. If you make non–library-qualified calls, such as CALL *LIBL/PROGRAM_1, and someone inserts another program named PROGRAM_1 into a library higher in the library list than your application library, your application calls the first PROGRAM_1. This is one way for an intruder to insert a Trojan horse or another impostor program to perform devious deeds. Making non–library-qualified calls is especially dangerous when your application adopts authority because an impostor program could exploit the adopted authority.

I know that making library-qualified calls is a controversial recommendation. Hard-coding library names in programs can be very limiting, especially when you're testing multiple language versions or releases of an application on the same system. To enable this type of testing, you can write your application to read the name of the library from a data area and then set the library variable to that value before making any calls. You'll also want to secure the data area (i.e., set public authority to *EXCLUDE) in the production version of the application. This way, when testing, the developer or tester can update the data area with the name of the library that contains the version of the application being tested.

Don't Store Passwords in Clear Text

Some applications require users to authenticate themselves to the application. Rather than storing the authentication information (typically a password) in clear text in a database file or in a configuration file in the IFS, as some applications do, you should use a validation list to store this information. For each application user, a validation list contains a user identifier (e.g., a user name) and authentication information (e.g., a password). You can use the validation list APIs to encrypt and verify passwords. IBM documents these APIs in the "Programming|application programming interfaces|Security APIs" section of the Information Center.

Testing, Testing

Now that I've addressed the design template questions and some implementation details, let's examine the final phase of the application development process: testing. Testing is usually necessary at several points in the development cycle, and you should consider security at every test point. This principle is especially true during the final test phase, which I call the *system test*. The system test is where it all comes together. Not only do you make sure the application works functionally, but you also verify that it meets security requirements.

Remember the requirements for the order-entry Web application — credit cards can't be compromised, and legacy systems can't be accessed from the Internet? System-test time is when you verify that the application meets such requirements. Some developers refer to this stage as "trying to hack the application." Whatever you call it, your test plans should include predefined test cases that specifically address your application's security requirements.

Last, be sure you test with the same authorities that will be given to actual application users. It does no good to test an application while the tester is signed on with a profile that has *ALLOBJ authority. Doing so only masks security issues and lets problems slip through to the user community. Ultimately, it creates a need for difficult and expensive debugging to fix security holes.

Moving Forward

I've introduced a lot of issues in this chapter, and you may have recognized that the applications on your systems have the "wrong" authorization scheme and the application data is exposed. You may be thinking, "Now what?" Keep reading. The next chapters will guide you through the process of redefining and implementing a new application security scheme so that it complies with your organization's security policy requirements.

Chapter 13

Role-based Access (RBAC)

Several laws, regulations, and standards — ISO27001, HIPAA, the Gramm-Leach-Bliley Act (GLBA), and the PCI DSS all have specific requirements when it comes to user access. The fundamental requirement is to grant users access to only the exact applications and capabilities needed to perform their job tasks — no more. To do so is to implement the concept of *least privilege access.*

While many organizations have taken the time to go through their applications and make sure users have access to only the menu options they need to perform their job functions, it's rare to see this principle extended to the user profile attributes and authorizations to application objects. Yet that is what implementing least privilege access demands.

How do you implement least privilege access in the IBM i world? Let's take a look.

Roles

The easiest way to set up least privilege access is to break down your user community into roles. In the security world, this approach is called *role-based access,* or RBAC. The premise of role-based access is that users are given capabilities and access to applications or system resources based on their job function, or "role." Separating role authorizations also enforces the separation of duties among roles, something that's important for companies that must comply with the Sarbanes-Oxley Act (SOX).

The type and number of applications you have will determine how many roles you have. As an example, consider an organization with the following roles: accounts payable clerk, accounts receivable clerk, finance manager, HR manager, and HR staff. Each role has a set of tasks its users are expected to perform. Therefore, access to specific application functions or to specific database files is permitted or denied based on the role in which a user has been defined.

Defining the Roles

To implement RBAC and the concept of least privilege access, you must first define what roles are required. Common sense will dictate how many roles to define. Too few, and you'll end up giving users more authority or access than they need to perform their job function; too many, and management of the roles becomes unwieldy.

The next step is to list the tasks that the roles perform. You might want to use a spreadsheet to keep the list organized. Table 13.1 shows an example.

Table 13.1
Sample Roles and Tasks

Role	Tasks
HR Staff	Accepts job applications, performs initial job interviews
HR Manager	Assists management with HR issues, oversees payroll operations
Finance Manager	Oversees all accounting functions, approves new accounts, approves check requests
AP Clerk	Processes invoices, submits check requests
AR Clerk	Generates invoices
Internal Audit	Audits system configurations, reviews compliance reports
Accounting Download	Downloads specific files within the accounting application to a spreadsheet for further processing

Notice that I've listed only the roles and the tasks they perform. I haven't yet identified which authorities are required — either within the application or for the IBM i user profiles. That's because you need to gain agreement about what tasks each role is supposed to perform. Don't be surprised if this step is a bit controversial. Different people may have different ideas about what tasks a particular role should perform.

After obtaining agreement about the tasks associated with each role, you can start listing the capabilities, application menu options, and specific authorities required for each role. In addition, you'll want to note any functions within System i Access for Windows or System i Access for Web that each role should be able to perform. Last, if a role is to have explicit authority to download or upload a specific file, list this fact as well.

For every role, document its purpose, capability (special authorities), and access requirements, as shown in Table 13.2.

Table 13.2
Sample Roles, Special Authorities, and Access Requirements

Role	Description	Special authorities	Access requirements
HR Staff	HR staff role	None	All HR application options except Payroll and Master
HR Manager	HR manager role	None	All HR application options
Finance Manager	Finance manager role	None	All accounting application options
AP Clerk	AP clerk role	None	All accounts payable options in the accounting application
AR Clerk	AR clerk role	None	All accounts receivable options in the accounting application
Internal Audit	Internal audit role	*AUDIT	Access to the WRKSYSVAL command, *RX authority to the /SkyView/Policy Minder directory to review compliance reports
Accounting Download	Users who need to download accounting data	None	*USE authority to the ACCTG authorization list securing the accounting files

Group Profiles

For IBM i, roles are implemented most easily through the use of group profiles. You create a group for each role and assign the group the special authorities required to perform the role's tasks. In addition, the group (role) is granted any private authorities required to application objects such as menus, files, authorization lists, commands, and so forth.

Here are a few tips to keep in mind when creating group profiles:

- When creating a group profile on IBM i, create the profile without a password (set the password parameter to *NONE) so the profile cannot be used for sign-on. Group profiles are supposed to represent a role, not an individual.
- To eliminate group profiles from appearing on reports that auditors may ask for:
    - Set limited capabilities to *YES so that the group doesn't appear on a report listing profiles with limited capabilities set to *NO.
    - Set the password expiration interval to *SYSVAL.
    - Set the initial menu to *SIGNOFF.
- If the role being represented isn't obvious from the group profile's name, add an explanation to the Text field.
- If the tasks that the role performs require IBM i special authorities, grant the special authorities to the group, not the individuals.

Table 13.3 shows the group profiles created to represent the roles in our example.

Table 13.3
Sample Roles and Group Profiles

Role	Group profile	Special authorities	Text description
HR Staff	GRPHRSTAFF	None	HR staff role
HR Manager	GRPHRMGR	None	HR manager role
Finance Manager	GRPFINMGR	None	Finance manager role
AP Clerk	GRPAPCLK	None	AP clerk role
AR Clerk	GRPARCLK	None	AR clerk role
Internal Audit	GRPAUDIT	*AUDIT	Internal audit role
Accounting Download	GRPACTDOWN	None	Users who need to download accounting data

Now you can go back to the role definition form and add the name of the group profile used to implement the role, as shown in Table 13.4.

Table 13.4
Role Definitions with Associated Group Profiles

Role	Description	Special authorities	Access requirements	Group profile
HR Staff	HR staff role	None	All HR application options except for Payroll and Master	GRPHRSTAFF
HR Manager	HR manager role	None	All HR application options	GRPHRMGR
Finance Manager	Finance manager role	None	All accounting application options	GRPFINMGR
AP Clerk	AP clerk role	None	All accounts payable options in the accounting application	GRPAPCLK
AR Clerk	AR clerk role	None	All accounts receivable options in the accounting application	GRPARCLK
Internal Audit	Internal audit role	*AUDIT	Access to the WRKSYSVAL command, *RX authority to the /SkyView/Policy Minder directory to review compliance reports	GRPAUDIT
Accounting Download	Users who need to download accounting data	None	*USE authority to the ACCTG authorization list securing the accounting files	GRPACTDOWN

Why Group Profiles?

Implementing roles via a group profile (whether within an application or for IBM i access) also provides an easy reporting and reviewing mechanism. Many laws and regulations, as well as internal and external auditors, require user access to be reviewed on a regular basis. If you cannot easily produce reports that list user access, the review process becomes confusing for all involved and, frankly, is often ineffective in reaching the objective (which is to remove access from users who no longer require it). When you implement roles, the appropriate manager reviews the capabilities and access associated with each role and then reviews the users in the role to determine 1) whether the role still requires the current access and capabilities and 2) whether each user should remain assigned to the role. Without roles, users and their individual access rights and capabilities must be reviewed individually, which can be a very tedious and time-consuming process.

Another reason to implement roles via group profiles is to keep a record of the access rights and capabilities of each role. I encourage you to provide access and capabilities only through roles (groups). This practice will ensure that if a user leaves the company, you have a clear and accurate picture of the access that whoever replaces that individual needs — simply place the new person in the former employee's role (group).

When helping our clients implement role-based access, I'm often asked the question, "If there's only one user in the role, do you still go ahead and create the group profile?" In my opinion, the answer is yes. That's because if you discipline yourself to only grant access to the group (role), the group will always have the authorities required to perform the role's functions. This benefit is very helpful when the one person in the role is on vacation, takes sick leave, is dismissed from the company, or otherwise is away from the office for an extended period. All you need to do to allow someone to take over the role is to add that person to the group. If you haven't implemented the capabilities and access at the group profile, you have to try to determine what authority the individual profile has been granted (or revoked) and replicate that for the person just placed into the role.

Implementation

To implement role-based access:

1. Create a group profile for each role using the recommended settings listed previously.
2. Assign the group profile the appropriate special authorities required by the tasks to be performed by the role.
3. Grant authority to the group for all the menu options, commands, authorization lists, and functions allowed in System i Access for Windows or Web.
4. For each user, assign the appropriate group profile (or group profiles, for users who belong to more than one role).

Table 13.5 shows an example of the final spreadsheet you'll work from to perform the actual implementation, including the users assigned to each role.

Table 13.5
Implementation Spreadsheet for Sample Roles

Role	Description	Special authorities	Access requirements	Group profile	Users
HR Staff	HR staff role	None	All HR application options except for Payroll and Master	GRPHRSTAFF	Ava Gracie
HR Manager	HR manager role	None	All HR application options	GRPHRMGR	Abigail Mareen
Finance Manager	Finance manager role	None	All accounting application options	GRPFINMGR	Mia Catherson
AP Clerk	AP clerk role	None	All accounts payable options in the accounting application	GRPAPCLK	Alan Walker
AR Clerk	AR clerk role	None	All accounts receivable options in the accounting application	GRPARCLK	Grant Thomas
Internal Audit	Internal audit role	*AUDIT	Access to the WRKSYSVAL command, *RX authority to the /SkyView/Policy Minder directory to review compliance reports	GRPAUDIT	Sam Johnson
Accounting Download	Users who need to download accounting data	None	*USE authority to the ACCTG authorization list securing the accounting files	GRPACTDOWN	Mia Catherson Alan Walker

Chapter 14

Role-based Access for IT

Until now, this book has focused primarily on end-user security. We've discussed how to identify user authorization roles and how to use system and application security to implement those roles. But one particular group of users requires further attention: the IT department. Included in this group are the IT professionals on staff — including operators, help desk personnel, network administrators for IBM i, administrators, and programmers — as well as software vendors who supply applications used on your system and IT consultants.

Security and Your IT Staff

IT is one of the most difficult departments to establish a security strategy for, yet IT personnel pose the greatest threat to your organization. As with all other employees, IT personnel should be authorized to just those objects and functions they need to perform their jobs. For example, you should no more let IT staff view or manipulate production data at will than you should any other employee. This simply makes good business sense.

Implementing role-based access is also a compliance requirement in many organizations. While many people loathe the extra burden the Sarbanes-Oxley Act (SOX) has placed on most companies, one good outcome is that many organizations have had to restrict how and when programmers access production. But implementing these new restrictions has been difficult for most. Why? Programmers — especially in the IBM i world — seem to take this restriction personally and feel their employers no longer trust them. They tend not to understand that the requirement has nothing to do with trust but everything to do with compliance with the law.

So, even though many in IT chafe over the restrictions, role-based access for IT is a requirement in many organizations. This chapter is devoted to helping you successfully implement role-based access across IT with the least amount of disruption possible.

Note: Although I describe the roles individually, I realize that in most organizations, the same person will perform some of these individual roles. While I think it would be unwieldy and highly impractical to create a separate user profile for each role that a user performs and expect that person to sign on using a different profile depending on the task, I do think it's wise to create a separate group profile that represents each role.

Identify the Roles

The first step toward achieving role-based access in the IT department is the same as I described in Chapter 13. That chapter presented a series of exercises designed to help you develop roles for non-IT users. You'll want to use the same method to define roles in IT. For most organizations, the roles are probably similar to those listed in Table 14.1. Depending on the size, some organizations will have more roles in IT, and some will have fewer, but this list serves a general starting point.

Table 14.1
Typical IT Authorization Roles

Role	Job functions	Required capabilities
Operator	Monitor for and answer messages, start and stop writers, submit batch jobs, monitor active jobs, save objects, restore objects upon request	• *SAVSYS, *JOBCTL special authorities • Authority to the RST*xxx* commands
Network Administrator	Creates lines, controllers, and devices; changes TCP/IP server configurations; manages TCP/IP configuration	• *IOSYSCFG special authority
Help Desk	Resets forgotten passwords, re-enables *DISABLED profiles	• No special authorities • Authority to utilities written to reset profiles and passwords
Programmer/Analyst	Designs and codes new applications, updates existing applications, debugs production application issues	• *JOBCTL special authority • Authority to the Application Administration feature that lets users with *JOBCTL see the job log of another user • Authority to debug commands
System Administrator	Manages all aspects of the system except the security configuration — performance, installation and configuration of new products, save/restore strategy, disaster recovery planning, hardware and operating system upgrades	• All special authorities except *AUDIT
Security Administrator	Manages the QSECOFR profile, creates (or manages the team that creates) user profiles; runs compliance checks to ensure the system, user profile, and application settings remain set to the appropriate settings; maintains the system's security architecture; reviews reports from the audit journal	• All special authorities (*AUDIT, *ALLOBJ, *JOBCTL, *IOSYSCFG, *SAVSYS, *SECADM, *SERVICE, *SPLCTL)

Define a Secure Environment for Each Business Function

After identifying the business authorization roles within IT, you need to determine what authorities are required to perform each job function — giving each role sufficient authority but not more authority or capabilities than it needs.

Operator

The first decision to make about the operator role is whether you should assign (or keep the assignment of) the QSYSOPR user profile (the IBM-supplied system operator user profile) as the operators' group profile. Doing so is certainly a handy approach because QSYSOPR is authorized to many system functions. However, it may surprise you to learn just how much power QSYSOPR has. So I recommend that you look at the objects QSYSOPR owns and is authorized to; any user who is given QSYSOPR as a group profile will own or be authorized to all these objects as well.

To determine which objects QSYSOPR owns, run this DSPUSRPRF (Display User Profile) command:

```
DSPUSRPRF  QSYSOPR *OBJOWN
```

To determine which objects QSYSOPR is authorized to, run this command:

```
DSPUSRPRF  QSYSOPR *OBJAUT
```

Given what you see after running the DSPUSRPRF command, decide whether QSYSOPR is an appropriate group assignment. If it provides too much power and it's already your operators' group, you may want to consider changing that assignment.

For organizations that have more than one system operator, it's essential for each operator to have his or her own user profile. I realize this may be difficult, but it is necessary to ensure that the history log, message queues, and audit journal entries provide an audit trail of actions taken by each individual, thus providing accountability. Although many operators are used to logging on to the system and leaving their workstations signed on all day to monitor for messages, this behavior may have to be altered.

After ensuring that all operators have individual profiles, you need to consider what special authorities and individual authorizations the operator role requires. Because the operator role's responsibilities usually include working with messages, devices, jobs, and printers, this role needs capabilities that allow it to work with these areas of the system. *JOBCTL is usually sufficient for this. I don't recommend giving operators *SPLCTL because that allows them to view any spooled file on the system. If you have any spooled file that needs to remain private, granting the role *SPLCTL defeats any controls you try to place on restricting access.

Another operator role function is performing save/restore operations. Once you've given the operator role *SAVSYS special authority, the operator will require no specific authority to any production or development libraries or directories to save or restore them. The default public authority for all restore (RST*xxx*) commands is *EXCLUDE. If you want all roles that have *SAVSYS special authority (including the operator role) to be able to restore anything to your system, you can change the public authority for these commands to *USE or make these users part of the QSYSOPR group. However, the approach I recommend is to secure all restore commands with an authorization list and grant *USE access to each role entrusted with this function.

Technical Note

A profile that has *SAVSYS special authority could access a secured object via the save/restore function. If the object is secured by public authorities but resides in a library secured by private authorities, a user with *SAVSYS authority could save the object and restore it to a library to which the user has authority. Because the object isn't secured by private authorities at the object level, the user could then access the restored object. This is why you want to ensure objects' *PUBLIC authority is set appropriately and also periodically review the actions of the operators.

Last, the operator role often has to submit batch jobs. Rather than authorizing the role to the production programs and files that these jobs work with, you can create CL-driven menus that contain the options the operator needs to submit batch jobs. These menu programs should adopt the application data owner's authority and should be authorized only to operators. When an operator selects an option, the program should submit the job using a job description that specifies a user profile to be used for the batch job. For instance, the job description for the batch job might be PRODJOBD, and the user profile defined for that job description might be PRODUSER. The SBMJOB (Submit Job) command would use the parameters JOBD(PRODJOBD) and USRPRF(*JOBD) to submit the job. The job would then run under the PRODUSER user profile.

This implementation has several key elements. First, it is critical that the menu options call programs that adopt. In many cases, it will be sufficient for the programs to be owned by the application's owning profile. However, depending on the task being performed, a more powerful profile may be required — for example, if the menu option is one taken when the operator has to restore an object and specify the restore command's ALWOBJDIF (Allow object differences) parameter. Because specifying this parameter requires the user to have *ALLOBJ authority, this menu option's program will have to be owned and adopt an *ALLOBJ profile.

In addition, because the menu options call programs that adopt, the operator role does not need authority to the application owner's profile or more powerful profiles. I've seen examples where organizations have given the operator role *USE authority to the application owner's profile so that the operators can submit jobs to run as this profile. This technique gives operators the ability to submit any task and run with these profiles' capabilities, including tasks not approved by an administrator and tasks that have not gone through a change management review. This is more capability than I prefer to give operators, so I recommend that you avoid this approach.

Implementing security for system operators requires a little work on your part, but the security and risk reduction you'll achieve is well worth the effort.

Network Administrator for IBM i

The network administrator role is responsible for the communications configuration of the system with the rest of the network. This includes all TCP/IP configuration, working with other network administrators to determine the appropriate routes and TCP/IP addresses, and so on. This role determines which TCP/IP servers should auto-start when the rest of the system starts. All tasks usually performed by this role on IBM i are available without granting private authorities, as long as the role has been given *IOSYSCFG special authority.

Help Desk

Most organizations have some type of help desk. In organizations where the help desk is a separate group from administrators, I prefer to assign these users to a separate group (role). Then, rather than give this role special authorities such as *SECADM to allow them to work with user profiles, I prefer to write utilities — typically short CL programs that adopt and are owned by a profile with sufficient authority to perform the task. Tasks this role usually performs include resetting user profiles when a user has disabled his or her profile after too many invalid sign-on attempts and assigning a new password when a user can't remember the current one.

If you give this role *SECADM and sufficient authority to work with all user profiles, help desk personnel end up with a lot of power. I prefer to provide the functionality through utilities and then assign these utilities to a menu specifically created to contain all the tasks the help desk role needs to perform. Using this approach also means that the profiles in this role can be set to be limited-capability users because they will be taking menu options, not running commands from a command line. This strategy further confines them to be able to run just the tasks required by their job function.

Programmer/Analyst

The programmer/analyst role is usually the most difficult to implement, from both a technical and an employee-relations point of view. Programmers typically resist security because they see it only as a limitation. Their jobs call for designing applications and writing, testing, and debugging the code to implement those applications. However, although programmers provide the physical framework for the data, the data itself belongs to those who administer it. For example, although the programmers may provide the application to manage human resources information, it's the HR department that owns the data and should be the one to determine who should be able to access it — not the programmer.

When creating the role (group profile) for the programmers — whether it's on a development or a production system — resist using the IBM-supplied profile QPGMR. QPGMR has authority to and owns many IBM i objects that programmers most likely don't need — and that you don't want them to have authority to. In addition, many vendor applications are owned by QPGMR. If your programmers are a member of QPGMR, they own these vendor applications. Last, QPGMR has been given *JOBCTL and *SAVSYS special authorities. Do you really want your programmers to have the ability to save any object off your system or to restore any object onto your system? I don't think so! But by making QPGMR your programmers' group profile, you give them these special authorities. Programmers may need *JOBCTL special authority, but they should not need *SAVSYS. And, regardless of what the programmers tell you, they do *not* need *ALLOBJ!

Let's talk about how to set up the development system, and then I'll go over the options for granting programmers access to production. On development, I usually put all programmers in one group. If you have a large programming staff, it may make sense to have more than one group and divide the programmers' group membership according to the project on which they're working. After creating the group (with, at most, *JOBCTL special authority), I set that as each programmer's first group profile and then specify the OWNER user profile attribute as *GRPPRF. This way, everything that a programmer creates is owned by the group, and the rest of the group is free to use these objects. I typically let programmers have "free reign" at this level of development. But, once again, this doesn't mean I give them *ALLOBJ. It's just that when they're developing code, they need to be able to test freely. That's why, on a development system, the programmer's group should not be the profile that owns the application. If it is, they own not just the application objects still in the development stage but the application objects that are in the "production" stage as well. If programmers own the production stage objects, they can bypass any change management software or process you might have in place. This, for most organizations, will constitute a SOX compliance issue and is just not a wise business practice.

Once you've created the programmer's group profile, you need to make sure programmers have access to all the programming and debug commands so that they can do their jobs. The easiest way to find out what those commands are is to find out what the QPGMR user profile is authorized to. Do that by running the following DSPUSRPRF command:

```
DSPUSRPRF  QPGMR *OBJAUT
```

Some of these commands are shipped by IBM as *PUBLIC *EXCLUDE. For the ones your programmers need, you'll want to grant their group a private authority of *USE.

Now, let's talk about programmer access to production. Although auditors would like to prevent programmers from ever accessing production, that approach is just not practical. Reality says that there are times when a programmer must go on to the production system to debug failures. So, how do you do that with the minimum risk to your production data? Here are some options. It should be noted, however, that none are completely without risk.

First, let me state right up front that I don't believe programmers should have continuous access to production. Programmers should be allowed on a production system only when there's a production issue. So one of the first issues you're going to have to resolve is how to give access to programmers. There are several approaches:

- Create a profile for each developer but leave it with a status of *DISABLED. When the programmer is on call, enable the profile.
- A more controlled approach, and the one I prefer, is to have the help desk enable the programmer's profile when the help desk receives a call that requires production access.

In either case, access is temporary. That is, the profile is set to *DISABLED once the programmer goes off call or when the problem is resolved. Another approach I've seen is that when the programmer signs on, a job is submitted that automatically disables the profile after two hours' use. However you implement it, make sure the programmer's access is not permanent.

The next issue you have to resolve is that of what authority the programmer has once he or she is allowed to sign on to production. Again, you have options. One option you do *not* have is to give the programmer *ALLOBJ. For the occasions when programmers might run a command that requires *ALLOBJ, you can write utilities that they can use to obtain the *ALLOBJ temporarily. And Application Administration can be used to let programmers see the job log of an *ALLOBJ user without having *ALLOBJ themselves. *Programmers do not need *ALLOBJ*— have I mentioned that recently?!? Okay, on to the real options:

- You can make the programmer a member of the profile that owns the application. The programmer doesn't have *ALLOBJ but still has all authority to all application objects. This is the most flexible but also the most risky approach (because the programmer, in effect, owns the application while signed on).
- If you've used authorization lists to secure the application data, you can add the programmer's group to the list with *CHANGE authority. This lets the group's members make data updates but does not give them the authority to clear or remove data or replace application objects. If you take this approach, you'll have to make sure that programmers have sufficient authority to debug programs; they need either *USE and *SERVICE special authority or *CHANGE to the programs to run debug against them. Although this approach requires a bit more setup, it is

less risky because your programmers only have *CHANGE authority to application data — not *ALL authority to all application objects.

Last, you'll want to consider logging programmers' activity while on production. Auditors would love for you to review these logs, but in reality most of you won't have time to do that. But the logs can provide valuable information should an investigation need to occur. You don't have to re-specify any value you've already specified in the QAUDLVL (Auditing Level) system value, but it's possible that you'll want to specify to log all commands that the programmers enter. To configure this, run the following CHGUSRAUD (Change User Audit) command for each developer's individual profile:

```
CHGUSRAUD  USRPRF(PGMR1 PGMR2 PGMR3) AUDLVL(*CMD)
```

(I'll explain more about IBM i auditing features in Chapter 15.)

Now, I've stated (several times now) that programmers don't need *ALLOBJ. There are some exceptions, however. In very small shops, programmers sometimes also play the administrator role — at least on occasion. In that case, it's likely the programmer will need to have *ALLOBJ — at least on occasion.

System and Security Administrators

Of all the IT roles in the average business, the system and security administrators might be the most poorly implemented. In most organizations, the security or system administrator is whoever possesses the QSECOFR profile and password, and that person often does all his or her work, whether security-related or not, under that profile. This is not appropriate.

Each administrator should have a separate profile from QSECOFR. QSECOFR should be used only in the situations in which it is required (by IBM's or a vendor's instructions). The security officer role will likely have all special authorities. The system administrator will have all special authorities but *AUDIT. This way, the system administrator cannot manipulate the auditing system values. My preference is to assign these special authorities to a group and to have the OWNER attribute be set to *GRPPRF so that anything created by the administrators (with the exception of IFS objects, as noted in Chapter 4) will be owned by their group, not the administrators themselves. This approach makes it much easier to delete an administrator's profile if the person leaves the organization.

Security Recommendation

Don't use the QSECOFR profile as an administrator profile. Instead, create a unique and distinct profile for each administrator.

Because *ALLOBJ provides for the opportunity to make accidental errors, I've seen some administrators have two profiles — one with and one without *ALLOBJ. This works well, especially when an individual performs multiple roles. In cases where *ALLOBJ isn't necessary, the administrator signs on with the "lower power" profile, using the profile with *ALLOBJ only when a task requires it. In larger shops, however, many administrators are constantly performing tasks that require *ALLOBJ, so this might not be practical in those situations.

Many auditors want to know what the administrator profiles are doing and so require more logging. I think turning on *CMD auditing is a good practice for all profiles assigned *ALLOBJ or for profiles that are members of a group that has *ALLOBJ. To configure this auditing, run the following command for all administrator profiles, including QSECOFR:

```
CHGUSRAUD  USRPRF(QSECOFR ADMIN1 SECADMIN1) AUDLVL(*CMD)
```

Security for Vendors and Consultants

Consultants and third-party software vendors are an integral part of most IT environments. All third-party packages bring a new element to your IT shop: menus, applications, and usually some form of security as part of the package. With these packages come some security risks. Not only do you need to be aware of these risks, but as the security administrator you must manage them, obtaining help from the vendors when necessary. Remember that although you've purchased software from a vendor, the data belongs to the enterprise, and that is where your responsibility lies. You can't afford to surrender the enterprise's security to a vendor or consultant.

A few basic weaknesses commonly occur in vendor applications and with consultant practices. I want to point out these areas and make a few recommendations for how to deal with these potential security threats. (Chapter 12 provides a vendor questionnaire that can help you assess the security impact of third-party applications before you make any new software selection.)

Vendor Support

Some vendors support their software packages by remotely accessing your system. I've found that they often claim to need more authority than they do. I see them usually requiring that they have at least *ALLOBJ special authority. Trust me, unless a vendor is managing your entire system, this "requirement" is out of convenience, not necessity. You should use the same approach you use for the programmer role for vendors accessing your systems. Make the access available (i.e., enable the profile) only after the vendor has called you and let you know it needs access and has explained why. Give only enough authority for the task being performed. For example, if a vendor is performing maintenance on its application, you could consider making the profile a member of the group that owns the application. With this approach, vendors should have sufficient authority to do their maintenance, but you've reduced the risk of their accessing your system by not giving them *ALLOBJ (even though that would be their preference). Remember, this is *your* system, not theirs. Vendors need to follow your security requirements — not the other way around. As with the programmers' profile used to access production, remove the access (i.e., set the profile status to *DISABLED) when the maintenance is complete.

Last, make sure vendors don't install programs that adopt a powerful profile and then put up a command line when called. I've seen vendors do this so they have a "back door" into the system in the event the organization attempts to place controls on them. Obviously, you'll want to configure auditing for these profiles. To ensure they are not gaining authority through adopted authority, I recommend you look for authority coming through adopted authority in addition to auditing the commands these profiles enter. To configure this, run the following command:

```
CHGUSRAUD  USRPRF(VENDOR1 VENDOR2) AUDLVL(*CMD *PGMADP)
```

Consultant Practices

Consultants provide a popular way to gain expertise in a specific area in which your current staff lacks proficiency. Consulting firms often offer specialists in particular areas who can help develop and maintain application software on your system. Be wary of consultants who claim to need or insist on having *ALLOBJ special authority. Consultants usually try to get as much authority as they can on your system — it makes their job easier. However, consultants should justify this requirement. In the case of my firm, the services — performing security assessments — require the consultant to have *ALLOBJ special authority because most of the IBM i functions being run require *ALLOBJ. However, this is not the case for most consultants. You should evaluate the business function of each consultant as you would for any IT staff position and then create specific user profiles, assigning them to the appropriate group (role) that reflects their job function.

If a consultant is creating programs on or restoring programs or objects to your system, require documentation for every program he or she creates or restores. Understand the purpose and function of each. If a consultant asks to load a tool kit on your system, obtain documentation that explains exactly which tools are being loaded, and be sure you know whether any of them adopt the authority of any powerful profiles (i.e., profiles that have *ALLOBJ special authority).

And it should go without saying that you should also turn on auditing for profiles assigned to consultants, just as those used for vendor access, with both *CMD and *PGMADP configured.

Role-based IT Access

Developing role-based access for IT will help you ensure that you are not assigning profiles more authority than is required to perform the relevant job function. Regular (I recommend at least annual) review of the duties performed by each role, the authorities required, and the users assigned to the roles will ensure users continue to have just enough authority to do their jobs — but no more.

Chapter 15

Auditing

The IBM i operating system provides extensive auditing features. Before I get into the details of how to implement these auditing features, let's discuss why you might need or want to enable them.

- Like the implementation of other security features, compliance with various laws and regulations often drives the requirement to enable auditing. For example, the Payment Card Industry requires auditing to be enabled and the resulting audit journal entries to be retained for at least a year, with the most recent 90 days' worth of activity available on the system or quickly and easily retrievable (should an investigation need to occur).
- Laws driving auditing requirements include the Sarbanes-Oxley Act (SOX) in the United States and its counterpart in Japan, the Financial Instruments and Exchange Law (JSOX). Many of you know that readying for one of these audits often requires the preparation of reports stemming from the activity logged in the audit journal.
- As I discussed in Chapter 14, you may want to log all the commands run by programmers who access production, as well as the actions of any user with *ALLOBJ authority and actions taken by your vendors and consultants.
- A proactive stance to security often drives the requirement to enable auditing. Although you may examine only a small set of the audit journal entries, you know that the information is available should you have to investigate an incident (more on this topic in Chapter 18).
- Last, even if you're not interested in the contents of the audit journal, your organization may have implemented a high availability (HA) solution that requires auditing to be enabled. Most HA solutions use the audit journal to know what and when to replicate to the target or failover system. If this is your situation, don't change your system's auditing configuration without first checking with your vendor (even after reading this chapter). Any change — but especially disabling some auditing — may produce significant problems in your HA solution.

I'd also like you to keep in mind a non–security-related benefit of turning on auditing, and that is the wealth of information auditing provides when debugging system problems. I can't tell you how many times I've been working with an administrator at a client site when someone walks into the office, calls, or sends an e-mail with a complaint or a plea for help that is resolved by examining information provided in the audit journal. Situations such as determining who or what process deleted a file or program are easily resolved by examining the deletion of objects (DO) audit journal entries. Unfortunately, most administrators don't use this information because they are unaware or forget that it's available! So I'm going to remind you several times throughout this chapter in hopes that this tip "sticks."

Security Recommendation

Once you understand what type of information is available in the audit journal, don't forget to use it as one of your sources of information when debugging problems.

The primary tool for auditing is the IBM i security audit journal. Before discussing the audit journal, though, let's briefly look at another useful tool: the history log.

The History Log

The history log records high-level activities that occur on your system, such as the start and completion of jobs, device status changes, system operator messages, and failed sign-on attempts. The history log records these events in the form of messages and stores the messages in system-created files.

Most security events are now logged only to the audit journal, but you can still learn from your system's history log. The history log helps you monitor system activities and reconstruct events to aid problem-determination and debugging efforts. The text description maintained by the system contains the beginning and ending date and time for the messages in the file — information that's useful when investigating events or activities.

You can display the contents of the history log by running the DSPLOG (Display Log) command:

```
DSPLOG  LOG(QHST)
```

The resulting display resembles the screen shown in Figure 15.1. Because system events such as job completions, invalid sign-on attempts, and users connecting via ODBC (known by their QZDASOINIT job names) are stored as messages in file QHST, you can place the cursor on a particular message and press the Help key to display second-level help text for the message.

Figure 15.1
Sample History Log

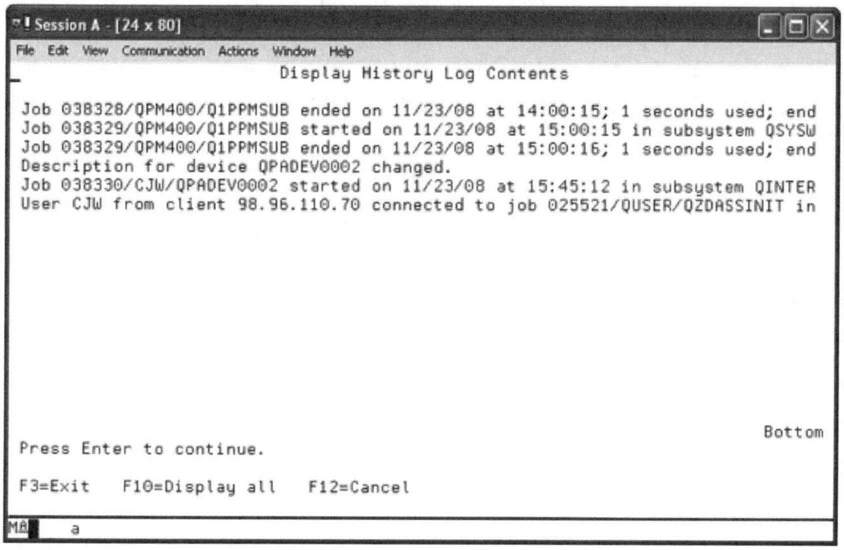

The DSPLOG command has several parameters that provide flexibility when you look through the log. These parameters, shown in Table 15.1, are fairly self-explanatory.

Table 15.1
DSPLOG Command Parameters

Parameter	Description
LOG	The default history log is QHST.
PERIOD	You can enter a specific time period for which log entries should be shown or accept the defaults for the beginning and ending dates and times. The default beginning time is the earliest available time, and the default beginning date is the current date.
JOB	You can use the JOB parameter to search for messages associated with a specific job name. You can enter just a job name, in which case the system might find several jobs with the same name that ran during a given time period. Or you can enter a specific job name, user name, and job number to retrieve history information for a specific job.
MSGID	Like the JOB parameter, the MSGID (Message ID) parameter helps narrow the search. You can specify one or more messages or a range of messages.
OUTPUT	The asterisk value (*) sends output to the screen. The *PRINT and *PRT*xxx* values give you a variety of options for sending information to a spooled file.

History Log Housekeeping

The history log consists of a message queue and system files that store history messages. The files reside in library QSYS and have names that follow the convention QHST*yydddn*, where *yyddd* stands for the Julian date on which the log was created and *n* is a sequence number (0–9 or A–Z) appended to the date. The system creates a new file whenever the existing file reaches its size limit, a number controlled by system value QHSTLOGSIZ (History Log Size). Another option is to specify *DAILY for this system value so that the system starts a new log each day. Because the system doesn't automatically delete history log files, you'll want to develop a strategy for deleting the files and for using the data before you delete the files.

It's wise to save previous versions of the history log to media so you can retrieve them if you need to investigate an event. I recommend making the QHST files part of your regular save strategy and retaining the files offsite for at least a year.

Inside Information

Just like the audit journal, the history log contains a lot of information that can be used to investigate a security event as well as help debug security or system administration issues. For example, a look through the log might surface a message such as "Receiver ACG0239 in JRNLIB never fully saved (I C)"; the second-level help text tells you which program was trying to delete the journal receiver. If you see the message "Password from device QPADEV007 not correct for user QSECOFR," you might want to investigate who is trying to sign on as the security officer.

One especially useful DSPLOG command option is the ability to select all security-related messages by specifying one parameter. All such messages (excluding some network and communications messages that might relate indirectly to security) fall within the message identifier range of CPF2200 to CPF2299. By entering the following command, you can view security-related messages for the specified period.

```
DSPLOG  LOG(QHST) MSGID(CPF2200) PERIOD(time_period)
```

As I mentioned earlier, the most complete repository of security event information is the security audit journal. Even so, the history log is one piece of the puzzle when you're reconstructing a scenario to allow investigation of an event.

The Security Audit Journal

The IBM i security audit journal is a system-supplied journaling function that records specific security-related events based on your configuration of auditing controls (e.g., system values, user profile parameters, object-auditing attributes). When you use the security audit journal, you can take advantage of many configuration options and features. To aid in your understanding and implementation of a successful auditing strategy, I'll discuss the following eight topics:

- The audit journal
- Auditing controls
- System-wide auditing
- User auditing
- Object auditing
- Event-auditing recommendations
- Working with the audit journal
- Displaying and printing audit journal entries

The Audit Journal

When an IBM i system is shipped, no audit journal exists. You must create the security audit journal before implementing any of its auditing features. IBM i requires you to create a journal named QAUDJRN, and it must reside in library QSYS. Creating this journal is actually a two-part process: First you create a journal receiver, and then you create the journal itself.

You can use the CRTJRNRCV (Create Journal Receiver) command to create a journal receiver (of any name) in a library of your choice. You then use the CRTJRN (Create Journal) command to create the QSYS/QAUDJRN journal and associate it with the journal receiver you defined in the previous step.

The QAUDJRN journal and the associated receiver function exactly like a normal database journal and receiver except that this journal is used only for audit journal entries. For more information about journals and managing journals and receivers, see the "Database|PDF files for Database|Journal management" topic in the Information Center.

The easiest way to configure auditing on your system is to use the system-supplied CHGSECSAUD (Change Security Auditing) command, whose prompt screen is shown in Figure 15.2. This command creates the audit journal and a receiver and sets the appropriate system values for you.

Auditing Controls

Now, let's discuss how you configure what is actually going to be audited. (Note that if you're following along on your own system, your user profile must be assigned *AUDIT special authority to configure auditing, display objects' auditing values, or work with audit entries.)

Several system values play a role in audit configuration. Recall from Chapter 3 that system value QAUDCTL (Auditing Control) is the on/off switch for IBM i auditing. The default QAUDCTL value of *NONE says that auditing is off. Specifying *AUDLVL in the QAUDCTL system value enables action auditing. Specifying *OBJAUD enables object auditing. The oddly named value *NOQTEMP tells the system not to log actions (e.g., creations, deletions) for objects in the QTEMP library. This value is useful because it eliminates entries that you really don't care about. You can specify any or all of these values for QAUDCTL, other than *NONE.

Figure 15.2

Sample CHGSECSAUD Prompt Screen

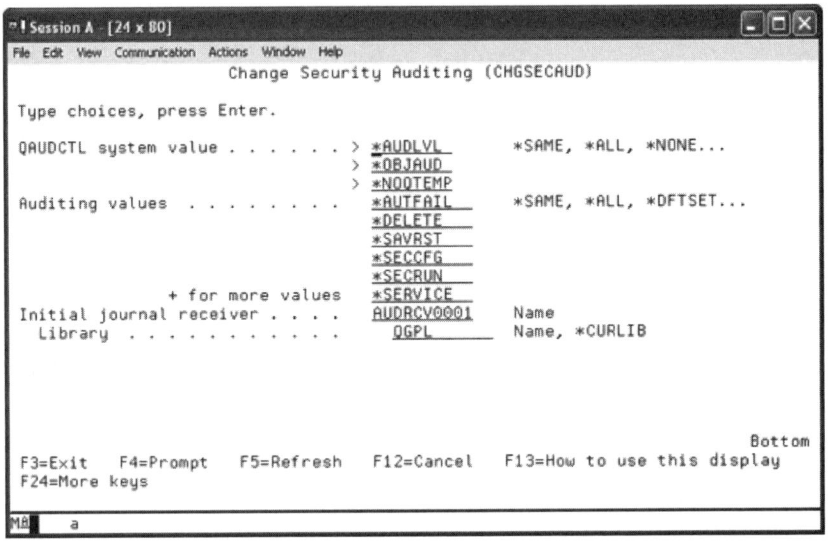

Two other system values control aspects of auditing, but let me say right up front that you don't want to change either of these settings. The only reason to change these values is if your organization requires the highest security and integrity settings. QAUDFRCLVL (Auditing Force Level) determines the maximum number of audit records the system will cache before forcing the records to disk. I recommend leaving the value as *SYS to maximize performance when auditing.

QAUDENDACN (Auditing End Action) determines the action the system will take when it's unable to continue auditing (i.e., when the system can't write another audit record to the journal). The default value *NOTIFY causes IBM i to send a message to the system operator as well as to the QSYS/QSYSMSG message queue if it exists. Specifying *PWRDWNSYS tells the system to power down immediately — in effect, this setting runs the PWRDWNSYS (Power Down System) command with the *IMMED option! In addition, the next IPL returns the system in a restricted state, and a user profile with *ALLOBJ and *AUDIT authority must sign on to restore auditing before the system can be restarted. Obviously, most installations will want to leave this system value at the default of *NOTIFY!

System-wide Auditing

The system value QAUDLVL (Auditing Level) — and its companion, QAUDLVL2 (Auditing Level Extension) — determines which actions are going to be audited. Table 15.2 lists the possible QAUDLVL/QAUDLVL2 values, with a notation as to those new in V6R1. The table also lists values for the AUDLVL (User action auditing) user profile attribute, which provides similar functions at the user profile level (we'll talk more about AUDLVL in a moment). You can specify one or more of the listed values (or *NONE, which causes other values to be ignored). Some values, such as *JOBDTA, *NETCMN, and *SECURITY, have subsets. Using a subset value rather than the full value lets you customize the audit entries so that only the ones you need are generated.

Table 15.2
QAUDLVL and AUDLVL Values

Value	Subsetted value	Description	QAUDLVL/QAUDLVL2	User profile attribute AUDLVL
*ATNEVT		Intrusion detection	X	
*AUTFAIL		Authority failure	X	X[1]
*CMD		Command strings		X
*CREATE		Creation of objects	X	X
*DELETE		Deletion of objects	X	X
*JOBDTA		Actions affecting jobs	X	X
	*JOBBAS[1]	Basic job changes	X	X
	*JOBCHGUSR[1]	Changes to job user	X	X
*NETCMN		Network communications	X	X[1]
	*NETBAS	Basic network functions	X	X[1]
	*NETCLU	Cluster functions	X	X[1]
	*NETFAIL	Network failures	X	X[1]
	*NETSCK	Socket tasks	X	X[1]
*OBJMGT		Object management	X	X
*OPTICAL		Optical	X	X
*PGMADP		Program adopt	X	X
*PGMFAIL		Program fail	X	X[1]
*PRTDTA		Print data	X	X[1]
*SAVRST		Restore operations	X	X
*SECURITY		Security operations	X	X
	*SECCFG	Security configuration	X	X[1]
	*SECDIRSRV	Directory services	X	X[1]
	*SECIPC	Interprocess communications	X	X[1]

Table 15.2 Continued

Value	Subsetted value	Description	QAUDLVL/QAUDLVL2	User profile attribute AUDLVL
	*SECNAS	Network authentication (Kerberos) actions	X	X[1]
	*SECRUN	Runtime security changes	X	X[1]
	*SECSCKD	Socket descriptors	X	X[1]
	*SECVFY	Verification functions	X	X[1]
	*SECVLDL	Validation list functions	X	X[1]
*SERVICE		Service functions	X	X
*SPLFDTA		Spooled file functions	X	X
*SYSMGT		System management	X	X

[1] *New in V6R1*

Let's look more closely at some of the most important QAUDLVL values: *AUTFAIL, *CREATE, *DELETE, *SAVRST, *SECCFG, *SECRUN, and *SERVICE. These seven values represent the minimum set of values that I recommend you specify for this system value. The thought behind specifying these values is that they will produce information that you need without generating so much data that it's overwhelming. My preference is to audit for the exceptions, not for what normally occurs every day.

The first value is *AUTFAIL. *AUTFAIL instructs the system to audit events related to insufficient authorities — events such as a user entering an incorrect user ID or password or attempts to access an object such as a database file without sufficient authority. Entries generated by the *AUTFAIL value can help you debug which object a user lacks authority to when a "Not authorized" message is issued. Applications can be written in ways that make it difficult to determine the actual object to which a user is not authorized. The details of this audit entry list the specific object as well as the profile that lacks sufficient authority. For example, I was at a client site once when a user signed on using the QSECOFR profile. A few screens into an application, the user received a "Not authorized" message. How could that be when the user had signed on as QSECOFR? When we examined the audit journal entry, we found that a profile swap had occurred and the user receiving the message was actually a different profile — not QSECOFR. The audit entry also listed the object for which the profile lacked authority.

*CREATE instructs the system to create a journal entry each time someone creates (e.g., compiles, duplicates) a new object on your system, including the replacement of objects that already exist on the system. Knowing about new objects on your system can help you manage the activities of programmers, consultants, and vendors.

*DELETE falls into the same category as *CREATE, only in reverse: The system creates an audit journal entry whenever an object is deleted. Such entries are vital when you're trying to determine the actions taken by a disgruntled employee on a production system. They're also extremely helpful in debugging errant programs and test cases that delete the wrong object.

Specify *SAVRST — especially on production systems — to monitor what objects are being restored onto your system.

The *SECURITY value provides a record of security-relevant changes to user profiles, changes to the authority of objects, and other security events. Fortunately, IBM has subsetted this value so that you don't have to generate the full scope of all the entries that *SECURITY produces. I recommend that you use the *SECCFG and *SECRUN values rather than the *SECURITY value, to monitor security configuration and runtime security changes, respectively.

*SERVICE provides a record of anyone who uses service tools or performs other service functions on your system. Given the power of these tools, you'll want to closely monitor their use.

Other Auditing Values

*PGMADP is an audit value you may want to specify occasionally. This value causes the system to generate an audit journal entry when a user's authority comes from adopted authority. You can specify this value when you suspect that someone is abusing adopted authority and using it to perform operations that he or she is not normally authorized to perform. But rather than turning on this value for QAUDLVL, I tend to enable it at the user level (more about user auditing shortly). (Recall that I recommended, in discussing role-based access for IT in Chapter 14, that you might want to enable *PGMADP for vendor and consultant profiles.) Specifying this value only for selected users is especially important when one or more applications are configured to provide application users' access to objects via adopted authority. Turning on *PGMADP in this case has the potential to produce massive numbers of audit journal entries, all of which reflect "normal" behavior.

The *JOBDTA, *PRTDTA, and *SPLFDTA values can provide useful information, but they generally produce large numbers of audit journal entries, most of which you won't often be interested in. To determine whether these values will provide information that will be useful for your organization, take a look at the security audit journal description in Chapter 9 of the *System i Security Reference* (which is available as a PDF file under the Information Center's Security topic).

User Auditing

In addition to configuring auditing for the entire system, you can configure auditing at the user level. Using the CHGUSRAUD (Change User Audit) command, you can turn on action auditing for an individual or configure auditing of users' access or modification of specific objects. Figure 15.3 shows the CHGUSRAUD prompt screen.

Figure 15.3
CHGUSRAUD Prompt Screen

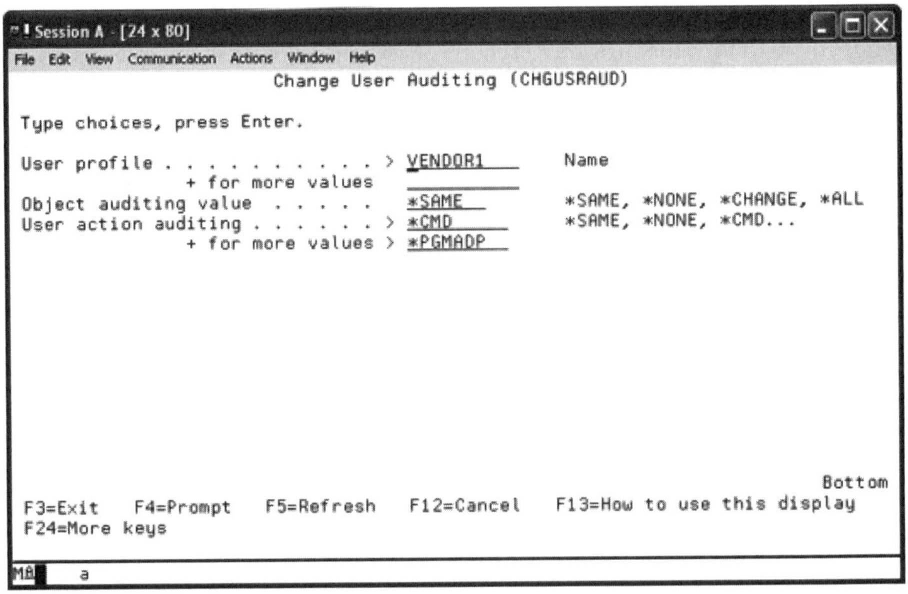

Note: Configuring auditing for a specific user is separate from the CRTUSRPRF (Create User Profile) and CHGUSRPRF (Change User Profile) commands to enable the separation of security duties from auditing duties.

The AUDLVL (User action auditing) attribute is similar to the QAUDLVL system value in that it enables actions to be audited. There's no need to repeat the values specified in the QAUDLVL system value in the AUDLVL parameter. Rather, use this parameter to specify additional actions to be audited. Also, there's one value, *CMD, that can be specified only at the user level. Specifying *CMD causes every command string a user enters or runs from a CL program to be audited (i.e., logged in the audit journal). This setting is especially useful if you must comply with a law or regulation that requires you to log or monitor administrators' activities. Given the volume of audit entries this setting generates, you'll want to assign this value only for user profiles that are required to be monitored — for example, users who have been assigned *ALLOBJ special authority and programmers accessing production (as I discussed in Chapter 14). You shouldn't turn on this setting for all users. That's why this value is available only at the user level and not at the system level.

Object Auditing

IBM i's object auditing lets you track which users read or modify a particular object. To configure auditing for a specific object, use the CHGOBJAUD (Change Object Auditing) command (Figure 15.4). Two types of actions cause an object-auditing audit journal entry to be generated: a user reads or uses an object or a user changes an object.

An OBJAUD (Object auditing) parameter value of *NONE means that use of the object will not be audited. *CHANGE tells IBM i to generate an audit journal entry when the object is modified. *ALL tells the system to audit all accesses (read or change) to the object. A final value, *USRPRF, tells IBM i to refer to the OBJAUD attribute of the user profile to determine what auditing to perform.

Figure 15.4

CHGOBJAUD Prompt Screen

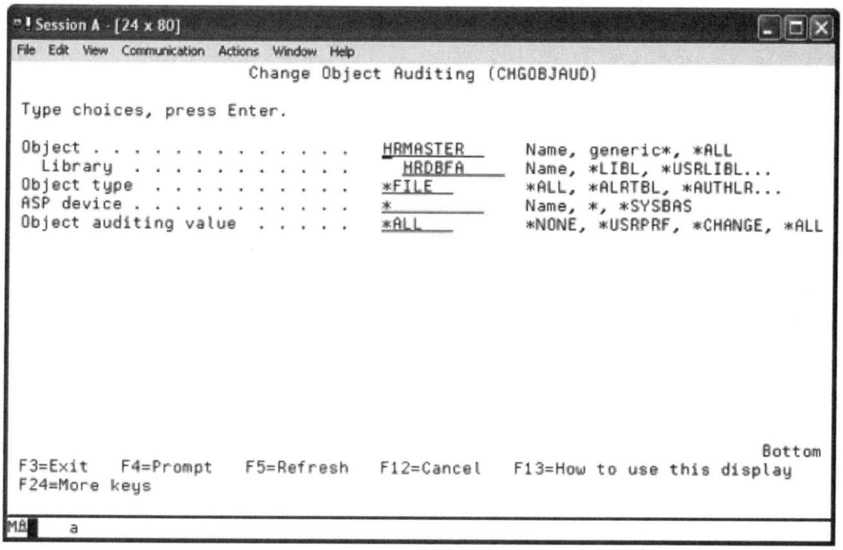

To change the auditing attribute of an object in one of the file systems (i.e., an object that's specified with a path name), use the CHGAUD (Change Auditing Value) command (Figure 15.5).

Figure 15.5

CHGAUD Prompt Screen

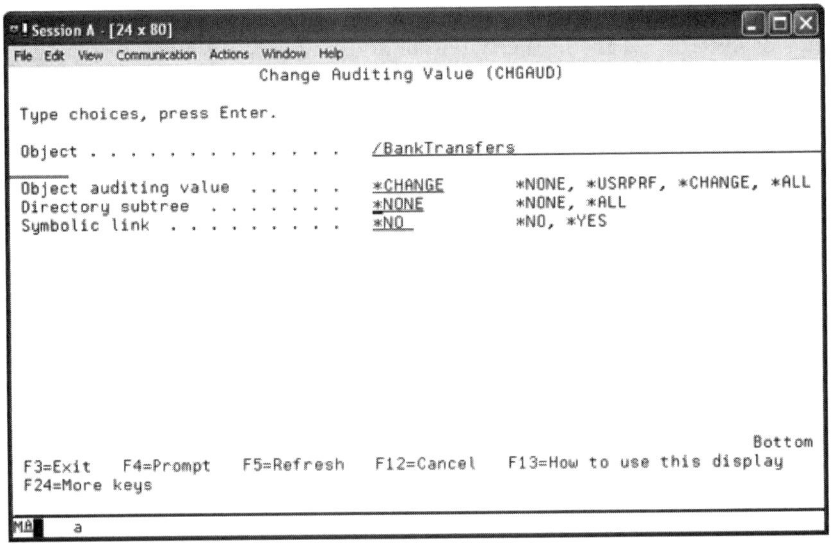

If you specify *USRPRF for the OBJAUD value on the CHGOBJAUD or CHGAUD command, IBM i looks at the object-auditing value specified in the user profile to determine whether to generate an object-auditing journal entry. A value of *NONE activates no object auditing for the user. *CHANGE tells the system to audit only the user's changes to objects. *ALL tells it to audit all accesses (read or change) of an object for that user.

The relationship between the object OBJAUD attribute and the user profile OBJAUD attribute is the key to understanding the control of object auditing on your system. Remember this rule: The OBJAUD value for an object takes precedence over the OBJAUD value of a user profile. This means that when you specify *NONE for the object OBJAUD attribute, IBM i doesn't generate an object-auditing entry for that object, regardless of the value specified for the OBJAUD attribute of the user profile accessing or changing the object. If you specify *CHANGE or *ALL for the object OBJAUD attribute, IBM i generates an object-auditing entry, regardless of the value of OBJAUD on the user profile. The operating system uses the user profile OBJAUD value only when you specify *USRPRF as the value for the object OBJAUD attribute. Table 15.3 lists the read and change operations that are audited when you configure the various combinations of OBJAUD values.

Table 15.3
OBJAUD Values

Object OBJAUD	*User profile OBJAUD*		
	***NONE**	***CHANGE**	***ALL**
***NONE**	None	None	None
***CHANGE**	Change	Change	Change
***ALL**	Change/Use	Change/Use	Change/Use
***USRPRF**	None	Change	Change/Use

The implications of this rule are simple. To implement user-specific auditing for object accesses, you must first specify OBJAUD(*USRPRF) for the objects you want to audit and then specify either OBJAUD(*CHANGE) or OBJAUD(*ALL) for the user profile for which you want to monitor activity. If you want to audit specific access to one or more objects regardless of which user profile is accessing the objects, simply modify the OBJAUD attribute of the objects. The diagram in Figure 15.6 summarizes your options as you use system value QAUDCTL and the OBJAUD attributes to control your auditing implementation.

Figure 15.6
How IBM i Determines Whether Auditing Should Occur

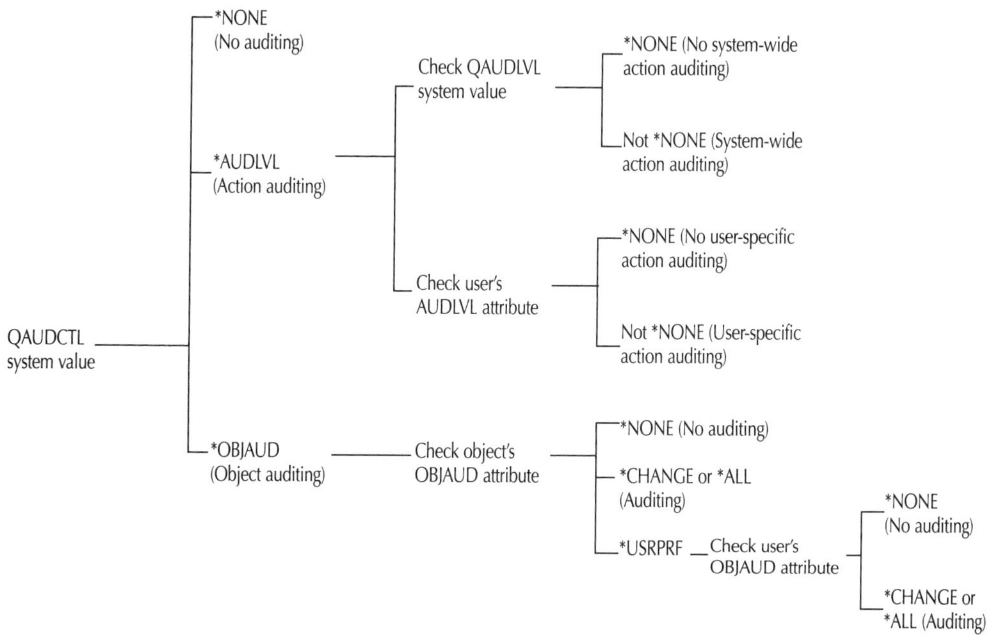

Object Auditing for New Objects

The auditing value for new objects is set based on the values in the CRTOBJAUD (Create object auditing) library attribute and the QCRTOBJAUD (Auditing for New Objects) system value. QCRTOBJAUD works much like the QCRTAUT (Create Authority) system value, which determines the default public authorities for new objects. When you create a new object, IBM i checks the CRTOBJAUD attribute for the library in which the new object will reside. If this value is *SYSVAL, IBM i takes the object's auditing value from the QCRTOBJAUD system value to determine which object-auditing value (*NONE, *CHANGE, *ALL, or *USRPRF) to assign to the OBJAUD attribute of the new object. Because the QCRTOBJAUD system value defaults to *NONE, newly created objects will be created with no object auditing enabled.

Before changing QCRTOBJAUD, I encourage you to think of the ramifications of doing so. Specifying *CHANGE for this system value will cause *all* newly created objects to be created with an object-auditing value of *CHANGE — this include user spaces, data queues, and data areas as well as programs and files. Sometimes, this is more auditing than is desirable. However, some high availability vendors require QCRTOBJAUD to be set to *CHANGE so that their software knows when objects have been changed.

Even more consideration needs to be made when setting QCRTOBJAUD to *ALL. With this setting, when newly created objects are changed or read, an audit journal entry is generated. Setting QCRTOBJAUD to *ALL may cause massive numbers of audit journal entries to be generated. So although you may think this setting is a good idea, you'll want to closely monitor your audit journal receivers and save them off quickly so as not to cause a storage problem.

My preferred approach for this system value, if you don't leave it at *NONE, is to set it to *USRPRF and turn on object auditing for selected user profiles. I typically turn on object auditing for the members of the administrator, programmer (on production), vendor, and consultant roles.

Event-Auditing Recommendations

Technology is generally worthless if you don't use it properly, and this maxim holds true for the auditing technologies available on IBM i. If you don't configure auditing, important information for debugging problems and valuable forensic information required if a breach occurs is lost. My company was called in to investigate a potential breach at a new client site. Unfortunately, the client had no auditing turned on, kept no job logs, and deleted its history logs. Needless to say, the investigation didn't take long because the data needed to make any type of investigation simply was not available.

The following sections list my recommendations concerning the use of the IBM i auditing functions.

Auditing Controls Security Recommendations

- For the QAUDCTL system value, specify both *AUDLVL and *OBJAUD so you can perform both types of auditing, and specify *NOQTEMP to eliminate unnecessary journal entries.
- Leave system values QAUDFRCLVL and QAUDENDACN at their respective default values of *SYS and *NOTIFY.

System and User Event-Auditing Security Recommendations

- For the QAUDLVL system value, specify *AUTFAIL, *CREATE, *DELETE, *SAVRST, *SECCFG, *SECRUN, and *SERVICE. You can specify additional values, but this group is my recommended "base set."
- To determine whether you have issues you need to resolve before going to security level 40 or 50, specify *AUTFAIL and *PGMFAIL for system value QAUDLVL. (For more details about moving to security level 40 or 50, see Chapter 3.)
- Use the AUDLVL attribute at the user profile level to monitor the activities of powerful profiles such as QSECOFR, profiles that have *ALLOBJ special authority, profiles used by programmers to access production systems, and profiles used by consultants and vendors who access your systems.

Object-Auditing Recommendations

- Turn on *ALL object auditing for files that contain private or critical data. Files containing cardholder data (encrypted credit cards), human resources information such as salaries, personal identifying numbers such as Social Security or social insurance numbers, or files containing restricted company information are all candidates for object auditing.

Technical Note

Some actions that you might think would cause an object read or change audit journal entry to be generated don't. If you're considering enabling object auditing on a specific object, make sure you understand which actions will cause audit journal entries to be generated for that object type. See Appendix E of the *System i Security Reference* for the actions that cause object auditing to occur on each object type.

Working with the Audit Journal

Once you've configured auditing, the next step is to examine the audit journal entries regularly for specific events, trends, or possible security breaches and exposures. The remainder of this chapter provides the information you need to accomplish this task.

First, we'll look at what a journal entry is and how it's formatted. Then, I'll describe methods for getting the information out of the audit journal.

For detailed information about and specifications for IBM i audit journal entries, I encourage you to become familiar with the information available in the *System i Security Reference*. Chapter 9 of this manual describes the values available for specification on the QAUDLVL system value as well as at the user level on the AUDLVL parameter of the CHGUSRAUD command. Appendix F includes the layout of the model outfiles provided for each audit journal entry type. These model files give you an easier way to use SQL or Query to find the exact information you're looking for without manually examining every entry in the journal. Last, the *Security Reference* gives you the details and specifications you need to perform some of the operations referred to below.

Understanding Journal Entry Formats

All journal entries have common fields (e.g., date, time, entry type, job name, IP address) followed by data specific to the type of entry. The standard journal entry fields are documented at the beginning of Appendix F in the *System i Security Reference*. These fields appear at the beginning of every security audit journal entry. Although the security manual documents the journal entries for previous releases, you need to use the most recent layout — that is, the layout for the *TYPE5 entry.

When examining the audit journal or narrowing your search of audit journal entries, you'll want to identify an *entry type*. Each journal entry type is identified by a two-character abbreviation. Table 15.4 lists the 75 types of journal entries possible in QAUDJRN. Each event action you might audit (e.g., *AUTFAIL, *CMD, *SECURITY) using the QAUDLVL system value, user-action auditing, or object-access auditing has one or more entry types that IBM i creates when adding a journal entry for those events.

Table 15.4
Audit Journal Entry Types

Entry type	Description
AD	Auditing changes
AF	Authority failure
AP	Obtaining adopted authority
AU	Enterprise Identity Mapping configuration change
CA	Authority changes
CD	Command string audit
CO	Create object
CP	User profile changed, created, or restored
CQ	Change of Change Request Description (*CRQD) object
CU	Cluster operations
CV	Connection verification

continued...

Entry type	Description
CY	Cryptographic configuration changes
DI	Directory services
DO	Delete object
DS	Dedicated Service Tools security password reset
EV	System environment variables
GR	Generic records
GS	Give descriptor
IM	Intrusion monitor
IP	Interprocess communication
IR	Actions on IP rules
IS	Internet security management
JD	Change to user parameter of a job description
JS	Actions that affect jobs
KF	Key ring file
LD	Link, unlink, or look up directory entry
ML	Office services mail actions
NA	Network attribute changed
ND	APPN directory search filter violation
NE	APPN endpoint filter violation
OM	Object move or rename
OR	Object restore
OW	Object ownership changed
O1	(Optical Access) Single file or directory
O2	(Optical Access) Dual file or directory
O3	(Optical Access) Volume
PA	Program changed to adopt authority
PG	Change of an object's primary group
PO	Printed output
PS	Profile swap
PW	Invalid password
RA	Authority change during restore
RJ	Restoring job description with user profile specified
RO	Change of object owner during restore
RP	Restoring adopted authority program
RQ	Restoring a *CRQD object
RU	Restoring user profile authority
RZ	Changing a primary group during restore

continued...

Entry type	Description
SD	Changes to system distribution directory
SE	Subsystem routing entry changed
SF	Actions to spooled files
SG	Asynchronous signals processed
SK	Secure sockets connections
SM	System management changes
SO	Server security user information actions
ST	Use of System Service Tools
SV	System value changed
VA	Changing an access control list
VC	Starting or ending a connection
VF	Closing server files
VL	Account limit exceeded
VN	Logging on and off the network
VO	Validation list actions
VP	Network password error
VR	Network resource access
VS	Starting or ending a server session
VU	Changing a network profile
VV	Changing service status
X0	Network authentication
X1	Identity token
XD	Directory server extension
YC	Document Library Object (DLO) accessed (change)
YR	DLO object accessed (read)
ZC	Object accessed (change)
ZR	Object accessed (read)

Displaying and Printing Audit Journal Entries

You have at least three options for displaying or printing QAUDJRN's contents. You can use the DSPAUDJRNE (Display Audit Journal Entries) command, the DSPJRN (Display Journal) command, or the CPYAUDJRNE (Copy Audit Journal Entries) command. DSPAUDJRNE is a simplified version of the DSPJRN command with parameters that pertain only to displaying audit journal entries out of QAUDJRN. DSPJRN provides a richer set of options for displaying and printing audit journal contents. CPYAUDJRNE makes it easy to retrieve audit journal entries and then run an SQL statement or query to refine information being displayed or printed.

Using the DSPAUDJRNE Command to Display Entries

To view journal entries, run the command DSPAUDJRNE. The command defaults to running a report of the AF (authority failure) entries. However, if you prompt the command, you can specify to see a list of the other types of entries (e.g., CP for changes to profiles, SV for changes to system values, and so forth.) Figure 15.7 shows an example of this command.

Figure 15.7
DSPAUDJRNE Prompt Screen

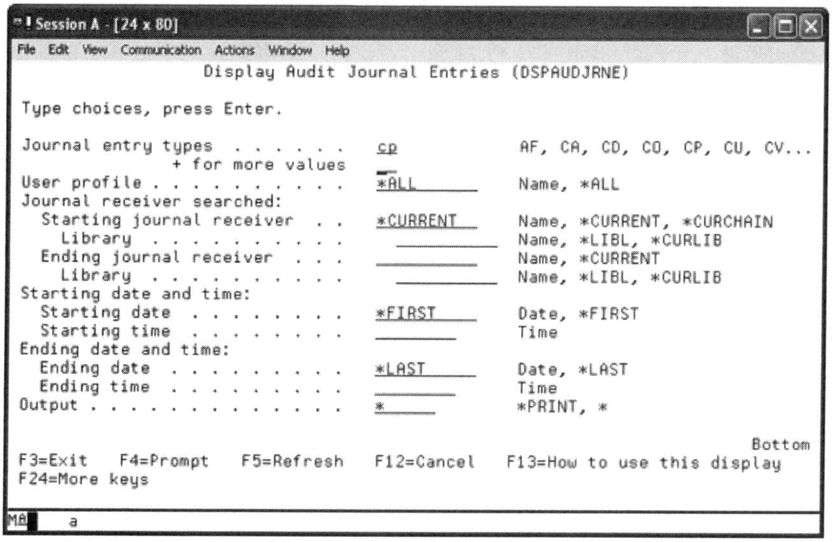

There are limitations with this option. First, there's no way to sort or query the results of running the command. You get either a spooled file or a list that appears on your display. Second, the command doesn't support all the audit types. Worse, it doesn't support integrated file system (IFS) objects. Figure 15.8 shows the output of the DSPAUDJRNE command. Notice the *N values in the Object name field. This indicates that these objects are IFS objects (whose path names are too long for the 10-character Object name field). To view an object's path name, you'll have to use the next technique.

Figure 15.8
DSPAUDJRNE Command Output

```
Session A - [24 x 80]
File  Edit  View  Communication  Actions  Window  Help
                              Display Report
Query . . . :   QSYS/QSECAF                 Report width . . . . . :     208
Position to line  . . . . .   ______      Shift to column  . . . . . .   _____
Line   ....+....1....+....2....+....3....+....4....+....5....+....6....+....7..
         Violation User      Object     Library    Object   Office     DLO
         type      profile   name       name       type     user       name

000155 AF    A     QUSER     *N         *N         *DIR
000156 AF    A     QUSER     *N         *N         *DIR
000157 AF    A     QUSER     *N         *N         *DIR
000158 AF    A     QUSER     *N         *N         *DIR
000159 AF    A     QUSER     *N         *N         *DIR
000160 AF    A     QUSER     *N         *N         *DIR
000161 AF    A     QPM400    *N         *N         *DIR
000162 AF    A     CWOODBURY SKYVIEWOUT R_SKYVWPMD *OUTQ
000163 AF    K     CWOODBURY QPOFEPFS   QSYS       *SRVPGM
000164 AF    A     QTMHHTTP  *N         *N         *DIR
000165 AF    K     CWOODBURY QPOFEPFS   QSYS       *SRVPGM
000166 AF    A     QTMHHTTP  *N         *N         *DIR
000167 AF    A     QPM400    *N         *N         *DIR
000168 AF    A     QPM400    *N         *N         *DIR
                                                                       More...
F3=Exit      F12=Cancel      F19=Left      F20=Right      F21=Split

MA    a
```

Figure 15.9
Sample DSPJRN Screen

```
Session A - [24 x 80]
File  Edit  View  Communication  Actions  Window  Help
                          Display Journal Entries

Journal  . . . . . . :   QAUDJRN        Library  . . . . . . :   QSYS
Largest sequence number on this screen  . . . . . . : 00000000000001234736
Type options, press Enter.
  5=Display entire entry


Opt    Sequence  Code  Type  Object      Library     Job         Time
 _      1234725   T     CA                           QGLDPUBA    11:04:12
 _      1234726   T     DO                           QGLDPUBA    11:04:12
 _      1234727   T     CD                           QPADEV0002  11:04:40
 _      1234728   T     CD                           QPADEV0002  11:04:40
 _      1234729   T     CD                           QPADEV0002  11:04:40
 _      1234730   T     CD                           QPADEV0002  11:05:35
 _      1234731   T     CD                           QPADEV0002  11:05:35
 _      1234732   T     CD                           QPADEV0002  11:05:54
 _      1234733   T     CP                           QPADEV0002  11:05:54
 _      1234734   T     SO                           QPADEV0002  11:05:54
 _      1234735   T     CD                           QPADEV0002  11:06:24
 _      1234736   T     ZC                           QPADEV0002  11:06:27
                                                                    More...
F3=Exit   F12=Cancel

MA    a
```

Using the DSPJRN Command to Display Entries

To view journal entries using the DSPJRN command, enter the command

```
DSPJRN  QAUDJRN
```

This command generates a screen similar to the one shown in Figure 15.9. Typing a 5 (Display entire entry) in the option column for an entry gives you a detail screen similar to the one shown in Figure 15.10. In this example, a profile (LINDAV) was changed. Pressing F10 (Display only entry details) here displays a series of detail screens similar to those shown in Figures 15.11A through 15.11C, providing further information about the event.

Figure 15.10
Display Journal Entry Screen

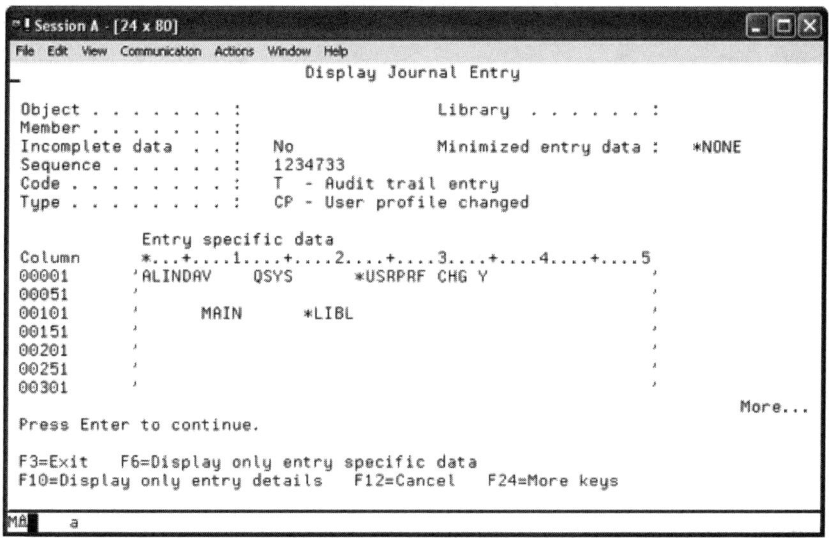

Figure 15.11A

Display Journal Entry Details Screen (Page 1 of 3)

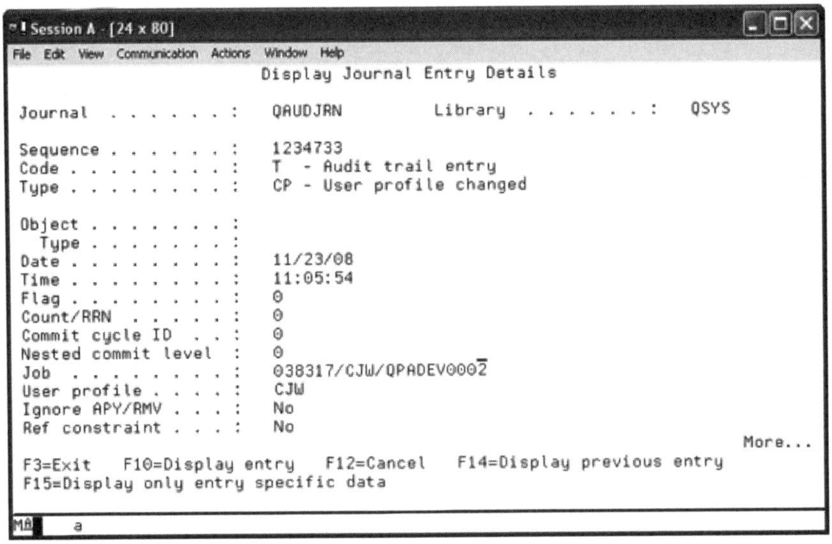

Figure 15.11B

Display Journal Entry Details Screen (Page 2 of 3)

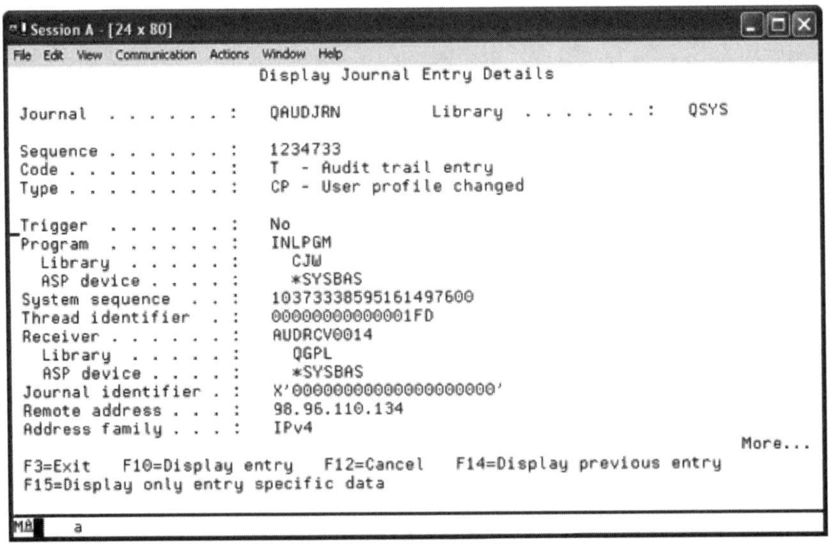

Figure 15.11C

Display Journal Entry Details Screen (Page 3 of 3)

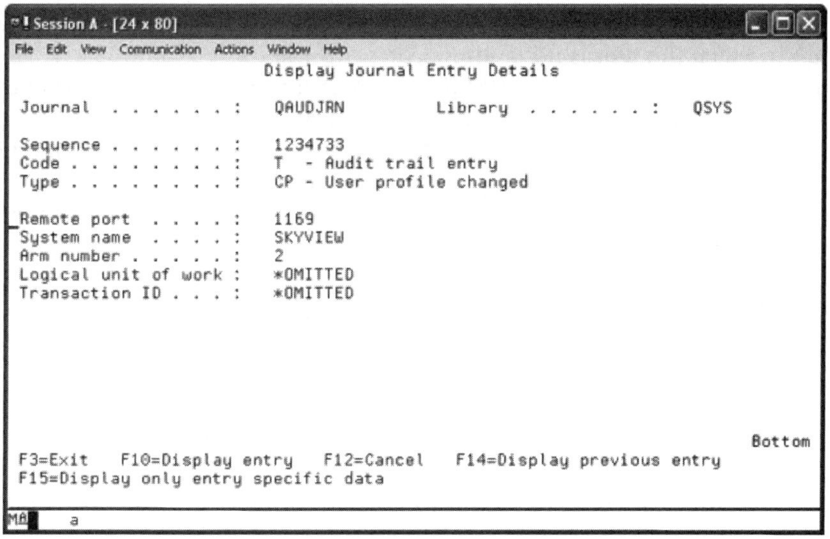

Figure 15.12

Using DSPJRN to Find the Path Name

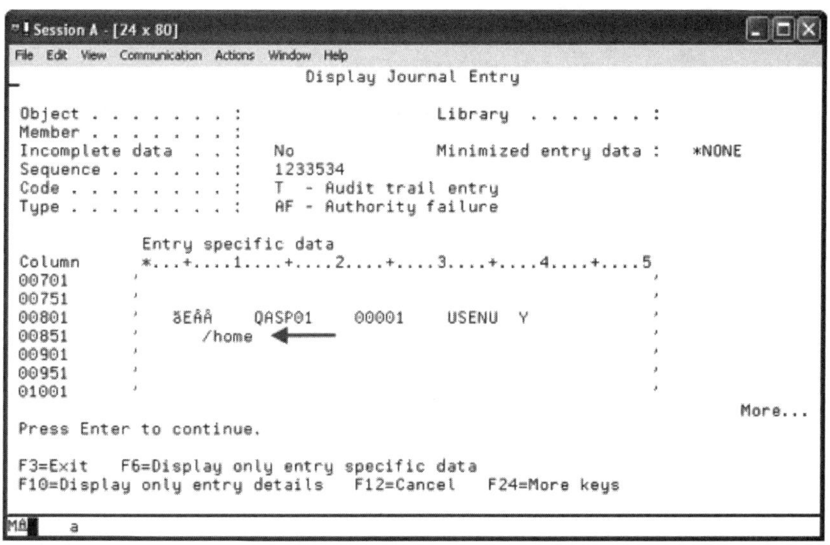

Going back to Figure 15.8, to find the path name of the object for which there was an authority failure, run the DSPJRN command, subsetting the list to see only the AF (authority failure) entries. When you find the entry you're interested in, type a 5 beside the entry and scroll to the last screen. Figure 15.12 (on page 261) shows the directory ('/home') to which the user is not authorized.

You can use parameters on the DSPJRN command to select only specific entry types or to narrow your search based on a specific time period, job, or — in some cases — user profile. If you know the exact time of the entry you're looking for, this method may be the easiest way to look at and find details about the entry. However, this approach provides no way to sort the entries or produce a report of the entries, so much of the time this method is not practical.

Using the CPYAUDJRNE Command

Your third option is to use the CPYAUDJRNE command. This command combines copying a model outfile for the specific type of audit journal entry being examined with the DSPJRN command and sending the results to the copy of the outfile. For example, if you're looking for all AF entries, the outfile that will contain the entries after CPYAUDJRNE is run will be based on the model outfile QASYAFJ5 that is shipped in library QSYS. Appendix F of the *System i Security Reference* lists the format of all outfiles provided with the system. Review this appendix to know what information is available for each audit journal entry type. After using the CPYAUDJRNE command, it's quite easy to run a query or SQL statement to find the exact entries you're looking for or to further narrow down the results and print a report of the activity.

Figure 15.13
CPYAUDJRNE Prompt Screen

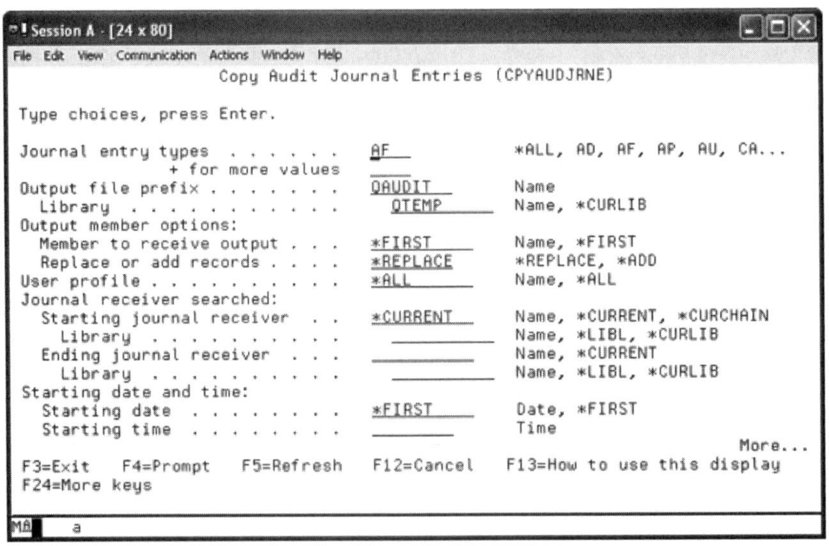

Figure 15.13 shows the prompt for the CPYAUDJRNE command. By default, the file containing the audit journal entries will be called QAUDIT*xx* where *xx* is the two-letter audit journal entry type. For example, the file name for the AF (authority failure) entries would be QAUDITAF. Also, the file will be in library QTEMP unless you specify otherwise.

Reporting on Activities from the Information in the Audit Journal

Although I would never recommend that someone examine every entry in the audit journal (unless I was trying to torture them!), there are some entries I think you should examine — or at least look for — on a regular basis. You can use any of the techniques described above to view these entries, but I tend to use the CPYAUDJRNE command and then write a query to narrow my search and produce a report that can be reviewed on a weekly basis. Keep in mind that the purpose of reviewing these reports is not to review known or approved activity. You're looking for exceptions and indications of inappropriate activity.

Here are the entries that I recommend you regularly review:

- AF
  - Authority failures to files containing private or confidential data
- CP
  - Password changes for powerful profiles such as QSECOFR and QSRV
  - Dedicated Service Tools (DST) QSECOFR reset
  - Creation of or changes to profiles by users not in the approved team
- PW
  - Invalid password attempts for powerful users (subtype P entries)
  - Invalid service tools ID (subtype Y entries)
  - Invalid service tools password (subtype Z entries)
- ST
  - Use of various service tools
- SV
  - Changes to system values (omitting changes to the performance system values because these are managed by IBM i itself)

I generate the following reports upon special request — typically when an issue is being investigated, an incident has occurred, or an auditor requests them:

- CD
    - Use of specific commands
    - All commands run by a specific user
- AP
    - Use of adopted authority by one or more specific users

When analyzing which users and processes are accessing or changing a specific object or set of objects, I generate reports for the following type:

- ZC
    - Object has been modified
- ZR
    - Object has been read

Benefits of the IBM i Architecture

Before I close this chapter, I would be remiss not to describe the benefits of the implementation of the IBM i audit features. At least one regulation — PCI DSS — requires you to use log integrity software that ensures that logs (audit journal receivers, in IBM i terms) — are preserved and that a mechanism is in place to detect changes to the log. While the history log is a file that can be modified, an audit journal receiver is not. The audit journal (QAUDJRN) determines what goes into the receivers, but it's the audit journal receivers that contain the actual entries. IBM specifically architected the audit function to be retained in journal receivers so that there is no possibility of an audit journal entry being modified or deleted. The inherent architecture of a journal receiver prevents this from happening. That's the reason that all security-critical messages were changed from being logged to the history log to being an audit journal entry — or they're logged to both the history log and the audit journal.

Even if someone deleted an entire journal receiver, you could detect that. The receivers are numbered sequentially, so you'd know one was missing. Also, when auditing is turned off, an entry is written in the journal to document this action. So, if you detect that auditing is not active, you can go back to the last entry written to determine who turned it off.

So, chalk up another benefit of running IBM i: Your audit journal entries are safe, and their integrity is assured.

Helpful Tools

Auditing security-related events daily is one of the most significant auditing tasks you can perform. Monitoring events as they occur gives you insight into the effectiveness of your security implementation and helps you manage the changes that naturally occur on your system. You'll find the four tools listed in Table 15.5 useful for performing auditing tasks.

Table 15.5
Tools for Performing Auditing Tasks

Security tool	Description
CHGSECAUD (Change Security Auditing)	Lets you set up auditing using a single command.
DSPAUDJRNE (Display Audit Journal Entries)	A simplified version of the DSPJRN command with parameters that pertain only to displaying audit journal entries out of QAUDJRN. Although IBM no longer updates this command (for example, the command provides only *TYPE4 information and doesn't support all the entry types currently available), DSPSECAUD remains a convenient command if you want a quick look at the most popular entries, such as authority failures or creations and deletions.
DSPSECAUD (Display Security Auditing)	Displays or prints a report of the current settings for system values QAUDCTL and QAUDLVL as well as the journal receiver currently attached to the audit journal.
CPYAUDJRNE (Copy Audit Journal Entries)	Combines the process of duplicating the model outfile for the type of audit journal entry you're looking for with running the results of the DSPJRN command to an outfile. Makes the gathering of audit journal entries substantially easier.

Navigator for i

You can use Navigator for i to change the auditing system values. At the Navigator window, double-click your system name, select Security and then Policies, and then double-click Audit Policy in the right panel. You'll see a screen similar to the one shown in Figure 15.14, where you can enter auditing system values.

Figure 15.14
Audit Policy Properties Dialog Box

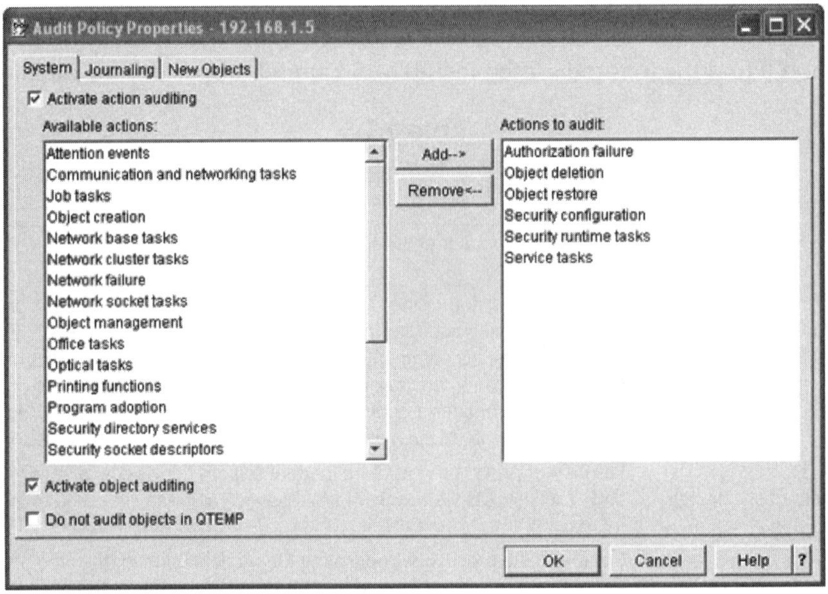

Navigator also lets you manage the auditing attributes of user profiles. To perform this task, double-click your system name, select Users and Groups, find the user profile for which you want to display or change the auditing attributes, and make the appropriate changes.

Technical Note

Although actions taken through Navigator for i are audited, as are objects accessed through Navigator, Navigator provides no command-string auditing.

Security Recommendation

One last and final reminder! Don't forget that a wealth of information is at your fingertips. All you have to do is get it out of the audit journal and use it to make your life as an administrator much easier when debugging system issues.

Chapter 16

Implementing Object-Level Security

Until now, we've been talking a lot of theory and a lot of facts, but we haven't gotten into much practical application. This is the chapter where all the information you've learned thus far in the book will be pulled together into practical implementation advice.

The focus of this chapter is implementing object-level security. Very few of you get the luxury of starting an application from "scratch" and designing its security scheme the right way from the beginning. Instead, most of you are stuck with one or more applications that use poorly architected security schemes. I've heard from many of you who have tried to implement object security, but the project has failed and you've had to back out your changes. This chapter's goal is to give you the information you need to successfully implement object-level security with as little "breakage" as possible. I'll describe the methodology I use with our clients to take a poorly implemented design and change it into one that meets the organization's security policy and compliance requirements.

So follow closely, and if you're tempted to skip a step or take steps out of order, don't. In many cases, the order in which you perform the steps is important to the success of the project. Also, be sure to read the entire chapter before making any changes. Although I cannot guarantee that your project will be implemented with no breakage, following this methodology should reduce the frequency of its occurrence.

Determine the Scope of Your Project

The first step in implementing object-level security is to determine what it is you're going to be changing and what security requirements you're trying to achieve. For example, you may only want to ensure that your employee master database is properly secured. Or perhaps your organization needs to comply with PCI DSS, so you need to make sure access to the files that contain cardholder data is "deny by default." In these cases, you're most likely required to set the *PUBLIC authority of the objects to *PUBLIC(*EXCLUDE). But that may not always be the case. In some situations, you might be concerned not with the confidentiality of the data but with the integrity of the data. Let me explain. We've had several customers whose security policy allowed any user on the system to view and download all the data on the system. But the customers wanted to make sure that no user was able to change the data from outside the application — through Database Utility (DBU) or File Transfer Protocol (FTP), by uploading a modified spreadsheet, or using any other means. That goal required that we architect a solution so that the *PUBLIC authority was *USE. This solution provided data integrity, but it did not keep the data private or away from the systems' users. Last, you might have one or more applications that use one of the common application security schemes described in Chapter 12 that requires users to have *ALLOBJ, be members of the group that owns the application, or have a wide-open *PUBLIC authority setting.

As you go through this chapter, it's helpful to have a specific project in mind. So, I'd like you to pause for a moment and decide which file or application you're going to address. Then, decide what the requirement for the data is — *PUBLIC authority of *EXCLUDE? *USE?

Now that you've thought about your own situation, here's the scenario I will use as an example for the rest of the chapter:

- Application to be remediated: Accounting, which includes accounts receivable (AR), accounts payable (AP), and general ledger (GL)
- Current security scheme:
    - The application is owned by the profile ACCTNG.
    - The ACCTNG profile has *JOBCTL, *SAVSYS, and *SECADM special authorities.
    - Each application user is a member of the ACCTNG profile.
    - The *PUBLIC authority of all application objects (including all database files) is *CHANGE.
    - Authority to the application directory is *PUBLIC DTAAUT(*RWX) OBJAUT(*ALL).
- Requirements of this project:
    - Security settings must enforce separation of duties.
    - PCI data must be deny by default.
    - Users must be able to rely on the integrity of the data.
    - No one should be able to change the data outside the protection of the application's logic (in other words, all changes must be made through an application interface, such as a menu option).
    - Only approved users should be able to view accounting data or download it.

High-Level Design of the Architecture

To remediate the flawed security scheme, we'll take the following approach, which accommodates all the requirements listed above.

- Application programs will be changed to adopt authority (programs will be set to USRPRF(*OWNER). It is through this adopted authority that application users will have sufficient authority to run the application.
- The ACCTNG profile will be changed to have only *JOBCTL special authority. Examining the function of the application reveals that it performs no tasks requiring either *SECADM or *SAVSYS. Therefore, these authorities will

be removed so that there's no possibility of application users exploiting the capabilities of adopting *SAVSYS or *SECADM special authority.

- The ACCTNG profile will also explicitly be set to PASSWORD(*NONE) so there's no possibility of someone using the profile for sign-on.
- All database files will be secured with an authorization list. There will be three authorization lists: AR, AP, and GL.
- All application objects, including authorization lists, will be owned by the ACCTNG profile.
- Application users will be removed from the ACCTNG group and placed into a different group profile.

Regardless of whether you intend to rework an entire application or secure one file, you'll want to follow the methodology outlined in this chapter. If you're only changing the authority on one or several files rather than an entire application, the amount of effort involved may be less, but you should work through each of these steps to be as certain as possible that your rollout will be successful.

Building the Big Picture

This phase of your project is the most critical. It is the information-gathering portion of the effort. This phase is where you determine what is currently happening to the application data — that is, what processes and which users are viewing or updating the data — not through application interfaces, but from outside the application. The access may come from another application that needs to read a file, a WebSphere application that updates a file via a Java Database Connectivity (JDBC) connection, a batch process that reads a file and creates a stream file in the integrated file system (IFS), or end users downloading data into spreadsheets. I call this process "building the big picture" (Figure 16.1). The more complete you make this picture, the greater the likelihood that you'll avoid breakage when you roll out the new security scheme.

Personally, I don't like it when things "break" — that is, when some aspect of an application has an error, such as a batch job failing, a user receiving an error when choosing a menu option, or a program receiving a function check. There's enough management attention on these types of projects that I want to avoid the negative attention such failures cause. That's why it's critical to document all known accesses of application data so that you can determine whether and how to accommodate that access in your new security architecture. Remember, the data is no longer going to be wide open. It will be either *PUBLIC *EXCLUDE or *USE. In either case, any process that needs to update the data will fail unless you do something to accommodate that process. It's unlikely your picture will be 100 percent complete — a process or two may fail or a user may experience a problem taking a menu option — but with this up-front investigation, the failure should not be so great so as to cause a rollback of the implementation or abandonment the project altogether. Also, with the architecture I use, recovery from failures is usually swift and relatively painless.

Figure 16.1
Building the Big Picture

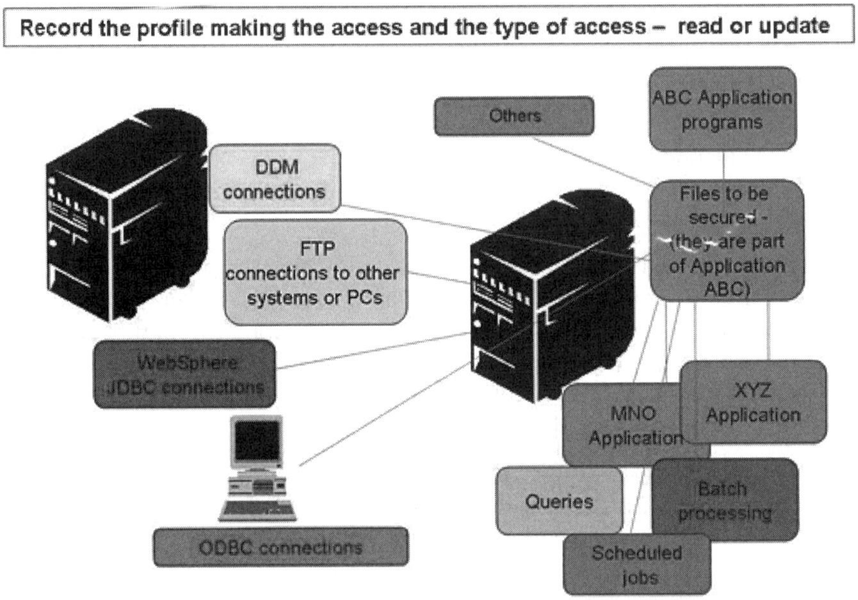

I'll talk about techniques for gathering information that will help you build the big picture in a moment. But first, let me describe the information you need to obtain. I usually use a spreadsheet, but the format you use doesn't matter. The important thing is to collect the following pieces of information:

- Name of the file being accessed (including the library or directory path)
- Name of the profile accessing the data
- A short description of the task so you can tell the tasks apart (make this description sufficient to know which task to run to test your settings during rollout of the new security architecture)
- Name of the program or type of access (e.g., APPLIB/PGMUPD2 if a program is accessing the data or DDM, FTP, System i Access data transfer, ODBC, and so on if it's an interface)
- Whether the access is a read or an update (if the process both reads and updates, just record the update)

You'll use this information during the rollout phase to accommodate these existing accesses so they don't break when you make the file's access more restrictive than its current setting. Although this process may seem a bit daunting, remember that you don't have to list the files being accessed by the application programs. You'll set these programs to adopt authority, so they will have sufficient authority to access the files. List only the tasks that are not calling the application's programs. Table 16.1 records this information for our sample scenario. (I'll talk about the authorization list column a bit later in this chapter.)

Table 16.1
Sample File Access Information

File library/name	Accessing profile	Description of task	Program or access type	Read or update	Authorization list name(s)
APLIB/MASTER	INVPROF	Inventory update	INVLIB/UPD	Update	AP_AUTL
APLIB/VENDOR	QEJBSRV	WebSphere Vendor AP app	JDBC	Read	AP_AUTL
GLLIB/BANK	FTPOUT	FTP bank process	FTP	Read	GL_AUTL

Collecting the Information

There are several ways to collect the required information. If you're just changing one or a handful of files, you can turn on object auditing on the file(s) and query the audit journal on a frequent basis to find the processes that are either reading (these will be ZR audit entries) or updating (ZC entries) the data.

We took this approach with one client that needed to change its cardholder data files from *PUBLIC *CHANGE to *PUBLIC *EXCLUDE. For each file, we ran the following command to enable auditing:

```
CHGOBJAUD  OBJ(ARLIB/AUTOPAY) OBJTYPE(*FILE) OBJAUD(*ALL)
```

Then, we ran the CPYAUDJRNE (Copy Audit Journal Entries) command for the ZR entries and ran it again for the ZC entries. We ran some queries against the data to eliminate application programs and processes being run by *ALLOBJ users. Next, we downloaded the results into two spreadsheets — one for processes updating the data and another for processes just reading the data. After sorting the results to list only unique entries, we had our list of profiles that needed to be authorized to the files' authorization list so that the processes they ran would continue to work.

If you're changing an entire application, gathering the information is more difficult. Turning on object auditing for all application files is too overwhelming — the sheer amount of generated data makes this approach impractical. If you have an exit point solution installed, you can use the logging feature to log the activity that occurs from outside the system (e.g., the DDM, FTP, and ODBC connections). If the application is a client/server or WebSphere application for which all transactions occur as the same IBM i user profile (usually as the owner of the database file), you'll want to configure the logging to ignore the "known" transactions (i.e., the application transactions) and log all others.

Regardless of whether you have another way to gather the information for the big picture, I recommend that you sit down with the application's developers or the application vendor as well as the system administrators, operators, and perhaps the application's power users and talk through the issue you're trying to solve. It's amazing how close you'll come to generating a complete list of tasks simply by getting the right people in the room.

After generating the initial list, you'll also want to look through the following items.

Job schedulers. Review the IBM and any vendor job scheduler you have installed, looking for jobs that access the application files. Remember that you can ignore any job running an application program (because adopted authority will be used to gain access to the application files). You can also ignore any job that runs as a profile with *ALLOBJ special authority.

Job descriptions. If you've decided to set libraries to *PUBLIC *EXCLUDE, you'll want to look through job descriptions to ensure that the libraries aren't listed in the library list for the job description or, if they are, that the users using the job description are authorized to the library. If the users aren't authorized to the library, the job won't start.

Queries. Look through your queries, specifically queries that generate files into the application's library. You may want to separate these out and create the files in a different library (more about queries later in the chapter).

The current configuration. You can use a vendor product such as Policy Minder or write your own routine to obtain a listing of the current security configuration on your application objects. At this phase, you'll want to pay specific attention to the private authorities that have been granted to the application objects. If there are individual or group profiles that have been granted a private authority to application objects — especially files — you'll want to investigate why the authority was granted. Perhaps it was granted to enable a specific feature or to resolve an issue with a failing process. If you can find out why the authority was granted, you can make a determination as to whether the access needs to be accommodated after you've changed the application's security configuration. I often find that most private authorities are removed in the process of implementing this type of project. Often, organizations determine that the individuals or groups really shouldn't be allowed to have direct access any more. Or they determine that the authority was granted in an effort to resolve an issue, but when it didn't solve the issue, the authority was never removed, or the object ownership was changed but the old owner's authority was never revoked.

Another area in the current configuration to examine is authorization lists. Again, you'll want to determine what problem the authorizations lists were intended to solve and whether that problem (of access control) still exists or will be accommodated with the new security scheme. I find that authorization lists are often attached to application objects and then never maintained. Many times, the current administrators don't even know why they're in place or what problem they were originally meant to address. As you can see, this type of project often results in the clean-up of extraneous private authorities and objects that are no longer used.

Dynamic SQL

Dynamic SQL presents an interesting challenge. For programs containing dynamic SQL, it's not sufficient to set the user profile attribute to *OWNER when compiling or changing the program. This setting causes the program to adopt, but it does not cause the dynamic SQL statement to adopt authority. To get the dynamic SQL statement to adopt, you have to set the DYNUSRPRF (Dynamic user profile) attribute to *OWNER when you compile the program. The only time you can specify this parameter is when the program is compiled.

For programs without dynamic SQL, run the CHGPGM (Change Program) command for all programs in the application libraries to configure the programs to adopt. But because CHGPGM has no dynamic user profile parameter, you'll need to identify the programs that contain dynamic SQL and recompile them before starting your tests. To know which programs to recompile, you must identify the programs containing dynamic SQL — a painful process. IBM provides the PRTSQLINF (Print SQL Information) command, but you have to run this command against a particular program, service program, or SQL package and then examine the resulting spooled file for dynamic SQL statements. You can either do this one at a time for each program in a library or write a CL program that gets the list of all the programs, service programs, and SQL packages in the library and loops through each of them to run the PRTSQLINF command. Then, you need to look through each generated spooled file to find the programs that contain dynamic SQL statements (note that you're looking for dynamic SQL statements, not embedded SQL statements). (This process is so time-consuming and painful that after going through it with one client we added a feature to the SkyView Policy Minder product to perform this analysis for you and generate a report of the programs containing dynamic SQL and the current user profile and dynamic user profile settings.) Once you determine which programs contain dynamic SQL, you need to recompile them, specifying *OWNER for the DYNUSRPRF parameter.

Security Recommendation

To avoid authority problems with programs containing dynamic SQL, consider changing the command default for all CRTSQL*xxx*I commands from DYNUSRPRF(*USER) to DYNUSRPRF(*OWNER).

Decision Points

This type of project involves several decisions, many of which require trade-offs — for example, reduced risk to your data versus less opportunity for organizational growth, or a more secure implementation versus the convenience and maintainability of the application. The following paragraphs discuss some of the decisions — and associated trade-offs — that this type of project requires you to make.

Owning All Application Objects Rather Than Being Authorized

The solution I've described for our sample application scenario assumes that all application objects are owned by the profile ACCTNG. This means that the ACCTNG profile has *ALL authority to all application objects. A slightly more secure method would be to separate ownership. In this approach, all data objects (*FILE, *DTAARA, and so on) are owned by one profile, and another profile owns the rest of the application objects, including *PGM objects. The programs are still configured to adopt. But rather than having the program owner possess all authority to the data (because it owns the files), the program owner is granted only enough authority to the files to run the application. This typically requires at least *CHANGE authority.

However, if the application performs any operations, such as adding or changing a physical file member or deleting a file and replacing it, the program owner will need not just *CHANGE authority to the file but *OBJMGT or *OBJEXIST (in the case of the delete and re-create) as well. Although possible, I've found that this configuration is usually not worth the slight exposure that accompanies it, given the fact that most program owners end up with close to the equivalent of *ALL anyway.

What Adopts

The debate with adopted authority is exactly which application programs adopt. One approach is to have all the application programs adopt authority — that is, have their USRPRF (User profile) attribute set to *OWNER. The concern is that if a programmer or other user with command-line access can call a program, they'll do so with adopted authority.

If this risk is unacceptable, one alternative is to have the "initial" programs adopt. You might think that there is only one initial program — the program that generates the application's menu. That's one of the programs, but the other initial programs are all the programs that are called on a SBMJOB (Submit Job) command or run in batch or through a job scheduler. All these programs must be configured to adopt because they all run in their own unique job; therefore, the adopted authority must be established by the first program called in that job. Finding all these programs can be time-consuming and is often a point of failure on rollout.

A slightly different approach is to have all programs adopt except for the program that generates the application menus. This way, the application's command line does not have adopted authority — something that reduces the risk of exploiting the authority of the application owner by users who have command-line access. (When all programs adopt, including the program that generates the menus, any command line that is on the menu usually also adopts. Therefore, you must ensure that all application users are set to limited capability *YES so they cannot exploit this authority.)

A potential problem with having all programs adopt except the program that generates the menu occurs when menu options are either commands or queries.

Now, this may be the exact effect you want. You may want commands to run with only the users' authority. Commands such as WRKSPLF (Work with Spooled Files) are generally best run with only the users' authority (and no adopted authority). The same issue exists if menu options contain queries. You may want them to run with only the users' authority. However, if you want either a command or a query to run with adopted authority, you'll need to place it in a CL program, have the application owner own the program, and create the program to adopt authority.

Finally, another variation is to have all programs adopt (i.e., have their user profile parameter set to *OWNER) and change their USEADPAUT (Use adopted authority) attribute from the default of *YES to *NO. With this setting, if a user creates a program with the same name as an application program and adds it to a library that's in the library list ahead of the application library, the user can't exploit or use the adopted authority from the previous application program. To completely prevent the use of adopted authority — that is, a program gaining adopted authority from a program higher in the call stack — you create an authorization list, specify it in the QUSEADPAUT (Use Adopted Authority) system value, and set its *PUBLIC authority to *EXCLUDE. Be sure to authorize any profile that is used to create or promote programs that need to set programs' USEADPAUT attribute to *YES.

Authorization Lists

I leave my customers no option but to secure physical and logical files with an authorization list. That's because an authorization list provides a "fail-safe" when rolling out the new security scheme. You can't change the owner or *PUBLIC authority or grant a private authority to a file that's in use. But if the file is secured with an authorization list, you can modify the authorization list. Therefore, if a user who needs direct access to a file is missed during your discovery or a new application is developed that needs access, you can add the access immediately without having to bring down the application so that the files won't be locked and you can modify the authorizations.

*PUBLIC Authority

When I use an authorization list, I also have the *PUBLIC authority of the objects that are secured by the list come from the authorization list. You do this by first securing the object with the authorization list and then specifying *AUTL for the object's *PUBLIC authority. The choice you need to make when securing your files with authorization lists is whether to set the *PUBLIC authority of the authorization lists to *USE or *EXCLUDE. This decision will, of course, depend on the type of data and your organization's security policy requirements.

How Many Files?

When securing only a handful of files, it probably makes sense to use the same authorization list to secure those files. However, when reworking an entire application, I've seen people get paralyzed over trying to decide how many authorization lists to use.

There is not a "right choice" as to how many authorization lists to use. Some will choose to use one authorization list across an entire application, acknowledging that users authorized to the list will have authority to all files secured by the authorization list, including some files to which they really don't need authority. Some will choose to have multiple authorization lists so that users have explicit authority to the files only if there is a business justification to give the authority. Most implementations end up somewhere in between. If you're having trouble deciding how many authorization lists to use, I encourage you to choose an approach and implement it, realizing that this architecture lets you easily change the number of authorization lists in the future. All you have to do is switch which authorization list is securing a particular file.

*PUBLIC Authority

*PUBLIC authority is another area where you have some choices to make in providing secure access. You must consider separately the *PUBLIC authority that should be granted to programs, libraries, and other application objects.

*PUBLIC Authority of Programs

If your programmers ever perform debug operations on programs on your production system, you're going to have to determine how they get authority to do so. Debugging a program requires the programmer to have either *CHANGE to the program or *USE to the program plus *SERVICE special authority. If you don't provide a service tools user ID to the programmer, giving him or her *SERVICE special authority is not as dangerous as it once was. However, if you're not comfortable with this, setting programs and service programs to *PUBLIC *CHANGE may be an option. About the only things you can do when you have *CHANGE authority to a program that *USE doesn't allow is debug the program and change the program's description. Given that end users shouldn't have authority to the debug commands, this alternative might be a viable option.

Securing Libraries

Although is may seem like the obvious thing to set the *PUBLIC authority of a library to *EXCLUDE to keep users out of the library, this setting may not be the best choice from a practical standpoint. If only a handful or small group of users need access to an application or library (e.g., an HR application), setting the libraries that make up this application to *EXCLUDE makes sense. However, if most users on the system use the libraries, it doesn't make sense. That's because all the users end up having authority to the library so that they can use the application. Also, if the libraries are in the users' library list, the users must have authority to the libraries or they can't sign on to the system. So it makes more sense to set the library to *PUBLIC *USE and appropriately secure the objects in the library (i.e., set all application objects — at least the files — to *PUBLIC *EXCLUDE).

*PUBLIC Authority of Other Application Objects

I've discussed the fact that application files' *PUBLIC authority is going to come from authorization lists, and I've talked about the *PUBLIC authority setting for libraries. But what about the rest of the application objects? One question you can ask is whether you're even going to bother with setting the *PUBLIC authority of the rest of the objects since it's really the data in the files that you're trying to protect. Some people will make sure to set the ownership of all application objects properly (so that the adopted authority is assured of working correctly) and leave it at that. My preference is to also address the *PUBLIC authority of the rest of the application objects. Here are some guidelines for you to follow:

- The following objects can generally be set to *PUBLIC *USE:
    - Libraries whose objects the general public can use
    - Directories — DTAAUT(*RX) OBJAUT(*NONE) (for directories containing objects with private data, use DTAAUT(*EXCLUDE) OBJAUT(*NONE))
    - Display files
    - Message files
    - Data areas
- The following objects may need to be *PUBLIC *CHANGE:
    - Libraries in which temporary objects (e.g., queries) are created
    - OUTQs, MSGQs
    - DEVDs

Queries

I've left the matter of queries to the end because this is a difficult area to manage. Why is securing queries so hard? In the case where a user runs a predefined query from a menu option that produces a report, it isn't. The trouble comes when users have the ability to modify the queries and view the files produced by the queries. When either of these activities can occur, that's when the difficulties start.

One of the options of a query is to produce a file rather than a printed report. If the same person always runs the query, things are fine. But if a different person runs the query on Monday morning and another person runs it on Wednesday afternoon, that's typically where problems arise. Why? Because the query is often written to delete and re-create the file produced by the query. Janet owns the file after running the query. If the *PUBLIC authority of the file isn't *ALL, when Alan then goes to run the query, he doesn't have sufficient authority to delete and re-create the file. Compounding the problem is that the query file is often being created into the same file as all production data. That's a problem when your new security architecture has set the production libraries to *PUBLIC *USE or *EXCLUDE.

I highly recommend moving these query files into a separate library from your production data. This approach gives you options for securing the files that you don't have if you leave the files in the same library. (It also allows production libraries to remain secured more appropriately.) I also recommend that you don't lump all your users' query files into the same library. That's because query files are often downloaded to a PC or sent to a server. When queries are downloaded to PCs, the user running a query to gather inventory information is most likely not the same person who runs a query related to your organization's financial data. For that reason, one user shouldn't have the ability to see — or download — the other person's data. If all query files reside in the same library, it becomes much more difficult to secure files to ensure only the appropriate people see the appropriate data. If you direct the query files to be created into separate libraries by area, you can easily allow the accounting group to have access to their own library and set *PUBLIC to *EXCLUDE, for example.

Another reason to separate the queries out of the production data file is because of the authority required to the library. If the file is being deleted and re-created, the users need *CHANGE authority to the library. You typically don't want to have your production data library set to *CHANGE. For query libraries, you can set the *PUBLIC authority to *EXCLUDE (to prevent unauthorized access of the data) and authorize the user or the group with *CHANGE authority.

Here are some methods I've implemented when securing queries:

- Option 1: The query files are owned by the application owner, the query is called from a program that adopts the application owner's authority, and the query specifies to replace the member rather than replace the entire file. This technique doesn't change the ownership of the file itself, and because the query is running with adopted authority of the file owner, there will be no authority issues.
- Option 2: If the users running a query are in the same group profile, consider configuring the users' profiles so that any new objects they create are owned by their group profile. In other words, set the OWNER parameter of the user profile to *GRPPRF. Note: This option will not work if the query is run by users in different group profiles.
- Option 3: Once you've separated the queries into different libraries, you have the option to specify the CRTAUT (Create authority) parameter on the library. In this case, you can specify an authorization list for that parameter so that any new object created into the library is secured with the authorization list. You typically grant the users or groups running the query *ALL authority and set *PUBLIC authority to *EXCLUDE. This option works well when users in different groups need access to the query files.

Making Changes to the Application

Now that you've done all your investigation, you're ready to make changes to the application configuration. Several of the changes you'll be making need to occur when the application objects aren't in use. You'll want to take advantage of down time to make these modifications. Fortunately, most of the changes can be done far in advance of the time when you actually start to lock down the files. Done in the right order, the following steps impose very little risk of affecting application processing when you make these changes ahead of the start of your project.

Step 1: Ownership of objects. The first change you'll want to make is to get all the application objects owned by the right profile. As you looked at the current configuration settings, you may have noticed that profiles other than the desired profile owned application objects. Because you want to have the authority come from adopted authority, it's critical that the program owner own the rest of the application objects.

Step 2: Program adoption. The next step is to change the application programs to adopt authority. You'll need to run the CHGPGM command against all the programs and service programs in your application libraries, specifying *OWNER for the USRPRF parameter.

If you haven't already done so, perform the analysis to determine whether any application programs contain dynamic SQL. Dynamic SQL is becoming more prevalent in the programming community, so this is not a step you should skip. If you have programs that contain dynamic SQL, recompile them and specify *OWNER for the DYNUSRPRF parameter.

Step 3: Create the authorization lists. Next, create the authorization lists that you'll use to secure your application files. I recommend using a naming convention that makes it obvious what each authorization list is being used for. If the naming convention doesn't reflect its purpose, make sure its text description does.

When creating the authorization lists, consider creating them with *PUBLIC authority set to *ALL. Although this setting might be slightly more open that the current application settings, it's not much more so, and it's a temporary setting you can use to ensure that no processes are affected until you're ready.

Step 4: Attach the authorization lists to the file. The next part of the setup that you can perform ahead of your scheduled rollout is to attach the authorization list to the appropriate files. You must perform this step when the application is not active because you can't modify the security attributes of a file that's open. The process involves two steps: First, associate the authorization list with the files. Then, set the *PUBLIC authority of the files to come from the authorization list. Again, assuming that you've created the authorization lists with a wide open security setting rather than the security setting you'll ultimately end up with, this step should not have any effect on your application.

Step 5: Set the *PUBLIC authority of the application objects. Last, you'll want to set the *PUBLIC authority of the rest of the application objects to their appropriate settings.

At this point, you've done nothing that should make any processes break or fail. Although nothing is foolproof, there is very little that could be failing at this point. The data continues to have wide-open access, so processes and users that need direct access should have sufficient authority. In addition, application users are still members of the group that owns the application or they have *ALLOBJ, so they, too, have sufficient authority.

Rolling Out the Changes

As you roll out your changes, order is once again important. If you follow the steps in the order given here, you'll limit which users or processes are affected by the changes, and you'll have a controlled rollout of your new security scheme.

Take the following steps if you're changing an application for which users must be members of the group that owns the application or must be assigned *ALLOBJ special authority.

Step 1: Authorize the outside processes and set the *PUBLIC authority of the authorization lists. Here's where you use Table 16.1. For every entry, grant the profile the required authority to the appropriate authorization list. In our example, you would add the authorities documented in the table *one at a time* to the appropriate authorization list. After adding each authority, you would then have the process tested or make sure the user can still access the data. After adding all the known required authorities to the authorization lists, you're at the point of making a configuration change that could have unknown effects. Up to this point, the only users or processes that could have been affected are the ones added to the authorization lists. The next step is to change the *PUBLIC authority setting of the authorization lists from wide-open to either *USE or *EXCLUDE. At this point, if you have missed any processes or users that need direct access to the application data, these will fail. After making this change, you may want to wait a couple of days before making other changes so that you can see whether anything is going to fail. Limiting the number of changes you make at once will allow you to more easily and quickly debug the failure because you'll have a better idea of where the error is originating.

Step 2: Remove excess special authorities of the owning profile. If you review the current capabilities of the application's owning profile and realize that it has more authorities than appropriate (in the example, the profile currently has *JOBCTL, *SAVSYS, and *SECADM but needs only *JOBCTL), now is the time to remove the excess special authorities. Again, you might want to wait a day or so to see whether there is any fallout from this step before proceeding.

Step 3: Modifying application users' access. It's now time to start removing users' *ALLOBJ authority or removing them from the group that owns the application. In both cases, I move the users into a new group. (This step is done to facilitate other security management practices, such as having the user's group own any objects that the user creates.) I usually find some power users scattered around the organization and move them into their new group (removing their *ALLOBJ if necessary). I tell these power users that some changes have been made, but that everything should work as it did before and to notify me if that's not the case. I let them run in their new configuration for a day or two before removing more users. Once the power users are working correctly, I start to move more and more users into the new group (and remove their *ALLOBJ special authority) until, by the end of the week, all users are in their new configuration.

If the application is using *PUBLIC authority to provide authority to application users, follow this additional step: Before you make any changes — even the setup changes

— make each application user a member of the group that owns the application. This is a temporary assignment, but it's done to ensure that application users aren't affected as you change the application's security configuration and lock down the authorization lists. So make this your very first step in the entire process, and then proceed as documented above.

Changes to Change Management

Don't forget to make sure that any changes you've made to the application's security scheme are reflected in the application's change management settings. For example, if you've changed the public authority settings of some types of application objects from *ALL to *CHANGE, you'll need to modify your change management software to create new objects of these types with the appropriate *PUBLIC authority setting. In addition, you'll want to make sure files continue to be secured with the appropriate authorization list, and so on. Last, you'll want to make sure the adopted authority attributes of programs are specified accurately so that programs are promoted with the correct attributes.

When Something Breaks: Debugging and Recovery Techniques

Hint 1. If you make a change and everything starts to fail and the phone starts to ring off the hook, you have several things you can do to recover and get things working again while you debug the issue. First, you can set the *PUBLIC authority of the authorization lists back to *ALL. Second, you can put the users back in the group that owns the application or give back their *ALLOBJ special authority. The users will have to sign off and sign back on to have the change to their profile take affect, but once they do, the profiles will work as they did before. Either or both of these steps can be done quickly and will stop the failures while you determine and correct the cause.

Hint 2. Use Appendix D of the *System i Security Reference* to determine the authority or special authorities required to perform a command or task.

Hint 3. Don't forget to look in the audit journal to obtain more details about the failure — or to prove that it's not a security issue. Changing the application's security scheme gets a lot of attention. Unfortunately, until the changes are complete and have been in place for a few weeks, people will attribute *every* (and I do mean every!) problem to the security project.

As I roll out the changes, I proactively use the DSPAUDJRNE (Display Audit Journal Entry) command to check the audit journal every 15 minutes or so. Often, I catch the failure before the user calls in to report a problem. I pay special attention to the user profile and object fields in the audit journal entry. Sometimes, the error that the user sees is not the real failure. The audit journal entry will report the actual user profile experiencing the failure and the exact object for which the failure occurred. If I need even more details from the audit journal entry, I look to the far right of the entry on DSPAUDJRNE to find the exact time. Then I use the DSPJRN (Display Journal) command to look for additional details, such as the program running. (For more information about using the audit journal, see Chapter 15.)

Hint 4. If the violation type (using DSPAUDJRNE) or the first character of the entry-specific data (using DSPJRN) is an A, the failure is an authority failure. If the character is a K, the user lacked some special authority. For example, if the first character is K and the object name is CRTUSRPRF (object type *CMD), look for the CRTUSRPRF command in Appendix D of the *System i Security Reference*; you'll see that *SECADM special authority is required to run the command.

Hint 5. If you see a "Not authorized. . ." message in the job log, look at the message details. If the message ID begins with CPF, the error is an authority failure; that is, a user doesn't have sufficient authority to some object. If the error message begins with SQL, the cause of the error is that a program that contains dynamic SQL has not been compiled with its DYNUSRPRF attribute set to *OWNER. In other words, a program has run that contains dynamic SQL, and the user doesn't have sufficient authority to something in the dynamic SQL statement. If the application's security scheme is to adopt authority, the program will need to be recompiled with the DYNUSRPRF attribute set to *OWNER so that the SQL adopts authority.

Making Sure the Changes "Stick"

Suppose a production issue arises and programmers need to promote an emergency fix. This emergency process bypasses change management, and the program ends up being owned by the programmer. How are you going to discover this?

Worse, imagine that a vendor updates its application. In doing so, the vendor replaces the entire application, and all the changes you've so carefully architected, tested, and implemented are gone. You need some method of documenting your changes and being able to re-implement them when things change. You can either use the CL programs that you used to implement the changes the first time or buy a product to do this for you. Be aware, however, that you cannot ignore this issue. Otherwise, you'll be doing this project all over again.

Gotchas

I try to do everything I can to avoid "breakage." This entire chapter is designed to help you do that as well. You can never identify everything before you start making changes. However, here are some of the issues to be sure you watch out for.

- Make the appropriate updates to your change management software and processes to reflect the applications' new security setup. Imagine forgetting this step, getting all application objects changed and the new configuration tested, and then having all new object promotions be done with the old security settings!

- Look in the audit journal before you make any configuration changes. Some AF (authority failure) entries occur "naturally" on your system. No one complains and processes don't fail, even though these failures occur. I've chased a problem for 20 minutes that I, in time, concluded was one of those naturally occurring audit entries and wasn't caused by the application changes that our client and I had just implemented.
- Don't forget to change the *PUBLIC authority of the authorization lists from their wide-open setting. A lot of testing time can be wasted testing with the wrong authority setting!
- Be sure to make the changes in the order described. Perform them in the wrong order, and you'll be affecting all users and all processes all at once. If you've covered everything, you'll be fine, but if not the phone will be ringing off the hook and your inbox will be flooded with incoming problems.
- Identify the processes that are difficult to re-run if they fail, and make sure you test these on a development or quality assurance machine that's configured with the new security scheme before running them with the new security scheme in production. I've seen multistep processes that can easily be re-run if they fail in step 3, but if the procedure gets all the way to step 7, the databases updated to that point have to be hand-corrected. These are the types of processes you'll want to make absolutely certain are tested before your rollout on production.
- Review the authority-checking algorithm before making any changes. When trying to debug why something isn't working correctly, it's imperative that you understand the order in which authorities are checked.
- When changing the security scheme of a vendor application, contact the vendor before making any changes. Some vendors have documents that will help you modify their application. And you want to make sure you won't be violating any license or maintenance agreements by making the changes.

Summary

I realize that this is an intense chapter with a lot of information. If you aren't comfortable or don't feel that you understand it, re-read it until you do before making any changes. Also, know that there are no guarantees. While this methodology has worked for me, I cannot promise it will work for you. But I hope I've provided enough information so that you'll be able to successfully implement an object-level security project of your own.

Chapter 17

Maintaining Compliance

This book has been about implementing IBM i security so that the system's configuration will be in compliance with your organization's security policy. The focus of this chapter is maintaining that configuration, or maintaining compliance. Compliance is not a one-time event. Compliance is a continuous process — or, as I like to describe it, a "lifestyle."

Unfortunately, unless you make an effort to keep your system's configuration in compliance, it's likely to drift out of compliance. How does that happen? You install a new application, and the installation process changes system values. Or, you install a new version of a vendor product and, because the vendor deletes the old libraries and restores the new product version, all the security changes you've made to the application objects are gone. Or, one or more systems are consolidated onto the same LPAR. Or, objects are promoted into production because an emergency fix was required, and now the ownership and object authorities aren't correct. A variety of occurrences can cause your security configuration to drift.

So, how do you maintain a compliance lifestyle? By putting processes in place that regularly check the system's settings. Now, I have to be honest. I designed my company's SkyView Policy Minder product to do this type of checking and automate compliance tasks. So I no longer monitor the compliance of a system "by hand." But I know that some of you would prefer to build your own processes. So let's look at what parts of the system need to be examined regularly.

Evaluating the Key Areas

Three areas of the system are key for maintaining compliance with your organization's security policy. Before we take a look at these areas, here are some reminders:

- The only way you can know you're in compliance is to compare the current settings with the values required by your security policy. If you don't know what these values are supposed to be, there's little point in examining the settings on a regular basis. My point is that you must first determine the appropriate settings — in other words, the values that support or implement your organization's security policy requirements.
- Document your exceptions (that is, write risk acceptance statements). I have never seen a system implemented entirely with best practices settings — there are always exceptions. You'll want to document those exceptions, explaining why the best practices setting can't be achieved and noting any mitigating factors (issues or

situations that cause the current setting to not be a significant risk), along with any compensating controls (other technologies, processes, or settings in place that reduce or eliminate the risk of the current setting). This documentation is useful in at least two situations:

- For auditors — Documenting your risk acceptance means that you have evaluated the situation and have an understanding of the risk the setting poses to your organization. In other words, you've thought through the situation.
- For future administrators or security officers — Documented risk acceptance statements provide important information for future employees or consultants when they examine the security configuration of the system. From experience, I can tell you that it's very helpful to know why a specific setting was chosen. With this information, you know what needs to be accommodated, if you're considering changing the value.

- Throughout this book, I've espoused best practices. If you're looking for guidance, see the chapters in which these topics have been discussed in detail to find the recommended best practices setting.

Now, let's take a look at what you'll want to examine on a regular basis — in other words, what you'll want to check to maintain a compliance lifestyle.

System Values

Although you may not want to look at *all* system values, you at least want to examine the security-relevant system values that I discussed back in Chapter 3 to make sure they don't change. There's a variety of ways that you can examine system value settings, the easiest being to run the DSPSYSVAL (Display System Value) command and compare the current settings to those required by your policy. If you want to guarantee that no one other than those with a service tools user ID can change system values, consider using the system value "lock-down" feature I described near the end of Chapter 3.

User Profiles

I've found that unless you have automated processes in place, user profiles are one of the first areas of the system to drift out of compliance. When examining user profiles for compliance, you'll want to look at several things. The first issue to look for is the configuration of the user profiles as they get created. In other words, as new profiles are created, do they have the correct special authorities, group assignment, initial program, initial menu, and limited-capability setting? Profiles are often created by copying an existing user's profile. Creating profiles with this method often results in profiles being created with excess special authorities and group assignments.

Next, look at the attributes of the user profile that could be issues during an audit — things such as profiles with default passwords, profiles with password expiration intervals set to *NOMAX (in other words, the users never have to change their passwords), and profiles with the limited-capability attribute set to *NO. In addition, auditors usually want you to keep a list of the users who have been assigned a special authority — especially those with *ALLOBJ — and to be aware if that list changes. Another list to keep track of is the users who have been made members of QSECOFR. Checking these attributes should not be an event that happens the week or two before the auditor appears. Not only are these attributes relevant to an audit; they're also relevant to the security of your systems. The easiest way to list the attributes of all profiles is to run the DSPUSRPRF *ALL command and send the results to an outfile, which you can then query or download to a spreadsheet for further examination.

Table 17. 1 lists some commands you can use to examine user profiles.

Table 17.1
Commands for Examining User Profiles

Command	Description
ANZDFTPWD (Analyze Default Passwords)	Lists the profiles with default passwords
DSPAUTUSR (Display Authorized Users)	Prints a report of all users and the groups they belong to (if any) or a report of all groups and their members
DSPUSRPRF (Display User Profile)	Can display an individual user profile or can send the details of all user profiles to an outfile
PRTUSRPRF (Print User Profile)	Sends the details of all user profiles to a spooled file

The user profile issue that I see affecting compliance most quickly is that of inactive profiles. Many laws and regulations require profiles to be disabled or removed from the system on a timely basis. For example, PCI DSS requires any inactive profile to be disabled or removed after 90 days of inactivity.

But aside from being a compliance issue, inactive profiles are a significant security issue. To leave profiles on the system in a configuration such that they are usable is opening your organization up for trouble. Many breaches have occurred by exploiting the accounts of individuals who have been dismissed from the company or who no longer need access. If you decide to manage inactive profiles with your own automated process, make sure you're looking at the right dates associated with the profile. You'll need to examine the profile's creation, last restore, and last used dates, with the last used date being the critical date. These dates are obtained by running the DSPOBJD (Display Object Description) command — not the DSPUSRPRF (Display User Profile) command. Many people have gotten "burned" by designing a process around the profile's last sign-on date (which is the date shown when you run DSPUSRPRF). The problem with this approach is that the last sign-on date is updated only when the profile signs on to the system via a 5250 sign-on screen. It is not updated when the profile is used only to run batch jobs or to sign on via a DDM, FTP, or ODBC connection; however, the last used date is updated in these cases.

IBM profiles are another area you need to monitor. I often see special authorities added to or removed from IBM profiles. Removing a special authority is typically not a problem, but adding a special authority (e.g., adding *ALLOBJ to QUSER) can create a significant exposure. Another thing to check is to ensure all IBM-supplied profiles (with the exception of QSECOFR) remain set at PASSWORD(*NONE). (All IBM profiles other than QSECOFR are shipped without a password.) You can check both of these items by looking through the report created by the PRTUSRPRF (Print User Profile) command.

Object Authority

Which object authorities you need to examine will depend on your security policy. Perhaps your examination can be limited to only a handful of files, such as your employee master database or the files containing cardholder (PCI) data. Or, it's possible you need to regularly check the authorities to files containing private data. Or, you may need to evaluate the ownership, adopted authority attributes, authorization list association, *PUBLIC authority, and private authorities associated with all objects in an entire application. Regardless of the scope of the requirement, having an automated process to evaluate whether the objects' security attributes are set correctly is key to compliance. I have seen well-meaning administrators try to keep objects secured appropriately, but without an automated process, other projects took priority and the authorities they meant to examine every Monday morning were forgotten or reprioritized. As an administrator, you have way too much to do to check these settings "by hand."

In addition to libraries and database files, make sure you don't neglect the directories and objects in the IFS. There are usually a handful of directories that need to be secured in a specific way so that private data is not exposed. You should examine these directories' security attributes along with those of the libraries and their objects.

Table 17.2 lists some commands you can use to examine object authorities.

Table 17.2

Commands for Examining Object Authorities

Command	Description
DSPOBJAUT (Display Object Authority)	Displays an object's owner, owner's authority, primary group and its authority, authorization list assignment, *PUBLIC authority setting, and any private authorities assigned to the object. You can specify to display the authorities of all objects of a particular type or all objects in a particular library and send the results to an outfile. You can then compare these results with the known or previously approved settings.
DSPAUT (Display Authority)	Displays the same attributes as DSPOBJAUT except for objects named using a path name (e.g., '/home/cjw').
PRTPUBAUT (Print Public Authority)	Prints a report of the *PUBLIC authority of all objects of the specified type.
PRTPVTAUT (Print Private Authorities)	Prints a report of the private authorities assigned to the objects of the specified type.
DSPAUTL (Display Authorization List)	Displays the users authorized to the authorization list.
DSPAUTLOBJ (Display Authorization List Objects)	Displays the objects secured by the authorization list.

An Annual Security Assessment

Just as we all should go to the doctor for an annual physical, our systems need an annual checkup as well. Several laws and regulations require a comprehensive security assessment. In addition to being a compliance requirement, it's a wise thing to do. An assessment is significantly broader than the compliance checking described earlier in this chapter. A security assessment looks for vulnerabilities across the system's configuration. Aside from the compliance requirements, why perform a security assessment? Because changes to your systems' configuration may have opened up vulnerabilities. These changes can occur when a new application is installed, the operating system is upgraded, systems are consolidated, or companies are merged.

Our SkyView Risk Assessor product performs an automated security assessment, so I no longer do this work "by hand." But if you're going to do the assessment yourself, I suggest you examine at least the following areas:

- All security-relevant system values (making sure to check for new system values after an operating system upgrade)
- User profiles settings (checking for extraneous private authorities, changes to IBM profiles, making IBM profiles groups, IBM profiles with passwords, and so on, in addition to the configuration items noted above)
- Service tools user IDs (making sure the passwords of all service tools user IDs, including IBM-supplied IDs, are changed regularly)
- Object authorities (including the create authority of libraries and the auditing values of both libraries and directory objects)
- File shares (making sure the appropriate directories and libraries are shared and no others)
- TCP/IP settings (checking auto-start and time-out values)
- Navigator for i settings (using Application Administration to limit what parts of Navigator and System i Access users are allowed to see on their desktops)
- NetServer settings (making sure no guest profile is assigned)
- Adopted authorities (especially examining programs and service programs that adopt an *ALLOBJ owner)
- Programs registered for exit points (making sure no programs have been removed or added)
- Commands that limited-capability users can run
- Job descriptions
- Subsystem descriptions
- Output queue and job queue descriptions
- Distributed Data Management (DDM) settings
- Web server configurations

You'll also want to run the CHKOBJITG (Check Object Integrity) command on at least an annual, if not a quarterly, basis. This command performs a "self-check" of the operating system. It will identify any altered operating system object as well as user-written programs that have been modified to run in system state. At the very least, you'll want to run this command for the objects owned by QSYS (if not all profiles on the system). The command can be CPU-consuming, however, so you'll want to submit it and run it in a low-priority subsystem so it doesn't affect other processing.

Regular Reviews

Even when items are known to be in compliance — that is, they match your security policy requirements — some areas of the system should be examined regularly to determine whether the settings are still appropriate. Just as you need to review your security policy to see whether it should be modified for changing business situations, so too do you need to review the following areas: group profile membership, special authority assignments, authorities to authorization lists, and the objects that authorization lists secure. Reviewing these items will ensure that your systems' configuration is in compliance with the requirement of "least privilege access" — that is, that users have only the capabilities and access required to perform their job functions. While an administrator or a security officer can facilitate these reviews, they typically should be performed by someone else because these roles may not be aware of changes that would cause these areas or assignments to be modified or removed. I recommend that these areas be reviewed at least annually, if not quarterly.

Group Profile Membership and Special Authority Assignments

If you have implemented role-based access (as described in Chapter 13), this review is fairly easy. Just print the configuration of the group profile (DSPUSRPRF GRP_NAME) to list the special authorities associated with the group. Then, print the members of the group (DSPUSRPRF GRP_NAME TYPE(*GRPMBR)). Give the appropriate manager a description or characterization of the types of functions allowed by each special authority assigned to the role (if any), along with the list of the users in the role. The manager should be able to quickly discern whether the list continues to be appropriate or adjustments need to be made. This is typically where you find out that users have changed positions within the organization. These users will often remain in their old group (role) until a review such as this is performed.

If you haven't implemented role-based access, you'll need to print every user's special authorities and group assignments and give all this information to each user's manager. This review is significantly more time-consuming for the manager and is far less likely to result in an accurate or comprehensive review (due to the overwhelming amount of information presented to each manager).

Authorization Lists

Although it's good to know that no additional authorities have been granted to an authorization list (in other words, it is in compliance), the present authorities need to be reviewed to ensure they are still appropriate for the current business needs. To perform this review, print the users authorized to the list (using the DSPAUTL, or Display Authorization List, command). You may also want to print the objects associated with the list (using the DSPAUTLOBJ, or Display Authorization List Objects, command). Present this information to the owner of the data that the authorization list secures. For example, if there's an authorization list that secures all HR data, give the list to the head of HR. It's likely the HR head will delegate the task of reviewing the users authorized to this information, and it's also likely that you (as an administrator) may have to explain the types of actions the authorities allow the user (for example, *USE permits viewing and downloading of data). But this is one way to get data owners to gain an understanding of the types of actions that can occur when they allow access to their data. After this, they tend to take the responsibility for approving users' access to the data more seriously.

Policies and Processes

Just as checking the compliance status of your system values, user profiles, and object authorities is not a one-time event, writing a security policy is not a one-time event. As I mentioned in Chapter 2, a security policy needs to be updated at least annually. And just like your security policy, your processes need to be up-to-date. Auditors will often take a process that implements a part of your organization's policy and follow it through to ensure it's being complied with. If what the auditor observes is different from the written policy, the auditor will know that something is wrong. However, the auditor won't know if the problem is out-of-date documentation or a person who is ignoring the written process; therefore, the matter will typically be written up as an audit finding. Here's the list of processes that I recommend you review and update at least annually:

- *Backup strategy* — Make sure you're backing up all pieces of your application as well as audit journal receivers and any other information needed for compliance.
- *Disaster recovery* — While your backup process is part of your disaster recovery documentation, your disaster recovery process is far more comprehensive, takes significant planning, and should be tested at least annually.
- *Patch strategy* — In IBM i terms, this is your strategy for obtaining and applying PTFs.
- *Change management* — This process should include the process of promoting application changes, enhancements, and fixes into production.
- *User account management*:
    - *Creation* — Documentation for this process should include the original request for access, proof that the request was approved, and confirmation that the request was fulfilled.

- *Removal* — This process documents the steps taken when you remove an account from the organization's system(s) and network. The process should have different steps, depending on why the account is being removed. For example, if the person has been let go (regardless of the reason), the steps taken will probably be implemented with a greater sense of urgency than if the person has retired or has taken an extended leave.
- *Requests for additional access* — This process documents the steps to be taken when users need to request access or capabilities in addition to their current settings.

Summary

While I highly recommend an automated approach to maintaining compliance, the important message in this chapter is that compliance is not a one-time event. Regardless of the method you use, it's important to verify compliance on a regular basis; otherwise, it's guaranteed that your system will fall out of compliance. In addition, both your organization's security policy and processes need to be reviewed for potential updates on an annual basis — more frequently with the addition of new technology, consolidation of systems, merging of companies, and so on.

Chapter 18

Preparing for the Worst: Creating a Security Incident Response Plan

No one wants to think about the fact that a breach could occur. In fact, most people are in total denial about the possibility. They use the excuse that their system is "secure" — never mind that the system is "secure-able" and that you actually have to use the features provided by IBM to make it secure. Or, they say they trust their employees and fellow workers. Are you telling me that you're willing to vouch for every employee, contractor, and vendor who has access to the system and that you'll guarantee they won't use the data available to them in an inappropriate way? If you're willing to guarantee this from now to eternity, then you don't have to read the rest of this chapter.

Still reading? That's what I thought, because when they're willing to let their minds go there, most people admit they need some help putting together a plan should the unthinkable happen: a breach that affects the organization or your IBM i system.

Be Prepared

Many laws and regulations (including PCI DSS) require that a section of your security policy address the need for a security incident response plan. But it simply makes good business sense to have plans in place in case a breach occurs. I've seen someone live through the aftermath of a significant breach. Even though plans were in place before the breach, the event was brutal — from the need to immediately lock down certain parts of the organization to dealing with various legal entities, law enforcement officials, the Federal Trade Commission (or FTC, the branch of the U.S. government that investigates breaches), and the board of directors to not sleeping or catching only naps for several weeks. Things would only have been worse had this organization not had a security incident response plan. Breaches can and do happen. It's best to be prepared.

Assembling the Incident Response Team

The first step is to assemble an incident response team. You'll need to include individuals that represent the following areas: information security, internal audit, risk management, fraud detection, and administrators from all system types — firewall, servers, IBM i, AIX, and so on. You'll also need to include selected members of management and perhaps the data owners. Appoint a team leader who is cool under pressure and can bring order in the midst of chaos. In fact, it's best to choose people who work well under pressure and can work effectively as a team.

Depending on the size of your organization, one person may fulfill several of the roles. That's not a problem as long as, in the case of an event, that person is physically able to fulfill all the roles in the stress-filled situation of investigating and handling an incident. Larger organizations may already have this team formed to some degree. Many larger organizations proactively monitor for network intruders or virus outbreaks. They are most familiar with responding to network-based attacks. They may not have thought to (nor may they be able to) address the theft or loss of data.

You'll want to seriously consider retaining a firm that specializes in forensics and investigating breaches. Having this contract in place prior to a breach will ensure the investigation starts as soon after discovering the breach as possible. And you'll be assured the investigation will be conducted in a thorough and legally submittable manner.

Two other critical members of the team are a representative from your organization's legal counsel and a public relations/media contact. Legal counsel is required to ensure that the company is complying with the law, as well as to protect the company and its contracts with customers and vendors. (In other words, you need to take service level agreements, or SLAs, into account if services must be stopped or access denied.) The media contact is vital if you have to make a public announcement about the breach. Depending on where your organization does business, many U.S. states and European countries require organizations to notify individuals if their private data or health information has been lost or stolen. Therefore, if this data has been breached, it's likely that a public announcement and individual notification will be required. This type of notification could significantly affect your organization's reputation and image; therefore, the situation must be handled carefully and professionally. The media representative may not be the person who actually appears in front of the media, but he or she would be the one to field all media inquiries, write any statement released to the press, and prepare the person who makes the public statements.

Responding to an Incident

One important aspect to responding to an incident is being aware of the laws and regulations with which your organization must comply when data is lost or stolen. For example, do you have to notify the individual of the loss? If the lost information was cardholder data, do you know who to contact? You might need to contact law enforcement, other government entities (e.g., the FTC in the United States if cardholder data was stolen), the bank your organization uses for processing credit cards, and/or the credit-card companies themselves. If private data or health information was breached, the individuals most likely need to be notified as well.

Another aspect to consider is where and how you start to investigate the incident. To determine which members of the incident response team should be notified and assembled, you'll need to make an initial determination of the type of incident. For example, if the incident is a virus, worm, or network breach, the team members you assemble will differ from those assembled if the IBM i audit journal shows that a programmer inappropriately accessed the routines that decrypt credit-card information.

Obviously, the media won't have to be notified if the incident is a virus, assuming that it is quickly contained and no loss of service to your customers occurs. However, if a database containing private data is thought to have been breached, everyone on the team will need to be gathered and previously defined action plans put into motion.

Last, it's important to assign the breach an "owner" — a point person who manages the overall incident response process and ensures nothing falls through the cracks.

Data Preservation

Critical to an incident investigation is data preservation. How you preserve data depends on the type of computer. For a PC, tools exist that let you obtain a mirror image of the contents of the PC's hard drive without affecting the files' dates. However, taking a "picture" of IBM i necessitates a system save and may mean bringing your organization's business to a screeching halt, kicking all users off the system, and going into restricted state. Before taking this step, however, you'll want to talk with your legal counsel to determine whether doing this will violate any service level agreements you might have with any of your customers or trading partners.

Because data preservation is critical when investigating breaches, it's best to gain an understanding of exactly what is required before a breach occurs and what will stand up in a court of law should your organization press charges or should charges be levied against your organization. For example, once a system save has been performed, it is critical to establish a "chain of possession" for the media containing the save. You don't want to leave the media lying around the machine room unattended because this might cause it to be inadmissible in a court of law. The best place to get the exact details about what data and system configuration information is required to perform a criminal investigation is the computer crime division of your local law enforcement office.

Performing the Investigation

IBM i provides many resources to aid in the investigation of an incident. Let's assume for the moment that data was inappropriately taken from your system. At a minimum, the resources you'll want to have available include job logs, the history log, journal receivers for both database files and the IBM i audit journal, and exit point logs if you've written or purchased exit programs that log network traffic. Because these sources of historical information are often cleared, deleted, or removed from the system during normal system maintenance, as part of planning for an incident you'll want to re-evaluate your processes to determine whether you need to alter your save strategy to preserve a copy of this information before performing system cleanup. The more complete a picture you can reconstruct, the more thorough the investigation can be.

Be Proactive

Before an incident occurs, assemble the team and role-play various incidents. Then create the policy, and document the incident reporting processes. It's likely to take several meetings before you've defined processes well enough to be followed.

Make Sure You're Saving the Right Information

Whether you're investigating a breach and need to reconstruct the system or you are in disaster recovery mode and need to restore your security information at a hot site or on a backup system, it's critical that you are saving the right pieces of information. The rest of this chapter focuses on saving and being able to accurately restore security data.

Saving Security Data

First, let me describe what I mean by "security data." Then I'll discuss how and when this data is saved. The security data that I'm about to discuss includes user profiles, passwords, private authorities, and authorization lists. How do you save this data? Either by running the SAVSECDTA (Save Security Data) command or by doing a SAVSYS (Save System). I'll also address the need for a save strategy for audit journal receivers.

How Often Should You Save Security Data?

How often you save your security data depends on how dynamic your system is. If you work for an organization that daily creates or changes numerous user profiles, grants new authorities to individuals or group profiles, or alters the authority to authorization lists, you probably want to save your security data on a nightly basis. If you are in the process of remediating your security architecture, you may want to save nightly or at least weekly to make sure you capture all the changes you're making. If your environment is not that dynamic and you have the "paper" trail of all requests for changes so that you could redo any changes not included in the save, you can consider saving this data less frequently. However, the overriding decision for save frequency should be made with the thought of rebuilding your system in the case of a breach.

How Often Should You Save Audit Journal Receivers?

I've found that, in most shops, the saving of audit journal receivers is not well-planned. In fact, some organizations remove them from the system without saving them at all. From a compliance point of view, not saving audit journal receivers may put you out of compliance. For example, PCI DSS requires that audit information be retained on your system for 90 days or at least be easily retrieved. In addition, audit information must be retained for at least one year. Some industries, such as the finance industry, have data retention requirements that may cause you to have to retain audit journal receivers for seven years. Given your current save strategy for audit journal receivers, would you be in compliance with the laws and regulations that govern your organization?

When you consider your save strategy, consider how easy or difficult it would be to retrieve a particular day's worth of audit information. Some organizations find it easiest to generate a new receiver and then save all the journal receivers from the previous day until the time of the save. That way, you know you have all the information for a particular day. Some organizations save their audit journal receivers on the same media as the rest of their save information. Others save their audit journal receivers on separate media. There is no "right way." The important thing is to make sure you're saving the journal receivers in a way that meets compliance requirements and lets you easily restore them should you need to investigate an issue.

Restoring Security Data

Now, let's turn our attention to the restoration of the security data.

Order Is Important

As with many processes, it's important to get the order right when restoring security information. If you're rebuilding a system to investigate a breach or rebuilding a system because you're in disaster recovery mode, the last thing you want to do is have to start the process over again because you didn't follow the directions. You *must* follow the instructions in the *System i Systems Management: Recovering Your System* manual (available in the Information Center under "Systems management|Backup and recovery|Recovering your system"). *Do not* deviate from the order described in this manual! In general, the manual will have you restore your security information using the RSTUSRPRF (Restore User Profile) command (RSTUSRPRF *ALL), then restore your objects, and then perform a RSTAUT (Restore Authority) operation. Restoring the user profiles first will make sure all the objects' owning profiles and authorization lists exist before the objects are restored to the system.

Performing the RSTUSRPRF before restoring your objects is critical. If you restore the objects before the user profiles, all objects owned by the non-IBM profiles will be owned by QDFTOWN. If you do these steps in reverse order, you'll have to start over. Once the ownership of the objects is established, it will stay that way unless you delete an object and restore it again — this time with all the profiles on the system before the objects are restored. It's a painful process, but it's the only way to get ownership corrected.

Restoring user profiles first also means that any authorization lists that secure the objects exist on the system. When you restore an object that is secured by an authorization list on the media but the authorization list doesn't exist on the system, IBM i changes the object so that it is no longer secured by an authorization list (i.e., it sets the authorization list attribute to *NONE) and changes *PUBLIC authority to *EXCLUDE. These may seem like harsh configuration changes, but to do anything else has the potential to open data up to public access when it was supposed to be tightly secured. Note: The only way to restore an authorization list is to restore all user profiles (RSTUSRPRF *ALL).

Once you've restored your user profiles (and authorization lists), you can restore your objects. Then, once your objects are restored, you'll want to perform one final step: Run the RSTAUT command for all profiles (RSTAUT *ALL). When user profiles are restored to the system, a temporary object is built that contains the profiles' private authorities. After the objects are restored and you run RSTAUT, IBM i reads that temporary object and grants all the private authorities that the user profile had to objects and authorization lists. If you don't perform this step, when you run the DSPUSRPRF (Display User Profile) command, you'll receive the message "CPD2206 Private authorities not restored," which is an indication that RSTAUT hasn't been run. Running RSTAUT is an important step that should not be overlooked. Skipping this step means that some users may not have the authority they need to certain objects to perform their job function. Or, users may be able to access objects you don't intend for them to access.

Security Data Restored with an Object

If you take a close look at the security data saved with SAVSECDTA and SAVSYS, you may realize that there are some other pieces of security data that we haven't talked about. Some security data is stored in an object itself and therefore is restored to the system when you restore the object. The following information is stored with an object:

- The name of the object's owner
- The owner's authority
- *PUBLIC authority
- The name of the object's primary group
- The primary group's authority
- The authorization list name
- The object's object audit setting

Seeing this list, you can probably understand why it is critical to have the profiles and authorization lists restored before restoring an object. Because the object keeps the names of the owner and authorization list (if the object is secured with an authorization list), those items need to be on the system when the object is restored.

Re-Creating the System After a Breach

Let's get back to the original topic of this chapter: breaches. Now that you know how restoring security data works, let's talk about re-creating the system after a breach.

If your organization suffers a breach, you may need to rebuild the system. Now that you've seen all the security parts that are required and how restoring this data works, you will want to consider what other "pieces" are required to reconstruct the entire picture. As I said earlier, you may need to be able to trace the activity through history logs, Web server logs, job logs, spooled files, and journal receivers associated with database files. The more completely you can re-create the system, the better. But this is one of those trade-off situations. Many of you don't save job logs, the history log, or spooled files so that your backup doesn't get to be huge and take forever to complete. This is one of the issues you'll want to discuss with your security incident response team to determine whether the business requirement of reducing the backup meets the recovery requirements of this team.

Chapter 19

Creating a Security Awareness Program

A security awareness program is an official or recognized project, typically under the "jurisdiction" of the Chief Security Officer (CSO), that is created for the purpose of educating individuals such as the organization's employees, contractors, vendors, and volunteers about the organization's security policies. The education should explain how the policies relate to each individual, and it should be performed on an ongoing basis.

A security awareness program is a requirement of many laws and regulations. But, beyond that, it simply makes good business sense to develop a security awareness program. By training your organization, you are mobilizing your entire workforce to help fight the security compliance battle and help you wage the war against inappropriate use of data.

Security awareness training typically starts the day an employee is hired, a contractor begins a project, or a volunteer starts. This is when the employee portion of your organization's security policy is explained and the person is required to read and sign it, acknowledging that he or she understands its contents and meaning. Although many organizations take advantage of this initial opportunity, that's often where the security awareness training stops. Everyone in the organization needs to develop a security "lifestyle." To ensure this happens, security awareness training should take place on a regular basis so that people receive periodic reminders of the organization's security requirements. They should also receive the appropriate training if they change positions and the requirements differ from the previous position.

Another reason for ongoing education is because new threats arise and new technology is developed. For example, when you first started working with e-mail, did you have to worry about users downloading and subsequently leaving confidential or private information on their mobile devices? Did you have to concern yourself with whether people were blogging during work hours or inappropriately posting statements as representatives of the organization? Without an ongoing security awareness program, you have no effective way of communicating the requirements of new or updated policies.

Another reason for periodic education is to update those in the organization about new or changed laws or regulations that affect the organization. For example, the purpose of the ID Theft Red Flag law (part of FACTA, the U.S. Fair and Accurate Credit Transactions Act of 2003) is fraud prevention. Much of the law talks about detecting different aspects of fraud, such as users reporting one address on a job application and the credit agencies reporting another address. The law is trying to ensure that organizations detect this type of situation and do something to investigate the discrepancy. While a lot can be done for fraud detection using technology, there's a lot that can be discovered by people simply paying attention to the details and questioning the situation when something doesn't look "right." But without the knowledge of what to look for and what to do if fraud is suspected, this opportunity for protecting your organization will be lost. Educating your workforce is key in helping detect fraud and other situations that can damage or otherwise put your organization out of compliance.

What Method Do I Use to Communicate?

Because we don't all learn in the same way, your security awareness program should include all types of learning. Giving security awareness training and tips using audio, graphics, and text will provide opportunities for everyone to absorb the information.

I've seen some of our clients use posters to communicate a "security slogan" of the month. The posters are often of high quality, featuring professional-looking images that capture the eye and draw you into the message. Other clients have used a "lighter" approach. Their posters feature a cartoon conveying the message of the month through humor. In both cases, the companies obviously know their "audience" and the type of messaging their workforce responds to. Because messages get "stale" and people start to ignore them, most organizations change their security message about once a month.

For the auditory learner, you can have podcasts with 30-second snippets of the latest security tip available from your organization's intranet home page. This method works well in the high-tech industry. For more "traditional" organizations, provide material for managers to present at their next department or team meeting.

As for the folks who read everything that comes their way, you can send out a monthly newsletter. Another approach is to change the network or IBM i sign-on screen to reflect the current month's security emphasis.

Finally, it's a good practice to have each employee annually review and re-sign the employee security policy. To accomplish this, some organizations use a secure Web site on their intranet where employees can read the policy and then digitally sign to indicate that the policy was reviewed. I've also seen organizations require management to present the policy to each employee individually or to the entire department during a department meeting. Regardless of the method, I highly recommend that you obtain some type of proof indicating that employees received the education and understand the updates and new requirements.

Getting Started

If you need help getting your security awareness program started, look to the Internet for help. There are companies that provide security awareness training packages — posters, videos, pamphlets, monthly security themes, and so on. In addition, organizations that focus on training for security professionals often have online security awareness training courses that you can buy. Both provide good places to start.

Whether you use one of these resources or create your own program, I encourage you to make your training "real." Provide scenarios that are meaningful to your organization so that the workforce can more easily relate to the concepts being explained. In addition, make sure you provide the right training for your audience. Talking about data classification and proper disposal of tapes containing private data will make the manufacturing worker's eyes glaze over! However, such employees do need to understand that the use of the Internet during their breaks should be for business purposes only — for example, to submit benefit claims or check out the company's internal bulletin board.

In addition to addressing topics that the entire organization will benefit from, it's important not to ignore the education aspects that should be targeted to a specific audience. For example, I believe it's vital to educate programmers and developers about the current and upcoming laws and regulations that affect your organization's data. Engaging these individuals as soon as you're aware of a new compliance requirement lets them better plan for application changes. In addition, if you see trends — such as the potential for Social Security numbers to be encrypted — you should warn them that this requirement might be coming. When equipped with this type of information, your programmers and developers can design their architecture to accommodate the requirement should it become law. In addition, engaging them early has the potential to make them feel like part of the information security team rather than having requirements foisted upon them without any input or warning.

If you want to begin a security awareness program and need a list of topics to get started, here are a few ideas:

- A reminder to keep confidential data confidential — even between co-workers (and even if they're good friends) when there is no business reason for another person to "need to know" the information.
- A reminder that family members are not employees and confidential data should not be shared with them.
- Reminders about what constitutes the "appropriate use of systems." For example, an organization's resources are not to be used to play games, gamble, or store personal information or data such as photos, music, or videos.
- Tips about how to recognize the signs of a security incident (or the potential for a security incident) and who to call if one is suspected.
- Information about how to recognize fraud and who to call if it's suspected.
- A reminder not to open e-mail attachments unless they are expected.
- A reminder about appropriate use of the Internet while at work (along with a reminder that user actions are logged).
- Education about blogging and the requirements (if any) of adding a disclaimer so that it doesn't look like an employee is speaking for the organization.

- Laptop security tips.
- A reminder about appropriate disposal methods for printed reports or media that contain private or confidential data.
- A reminder that employees should wear badges at all times while on company property.
- A reminder or instructions about invoking a screen saver when employees leave their desks.
- Tips addressing safety and security when traveling.
- Details about what type of data is allowed (or not allowed) to be stored on mobile devices.
- An explanation of what constitutes spam, what the organization is doing to control spam, and what action users should take if spam still ends up in their inboxes.

Index

#

11111111 service tools user ID, 93, 96, 104
22222222 service tools user ID, 93, 95, 96

A

Access
- application-only, 214–215
- control, 5–6
- data, 11
- least privilege, 6, 203, 223, 290
- menu, 212–214
- object, in IFS, 146–148
- production, 235
- role-based, 223–228
- SST panel, 97
- user, modifying, 280–281
- user profile, 37

*ADD authority, 81, 114
ADDSVRAUTE (Add Server Authentication Entry) command, 183, 184, 185
Adopted authority
- concepts, 127
- defined, 127
- example, 128–129
- guidelines, 129
- IFS and, 145–146
- program attributes controlling, 128
- security recommendation, 129
- *See also* Authorities

*ALL authority, 274
*ALLOBJ special authority
- assignment, 72
- consultants and, 238
- defined, 72
- encryption keys and, 169
- guidelines, 2
- menu access control, 214
- profiles with/without, 236
- red flags, 6
- security level 20, 20
- security recommendation, 73
- system/security administrators and, 236–237
- vendor application requirement, 219
- vendor use of, 6, 237
- *See also* Authorities; Special authorities

ALWOBJDIF (Allow Object Differences) command, 23
Annual security assessment, 289–290
Anonymous FTP, 179
ANZDFTPWD (Analyze Default Passwords) command, 80, 287
ANZPRFACT (Analyze Profile Activity) command, 80
Application Administration window (Navigator for i), 134
Application design, security policy, 15
Application programming interfaces (APIs), 22
- GID, 215–216
- IFS, 149
- profile-swap, 215
- profile-token, 215
- UID, 215–216
- validation list, 221

Application-only access, 214–215
Applications
- *PUBLIC authority definition, 219
- application ownership, 216
- audit trail requirement, 218
- authorization schemes, 209, 212–216
- change management settings, 281
- changing, 278–279
- client/server, 217
- design considerations, 210–218
- design template questions, 211
- IBM i implementation, 218–221
- job descriptions, 219
- library list management, 219–221
- library-qualified calls, 221
- logging, 217
- ownership, 216
- password storage, 221
- profile requirement, 217
- profile running, 217
- questions, 210
- role usage, 211–212
- security principles, 209
- testing, 221–222
- user access, modifying, 280–281
- WebSphere, 217

Asymmetric key algorithms, 164–165
Audit journal
- change management and, 283
- creating, 243
- defined, 242
- for failure details, 281
- receiver architecture, 264
- receiver contents determination, 264
- receiver save frequency, 296–297
- reporting on activities from, 263–264
- working with, 253–256

Audit journal entries
- deletion of objects (DO), 240
- displaying (DSPAUDJRNE), 257–258
- displaying (DSPJRN), 258, 259–262
- examining, 253–254
- formats, 254–256
- printing (CPYAUDJRNE), 262
- QAUDLVL values, 58–60
- regular viewing of, 263
- specifications, 254
- types, 254–256

Audit Policy Properties dialog box (Navigator for i), 266
Audit-related system values, 56–62
- defined, 56
- QAUDCTL, 56, 57, 243
- QAUDENDACN, 56, 57
- QAUDFRCLVL, 56, 58
- QAUDLVL, 56, 58–60
- QAUDLVL2, 56, 61
- QCRTOBJAUD, 56, 61–62, 252
- summary, 56
- *See also* System values

Auditing, 239–266
  Audit Policy Properties dialog box, 266
  configuring, 243
  controls, 243–244
  controls, security recommendations, 253
  force level, 58
  history log, 240–242
  initial configuration, 60
  level, 58–60
  level extension, 61
  Navigator for i, 265–266
  need for, 239–240
  objects, 249–252, 253
  objects (IFS), 146
  objects, new, 61–62
  on/off switch, 57
  recommendations, 252–253
  security audit journal, 242–243
  system values, 56–62
  system-wide, 245–247
  tools, 265
  unable to continue, 57
  user, 247–248, 253
  user-specific, 251
Audits
  changes, 59–60
  system management, 60
  violations, 59
AUDLVL (User action auditing)
  user profile attribute, 245
  defined, 248
  values, 245–246
Authorities
  *ADD, 81, 114
  *ALL, 274
  *AUTLMGT, 113
  *CHANGE, 81, 121, 130, 274
  *DLT, 81, 114
  *EXCLUDE, 116, 121
  *EXECUTE, 81, 114
  *OBJALTER, 113
  *OBJEXIST, 113
  *OBJMGT, 81, 112
  *OBJOPR, 81, 112
  *OBJREF, 113
  *PUBLIC, 15, 79, 117–121, 124–125, 276–277
  *READ, 81, 114
  *UPDATE, 81, 114
  adopted, 127–129
  assigning, 79
  authorization search progression, 124
  checking example, 125–126
  in compliance evaluation, 288–289
  data, 111, 113–114
  dynamic SQL and, 273
  examination commands, 288
  group profile, 116
  groupings, 115–116
  IBM i checking of, 123–126
  IFS, 140–141
  object, 111, 112–113
  for printing functions, 158
  relationships, 114–115
  save/restore functions, 130–131
  setting, 218–219
  user profiles and, 81
  See also Special authorities
Authority cache, 126
Authority-checking algorithm, 283
Authorization lists, 121–123
  *PUBLIC authority, 275, 280
  advantages, 122
  attached to application objects, 272
  attaching, 279
  creating, 279
  decision point, 275–276
  defined, 121
  group profiles comparison, 123
  number of files and, 275–276
  QPWFSERVER, 153
  reviewing, 291
  securing objects with, 121
  security recommendation, 122
  specifying in QUSEADPAUT, 275
Authorization schemes, 209, 212–216
  application-only access, 214
  application-only access with a twist, 215
  menu access control, 212–214
  process change, 215–216
  secure, 214–216
  See also Applications
Authorization search progression, 124
*AUTLMGT authority, 113
Auxiliary storage pools (ASPs), 171
Availability, 4

B

Backup media, encrypting, 170–171
Backup, Recovery and Media Services (BRMS), 133, 171
Backup strategy, reviewing, 291
Breach, security
  being prepared for, 293
  incident response team, 293–294, 295
  investigating, 294, 295
  possibility of, 293
  response to, 294
  See also Incidents; Security incident response plan
"Breakage," avoiding, 282–283
Business contingency plan, 208
Business events/procedures, 17

C

CGI-BIN programs, 204
*CHANGE authority, 81, 274, 278
Change management
  audit journal and, 283
  authority-checking algorithm, 283
  authorization list *PUBLIC authority, 283
  change order, 283
  effect of application security changes on, 281, 282
  reviewing, 291
  security scheme, 283
  settings, 281
  software updates, 282
Change Service Tools User Privileges screens, 98–99
  Debug, 102
  Dedicated Service Tools (DST) environment, 102
  Disk units – administration, 100
  Disk units – operations, 100
  Disk units – read only, 100
  Display/Alter/Dump, 102
  Hardware service manager, 102
  Install, 102
  Licensed Internal Code fixes, 102
  Licensed Internal Code log, 102
  Main storage dump, 102
  Operating system initial program load (IPL), 102
  Operator panel functions, 102

Partition remote panel key *XXXXXXXX* (0-7), 101
Performance data collector, 102
Product activity log, 102
Remote service support, 102
Service tool save and restore, 102
Service tool security, 102
Start Service Tools, 102
System capacity – administrator, 102
System capacity – operations, 102
System partitions – administration, 101
System partitions – operations, 100–101
System security, 102
Take over console, 102
Trace privilege, 102
See also Service tools functional privileges
Characters, password
adjacent, 47, 53–54
limiting use of, 47–48, 53–54
maximum number of, 48–49, 54
minimum number of, 48–49, 54
prevention, 49, 54
repeating, 48, 53–54
special, 54
CHGACTPRFL (Change Active Profiles List) command, 80
CHGACTSCDE (Change Activation Schedule Entry) command, 81, 203
CHGAUD (Change Auditing Value) command, 250
CHGAUT (Change Authority) command, 141, 143, 150
issuing, 141
managing authorities with, 144
CHGCMDDFT (Change Command Default) command, 176
CHGDDMTCPA (Change DDM TCP/IP Attributes) command, 184
CHGEXPSCDE (Change Expiration Schedule Entry) command, 80
CHGFCNUSG (Change Function Usage) command, 134, 135
CHGJOBD (Change Job Description) command, 23
CHGOBJAUD (Change Object Auditing) command, 249
CHGOBJPGP (Change Object Primary Group) command, 117
CHGOUTQ (Change Output Queue) command, 155–156
CHGOWN (Change Owner) command, 150
CHGPGM (Change Program) command, 273
CHGPWD (Change Password) command, 68
CHGSECAUD (Change Security Auditing) command, 243, 244, 265
CHGSMTPA (Change SMTP Attributes) command, 180
CHGSYSVAL (Change System Value) command, 19, 62
CHGUSRAUD (Change User Audit) command, 247–248
defined, 247
prompt screen, 248
CHGUSRPRF (Change User Profile) command, 64, 128
*STATUS parameter, 39, 217
auditing duties and, 248
Chief Security Officer (CSO), 299
CHKOBJITG (Check Object Integrity) command, 289
Client-side authentication, 192
Client/server applications, 217
Coded character set ID (CCSID), 84
Communications security, 191–194
Digital Certificate Manager, 193
digital certificates, 191
Secure Shell, 194
SSL, 191–193
VPNs, 193
wireless, 194–195
Communications
process, 17
in security awareness program, 300
security policy, 17
Compliance
exceptions, documenting, 285–286
key area evaluation, 285–290
knowledge, 285
lifestyle, 285
maintaining, 6–7, 285–292
object authority, 288–290
policies and processes, 291–292
reviews, 290–291
summary, 292
system values, 286
user profiles, 286–288
Confidential data, 12
Confidentiality, 2–3
Consultants
*ALLOBJ special authority and, 238
practices, 238
security for, 237, 238
Control-block modification, preventing, 25
COPY command, 149, 150
CPY command, 149, 150
CPYAUDJRNE (Copy Audit Journal Entries) command, 109, 256, 265
defined, 256, 262
prompt screen, 262
using, 262–263
for ZC entries, 271
for ZR entries, 271
CPYSPLF (Copy Spooled File) command, 158
CPYTOIMPF (Copy to Import File) command, 150
CPYTOSTMF (Copy to Stream File) command, 150
Create Service Tools User ID panel, 95
CRTAUT (Create authority) parameter, 278
CRTAUTL (Create Authorization List) command, 121
CRTCLPGM (Create CL Program) command, 129
CRTJOBD (Create Job Description) command, 23
CRTJRNRCV (Create Journal Receiver) command, 243
CRTOBJAUD (Create object auditing) attribute, 252
CRTOUTQ (Create Output Queue) command, 155
AUT (Authority) parameter, 157
AUTCHK (Authority check) parameter, 157

DSPDTA (Display data) parameter, 156
OPRCTL (Operator control) parameter, 156–157
parameters, 156
CRTUSRPRF (Create User Profile) command, 64, 248
AUT (Authority) parameter, 79
CURLIB (Current library) parameter, 74
DSPSGNINF (Display sign-on information) parameter, 76
GRPAUT (Group authority) parameter, 76, 78, 79
GRPAUTTYP (Group authority type) parameter, 76, 78, 79
GRPPRF (Group profile) parameter, 76, 77, 79
INLMNU (Initial menu) parameter, 74
INLPGM (Initial program to call) parameter, 74
LCLPWDMGT (Local password management) parameter, 68
LMTCPB (Limit capabilities) parameter, 75–76, 77
LMTDEVSSN (Limit device sessions) parameter, 76
OWNER (Owner) parameter, 76
parameters, 64–65
PASSWORD (User password) parameter, 67–68, 77
PWDCHGBLK (Block password change) parameter, 69–70
PWDEXP (Set password to expired) parameter, 68, 82
PWDEXPITV (Password Expiration Interval) parameter, 69, 76, 77
SPCAUT (Special Authority) parameter, 70, 72–74, 77
STATUS (Profile status) parameter, 70
SUPGRPPRF (Supplemental groups) parameter, 76
TEXT parameter, 77
USRCLS (User Class) parameter, 70–71
USRPRF (User profile) parameter, 65–67, 77, 128, 274
Crypto card, 165–166

D

Data
access, 11
categories, 12
disposal methods, 11
encrypting, 12, 165–170
IT role as custodian, 14
ownership, 13
preservation, 295
retention period, 11
security, restoring, 297–298
security, saving, 296–297
storage location, 166–167
use, 12
Data area, *PUBLIC authority, 277
Data authorities, 113–114
application, 113
defined, 111
types of, 114
Data classification, 11–14
areas addressed by, 11–12
ownership and, 13
responsibility for, 13
by title, 12
Database Utility (DBU), 267
DDMACC attribute, 175
Debugging techniques, 281–282
Decision points, 273–278
*PUBLIC authority, 276–277
authorization lists, 275–276
queries, 277–278
trade-offs, 273
See also Object-level security
Dedicated Service Tools (DST), 63, 93, 97, 219
environment privilege, 108
panel access, 97
Deny by default, 11, 112, 116, 203, 268
Device profiles, 105–106
defined, 105
information about, 105–106
management, 106
QCONSOLE, 106
Digital Certificate Manager (DCM), 192, 193
Digital certificates, 191
Digital signatures, 43
Digits, password
limiting, 47, 53
preventing, 53
repeated, 53
requiring, 50, 54
Directories
*PUBLIC authority, 277
application, 151
home, 152
IBM-supplied, 151–152
locked down, user access, 146
object ownership in, 154
user, 151
Disaster recovery, reviewing, 291
Display files, *PUBLIC authority, 277
Display Station Passthrough (DSPT), 31
Distributed Data Management (DDM), 6, 153, 214
configuring, 183
request processing, 175
security, 183–184
Distributed Relational Database Architecture (DRDA), 175
defined, 183
security, 183, 185
*DLT authority, 81, 114
DLTUSRPRF (Delete User Profile) command, 64
Domain Name Service (DNS), 182
Domains, 22
DSPACTPRFL (Display Active Profiles List) command, 81
DSPACTSCD (Display Activation Schedule) command, 80
DSPAUDJRNE (Display Audit Journal Entries) command, 256, 265, 281–282
defined, 256
failures and, 281–282
output, 258
prompt screen, 257
using, 257–258
DSPAUT (Display Authority) command, 141, 288
DSPAUTL (Display Authorization List) command, 288, 291
DSPAUTLOBJ (Display Authorization List Objects) command, 288, 291
DSPAUTUSR (Display Authorized Users) command, 287
DSPEXPSCD (Display Expiration Schedule) command, 80

DSPFCNUSG (Display Function Usage) command, 134
DSPJRN (Display Journal)
command, 256
defined, 256
details screen, 260–261
to find path name, 261
parameters, 262
sample screen, 258
screen, 259
using, 259–262
DSPLOG (Display Log) command, 240
message selection, 242
parameters, 241
DSPOBJAUT (Display Object Authority) command, 117
AUT (Authority) parameter, 117–118
CRTAUT (Create authority) parameter, 118
in examining object authorities, 288
DSPOBJD (Display Object Description) command, 287
DSPSECA (Display Security Attributes) command, 19
DSPSECAUD (Display Security Auditing) command, 265
DSPSPLF (Display Spooled File) command, 158
DSPSSTUSR (Display Service Tools User ID) command, 104–105
DSPSYSVAL (Display System Value) command, 19
DSPUSRPRF (Display User Profile) command, 231, 287, 298
Dynamic SQL, 273, 279, 282
DYNUSRPRF (Dynamic user profile) attribute, 273

E

EDTAUTL (Edit Authorization List) command, 121
EDTOBJAUT (Edit Object Authority) command, 111, 117, 157
AUT (Authority) parameter, 117–118
CRTAUT (Create authority) parameter, 118
Employee security policy, 15–16, 300
Encryption, 163–172
algorithm, 163
architecting applications for, 167–168
asymmetric key algorithms, 164–165
auxiliary storage pools (ASPs), 171
backup media, 170–171
basics, 163–165
components, 163
disaster recovery and, 172
file data, 165–170
hash algorithms, 165
independent storage pools (IASPs), 171
PKI, 164–165
resources, 172
SHA, 95
transmission of data, 165
Encryption keys, 163
*ALLOBJ special authority and, 169
backing up, 169
changing, 170
controlling access to, 169
generation, 168
managing, 168–170
storing, 169
ENDHOSTSVR (End Host Server) command, 186
ENDTCP (End TCP/IP) command, 186
Evaluating risks
availability, 4
confidentiality, 2–3
integrity, 3
privacy, 4–5
Evaluating threats, 5
*EXCLUDE authority, 116
*EXECUTE authority, 81, 114
Exit points
defined, 189
using, 189–190
Exit programs
defined, 189
Internet security, 206
in security configurations, 190
use of, 189–190
writing, 190
Extranets, in security policy, 14

F

Federal Information Processing Standard (FIPS), 165
File attributes, 144–145
defined, 144
setting, 145
types of, 144
File ID (FID), 146
File shares, 146–148
creating, 148
defined, 146
IBM-defined, 147
properties, 148
reviewing, 153–154
viewing, 147
File Transfer Protocol (FTP), 6, 146, 153, 178–180, 214, 267
anonymous, 179
INACTTIMO parameter and, 179
QMAXSIGN system value and, 178
RCMD (Remote Command Facility), 179
scripts, 63
Firewalls, 200–201
configuration, 200–201
features, 201
technology, 200
See also Internet security

G

General security system values, 27–44
defined, 27
details, 29–44
summary, 28
V6R1, 44
See also System values
Gramm-Leach-Bliley Act (GLBA), 223
Group authority, menu access control, 213–214
Group Identification Number (GID), 79, 215–216
Group policy objects (GPOs), 186, 187
Group profiles, 116–117
assigning users to, 77–78
authorization lists comparison, 123
creating, 77, 225

creation guidelines, 225
defined, 76
GID, 79
granting authority to, 117
membership review, 290
multiple, 116
names, 77
parameters, 76
passwords, 77
private authority, 79
in role implementation, 225–227
sample roles and, 226
GRTOBJAUT (Grant Object Authority) command, 111, 157
AUT parameter, 121
use example, 116

H

Hash algorithms, 165
Header, 125
Health Insurance Portability and Accountability Act (HIPAA), 207, 223
Help desk role
*SECADM special authority, 233
job functions, 230
required capabilities, 230
as separate from administrators, 233
High availability (HA) software, 60
High-level architecture design, 268–269
History log, 240–242
defined, 240
housekeeping, 242
information, 242
sample, 241
See also Auditing
Home directory, 152
HTTP Server, 204
Hypertext Markup Language (HTML), 215

I

i5/OS, 7
IBM Disk Sanitizer, 12
integrated encryption key management, 166
security best practices, 7
IBM Disk Sanitizer, 12
IBM i
application implementation, 218–221
architecture benefits, 264
authorities, setting, 218–219
authority cache, 126
authority checking, 123–126
group profiles, 76
header, 125
SECTOOLS menu, 154
security audit journal, 242–243
security best practices, 1–2, 7
user identities, 63
IBM profiles, 287–288
IBM-supplied directories, *PUBLIC authority for, 151–152
Incident response team, 293–294
Incidents
being proactive, 295
investigations, 295
response to, 294–295
See also Security incident response plan
Independent auxiliary storage pools (IASPs), 171
Information collection, 271–272
Initial Program Load (IPL), 19
Integrated file system (IFS), 269
accessing objects in, 146–148
adopted authority and, 145–146
APIs, 149
auditing objects in, 146
authorities, 140–141
defined, 139
file attributes for, 145
file systems in, 139
objects, authority management, 141–146
POP e-mail storage, 181
security for, 139–154
security recommendations, 151–154
virus scanning, 150–151
Integrity, 3
Internet Assigned Numbers Authority (IANA), 177
Internet security, 197–208
business contingency plan, 208
configuration, 207
control, 203–204
corporate policy, 198–199
exit programs, 206
firewalls, 200–201
intrusion detection, 206–207
ISP, 199
monitoring, 206–207
policy, 14
process, 197–198
resource, 203
risk determination, 197
system values, 201–202
testing and evaluation, 207
user profiles, 202–203
Web applications, 204–206
Internet Service Providers (ISPs), 199
Intrusion detection, 206–207
*IOSYSCFG special authority
defined, 73
power, 174
requirement, 174
See also Authorities; Special authorities
iSeries Access for Web. *See* System i Access for Web
iSeries Access for Windows. *See* System i Access for Windows
iSeries Navigator. *See* Navigator for i
IT staff
authorization roles, 230
role-based access, 229–238
security and, 229

J

Java Database Connectivity (JDBC), 187, 214, 269
Job descriptions, 272
libraries and, 272
securing, 219
Job schedulers, 272
JOBACN attribute
*FILE, 175
*REJECT, 174
*SEARCH, 175
defined, 174
*JOBCTL special authority, 135–136
assigning, 157
in output queue control, 156
See also Authorities; Special authorities

K

Key management, 168–170
access control, 169
backing up, 169
capabilities, 168
changing, 170

generation, 168
separation of duties, 169
storage, 169

L

Least privilege access, 6, 203, 223, 290
Legal review, 18
Libraries
*CHANGE authority, 278
*PUBLIC authority, 272, 276, 277
job descriptions and, 272
QALWOBJRST and, 30
securing, 276
Library lists
for job description, 272
managing, 219–221
Library-qualified calls, 221
Licensed Internal Code (LIC), 93
Lightweight Directory Access Protocol (LDAP), 192
Line Printer Daemon (LPD), 203

M

Mail-bombing, 180
Management Central (Navigator for i), 83, 190–191
Menu access control, 212–214
*ALLOBJ, 214
*PUBLIC authority, 213
authorization model, 212
group authority, 213–214
Message files, *PUBLIC authority, 277
Monitoring, 206–207
intrusion detection, 206–207
security configuration, 207

N

Navigator for i, 7, 63
Application Administration window, 134
disk unit management, 93
logical partitioning, 93
Management Central, 83, 190–191
New User interface, 83
Permissions window, 136–137, 143
for spooled files, 161
TCP/IP management, 195–196
user capabilities, 135
in user profile creation, 81–83
Network Address Translation (NAT), 178
Network administrator role
job functions, 230
required capabilities, 230
TCP/IP configuration, 233
Network connections, 14
Network security attributes
DDMACC, 175
JOBACN, 174–175
PCSACC, 175
New Group window (Navigator for i), 82
New User window (Navigator for i), 81

O

*OBJALTER authority, 113
OBJAUD (Object auditing)
attribute, 249, 250
object/user profile relationship, 250
values, 251
Object auditing, 249–252
*ALL, 253
IBM i determination, 251
for new objects, 252
OBJAUD attribute and, 249, 250, 251
recommendations, 253
See also Auditing
Object authorities
defined, 111
granting, 111
revoking, 111
types of, 112
See also Authorities
Object-level security, 111–137
adopted authority, 127–129
authority checking, 123–126
authorization lists, 121–123
decision points, 273–278
high-level architecture, 268–269
implementing, 267–283
Navigator for i, 136–137
object ownership, 131–132
private authorities, 111–117
project scope determination, 267–268
public authority, 117–132
tools, 136
user function limit, 133–136
See also Security
Objects
accessing in IFS, 146–148
auditing, 61–62, 146
copying, 149
creating, 149
ownership, 131–132, 279
owning all, 274
securing with authorization lists, 121
security data restored with, 298
validation list, 83
*OBJEXIST authority, 112, 113
*OBJMGT authority, 81
*OBJOPR authority, 81
*OBJREF authority, 113
Open data path (ODP), 25
Open Database Connectivity (ODBC), 6, 185, 214
QPWFSERVER authorization list and, 153
security considerations, 187
user access, 146
OpenSSH, 194
Operator role, 231–233
*SAVSYS special authority, 232
*SPLCTL special authority, 231
batch job submission, 232
decisions, 231
job functions, 230
required capabilities, 230
OS/400, 7
Output queue
attributes, 155–158
attributes, changing, 155–156
creating, 155–156
as functional necessity, 155
Navigator for i and, 161
ownership, 157, 159
sample security implementation, 159–160
securing, 155
tools, 160

P

Parameter-passing validation, 24
Parking, 203
Password-related system values, 44–56
defined, 44
details, 45–56
list of, 45
V6R1, 52–56
See also System values

Passwords
- adjacent characters, 47, 53–54
- adjacent digits, 47, 53
- block change, 55, 69
- characters, 47–49, 53–54
- checking rules, 53–55
- in clear text, 221
- different, requiring, 50–51
- digits, 47, 53–54
- expiration interval, 46–47, 69
- expiration warning messages, 56
- group profile, 77
- IBM-supplied profiles and, 68
- levels, 46
- local management, 68
- maximum number of, 48–49
- minimum number of, 48–49
- non-allowed characters, 47–48
- policies, 97
- possible values, 45
- repeating characters, 48, 53
- repeating digits, 53
- security level 20, 20
- service tools user ID, 95–97
- setting to expired, 68
- special characters, 54
- time change interval, 55
- user profile, 67
- validation program, 51–52

Patch strategy, reviewing, 291

Payment Card Industry Data Security Standard (PCI DSS), 166
- audit information requirement, 296
- compliance, 267
- security incident response plan requirement, 293
- wireless networks requirements, 195

PCs
- ODBC, 187
- security for, 185–189
- System i Access for Web, 188–189
- System i Access for Windows, 185–187

PCSACC attribute, 175

Permissions, 136

Permissions window (Navigator for i), 136–37, 143

Physical security, 173
- precautions, 173
- security policies, 10
- *See also* Security

Platform-specific issues, 15

Portable Application Solutions Environment (PASE), 29, 150, 215

Post Office Protocol (POP), 181

Privacy, 4–5

Private authorities. *See* Authorities

Private data, 12

Processes, reviewing, 291–292

Profile-swap APIs, 215

Profile-token APIs, 215

Program adoption, 279

Programmer/analyst role, 233–236
- activity logging, 236
- creating, 234
- implementation challenge, 233
- job functions, 230
- production access, 235
- production authority, 235–236
- required capabilities, 230
- *See also* Role-based access; Roles

PRTADPOBJ (Print Adopting Objects) command, 136

PRTDIRINF (Print Directory Information) command, 154

PRTPRFINT (Print Profile Internals) command, 80

PRTPUBAUT (Print Public Authority) command, 136, 154, 288

PRTPVTAUT (Print Private Authority) command, 136, 154, 288

PRTQAUT (Print Queue Authority Report) security tool, 160

PRTSQLINF (Print SQL Information) command, 273

PRTSYSSECA (Print System Security Attributes) command, 62

PRTUSRPRF (Print User Profile) command, 68, 80, 287, 288

PRTVLDL (Print Validation List) command
- code, 84
- DDS, 85
- defined, 84
- processing program, 88–89
- report, 90
- RPG IV prototypes, 85–87
- screen, 90

*PUBLIC authority, 15, 79, 117–121
- *USE, 267–268
- for application and user directories, 151
- application objects, 277
- in authority check, 124–125
- authorization lists, 275
- authorization lists, setting, 280
- commands using, 117
- in decision process, 276–277
- default, using, 119–121
- defined, 117
- defining, 219
- determination, 119
- establishing, 117–119
- for IBM-supplied directories, 151–152
- menu access control, 213
- programs, 276
- save/restore functions, 130–131
- setting, 279

Public data, 12

Public key infrastructure (PKI), 164–165

Public/private key-pairs, 165

Q

QALWOBJRST system value, 29–30
- defined, 28, 29
- security recommendation, 30

QALWUSRDMN system value
- defined, 28, 30
- security recommendation, 30

QAUDCTL system value, 56, 57, 243

QAUDENDACN system value, 56, 57, 244

QAUDFRCLVL system value, 56, 58, 244

QAUDJRN journal, 243

QAUDLVL (Auditing Level) system value, 26, 56, 58–60
- *AUTFAIL value, 246
- *CREATE value, 246
- *DELETE value, 246–247
- *JOBDTA value, 247
- *PGMADP value, 247
- *PRTDTA value, 247
- *SAVRST value, 247
- *SECURITY value, 247
- *SERVICE value, 109, 247
- *SPLFDTA value, 247

defined, 58, 245
security recommendation, 60
specifying, 253
values, 58–60, 245–246
QAUDLVL2 system value, 56, 61
defined, 245
values, 245
QAUTOCFG system value, 28, 31, 173
QAUTOVRT system value, 28, 31, 174
QCONSOLE device profile, 106
QCRTAUT system value
*ALL value, 120
*CHANGE value, 119–120
*EXCLUDE value, 120
defined, 28, 32
security recommendation, 32
values, 32
QCRTOBJAUD system value, 56, 61–62
changing, 252
defined, 61
security recommendation, 61
QDSCJOBITV system value, 28, 35–36
QDSPGNINF system value, 28, 33
QFRCCVNRST system value, 33–35
behavior, 33
defined, 28, 33
security recommendation, 35
values, 34–35
QHST file, 240, 242
QINACTITV system value, 28, 35–36
changing, 36
defined, 28, 35
in Internet security, 202
security recommendation, 36
session application, 36
QINACTMSGQ system value, 28, 35–36
QLMTDEVSSN system value, 28, 36–37
concurrent device session control, 37
defined, 28, 36
security recommendation, 37
QLMTSECOFR system value, 28, 37–38
defined, 28, 37
security recommendation, 38
usefulness, 38
QMAXSGNACN system value, 28, 38–39
defined, 28, 38
security recommendation, 39
values, 39
QMAXSIGN system value, 28, 38
QPGMR user profile, 234
QPWDCHGBLK system value, 55
QPWDEXPITV system value, 45, 46–47
*NOMAX value, 69
defined, 45, 46
password expiration control, 47
security recommendation, 47
QPWDEXPWRN system value, 45, 56
QPWDLMTAJC system value, 45, 47
QPWDLMTCHR system value, 45, 47–48
QPWDLMTREP system value, 45, 48, 69
QPWDLVL system value, 45–46
defined, 45
in Internet security, 202
security recommendation, 46
QPWDMAXLEN system value, 45, 48–49
QPWDMINLEN system value, 45, 48–49
QPWDPOSDIF system value, 45, 49
QPWDRQDDGT system value, 45, 50
QPWDRQDDIF system value, 45, 50–51
defined, 45, 50
security recommendation, 51
values, 50
QPWDRULES system value, 45, 53–55
default, 53, 54
defined, 53
security recommendation, 55
values, 53–54
QPWDVLDPGM system value, 45, 51
defined, 45, 51
security recommendation, 52
validity-checking program, 51–52
QPWFSERVER authorization list, 153
QRETSVRSEC system value, 28, 39–40
defined, 28, 39
remote client authentication, 184
security recommendation, 40
for validation list users, 90–91
QRMTSIGN system value, 28, 40
QSCANFS system value, 28, 41–42
defined, 28, 41
IFS and, 151
security recommendation, 42
QSCANFSCTL system value, 28, 41–42
defined, 28, 41
IFS and, 151
security recommendation, 42
QSECOFR (security officer profile), 22, 64
default password, 68
disabled, 70
in Internet security, 202
production object ownership and, 132
QSECOFR (service tools user ID), 93, 97, 104, 110
QSECURITY (System Security Level) system value, 19–20
changing, 20
defined, 19
displaying, 19–20
security level 20, 20
security level 30, 21
security level 40, 22–24
security level 50, 24–26
security levels 40/50, moving to, 26–27
security recommendation, 21
QSSLCSL system value, 44
QSSLCSLCTL system value, 44
QSSLPCL system value, 44
QSYCHGPW (Change Password) API, 44, 70
QSYS, 22, 23, 143, 290
QSYSOPR user profile, 231
QTEMP library
maintenance, 25–26
objects, 25
Queries, 272
options, 278
securing, 277–278
separating, 277–278
QUSEADPAUT system value, 28, 42–43
defined, 28, 42

for production system, 42
security recommendation, 43
specifying authorization list in, 275
QVFYOBJRST system value, 28, 43–44
defined, 28, 43
security recommendation, 44
values, 43
QZLSADFS (Add File Share) API, 154
QZLSCHRS (Change File Share) API, 154

R

RCLSTG (Reclaim Storage) command, 26
Re-creating system, 298
*READ authority, 81, 114
Recovery techniques, 281–282
Remote Execution (REXEC) facility, 182
Reports, audit, 264
Requirements, reviewing, 6–7
Resource security
defined, 20
Internet, 203
Responsible parties
data classification, 13
security policies, 10
Restoring
modified programs, 23–24
security data, 297–298
Restricted data, 12
Restricted MI instructions, 22–23
Reviews, 290–291
authorization lists, 291
group profile membership, 290
need for, 290
process, 291–292
requirements, 6–7
special authority assignments, 290
See also Compliance
Risks
availability and, 4
confidentiality and, 2–3
evaluating, 2–5
integrity and, 3
privacy and, 4–5
Role-based access (RBAC), 223–228
defined, 223
group profiles, 225–227
implementation, 227–228
for IT, 229–238
premise, 223
Roles
defined, 223
definitions, 223–224
definitions with associated group profiles, 226
help desk, 230, 233
identifying, 230
implementation spreadsheet for, 228
implementation with group profiles, 225–227
IT authorization, 230
network administrator, 230, 233
operator, 230, 231–233
programmer/analyst, 230, 233–236
sample, 224–225
security administrator, 230, 236–237
system administrator, 230, 236–237
RSTAUT (Restore Authority) command, 117, 130
RSTOBJ (Restore Object) command, 23
RSTUSRPRF (Restore User Profile) command, 297
RTVDIRINF (Retrieve Directory Information) command, 154
RVKOBJAUT (Revoke Object Authority) command, 111, 157

S

Sarbanes-Oxley Act (SOX), 223, 229
Save files (*SAVF), in encrypting backup media, 170
SAVSECDTA (Save Security Data) command, 117, 190, 202, 296
*SAVSYS special authority
defined, 73
operators, 232
risks, 73
for system operators, 130
See also Authorities; Special authorities
SAVSYS (Save System) command, 296
SBMJOB (Submit Job) command, 232, 274
*SECADM special authority
for creating/maintaining user profiles, 64
defined, 73
help desk, 233
See also Authorities; Special authorities
SECTOOLS menu, 154
Secure Hash Algorithm (SHA) encryption, 95
Secure HTTP (HTTPS), 205
Secure Shell, 194
Secure Sockets Layer (SSL), 105, 191–193
client-side authentication, 192, 205
defined, 191
servers enabled for, 192
Security
annual assessment, 289–290
availability and, 4
as business function, 8
confidentiality and, 2–3
for IFS, 139–154
integrity and, 3
Internet, 197–208
IT staff and, 229
as lifestyle, 8
object-level, 111–137, 267–283
password, 20
PCs, 185–189
physical, 10, 173
plan, maintaining, 7
printed output, 155–161
privacy and, 4–5
questions for vendors, 220
resource, 20, 203
service tools, 93–110
strategic issue management, 5–7
system level, 19–62
TCP/IP, 175–185
for vendors/consultants, 237–238
Security administrator role, 236–237
*ALLOBJ special authority and, 236–237
job functions, 230
required capabilities, 230
Security awareness program, 299–302

communication methods, 300
creating, 301–302
employee security policy, 300
packages, 300
reasons for, 299–300
as requirement, 299
starting, 300–302
topics, 301–302
training, 299
Security data, 296–298
restored with an object, 298
restoring, 297–298
save frequency, 296
saving, 296–297
Security incident response plan, 293–298
data preservation, 295
incident response, 294–295
incident response team, 293–294
investigation, 295
preparing, 293–295
security data, restoring, 297–298
security data, saving, 296–297
system recreation, 298
Security level 20
*ALLOBJ special authority and, 20
defined, 20
moving to level 30, 21
program testing, 22
special authorities, 20
Security level 30
defined, 21
moving from level 20 to, 21
program testing, 22
special authorities, 21
Security level 40, 22–24
defined, 22
job initiation validation, 23
modified program restoration, 23–24
moving to, 26–27
parameter-passing validation, 24
reasons for using, 24
restricted MI instructions, 22–23
state and domain restrictions, 22
Security level 50, 24–26
control-block modification prevention, 25
defined, 24
message restrictions, 24–25
moving to, 26–27
pointer removal, 25
QTEMP library maintenance, 25–26
reasons for using, 26
Security policies, 9–17
application design, 15
benefits, 9
communicating, 17
compliance, 16
data classification, 11–14
defined, 9
employee, 300
employee guidelines, 15–16
enforcement, 16–17
functions, 9
Internet, 198–199
legal review, 18
need for, 9
network connections, 14
notification, 16
notification, enforcement, compliance, 16–17
overview, 9–10
physical security, 10
platform-specific issues, 15
responsible parties, 10
security incident response plan, 293
starting, 18
Security scheme
change management, 283
implementation, 9
initial steps, 7
starting, 7
*SERVICE special authority
defined, 73
selective function, 73–74
See also Authorities; Special authorities
Service functions, 93
Service tools
defined, 93
security, 93–110
security log, 108
security recommendations, 109–110
use, monitoring, 108–109
V6R1 features, 104
Service tools functional privileges, 97–105
changing, 98–99
default, 103
defined, 97
list of, 100–102
Service tools user IDs, 93–97
11111111, 93, 96, 104
22222222, 93, 95, 96
alternate password policy details, 96–97
creating, 94
default password policy details, 96
defined, 63
disabling, 109
expired and defaulted passwords, 109, 110
IBM-supplied, 93, 95, 96, 104, 109, 110
minimum number of, 109
password policy selection, 97
passwords, 95–97
QSECOFR, 93, 97, 104, 110
security recommendations, 109–110
Simple Mail Transfer Protocol (SMTP), 180
Simple Network Management Protocol (SNMP), 182–183
Simple Network Time Protocol (SNTP), 183
SNDNETSPLF (Send Network Spooled File) command, 156
SNDTCPSPLF (Send TCP/IP Spooled File) command, 156
SPCAUT (Special authority) parameter, 71–74
in creating group profiles, 77
overriding, 72
values, 71–72
Special authorities
*ALLOBJ, 2, 6, 72–73, 237, 238
*IOSYSCFG, 73, 174
*JOBCTL, 135–36, 156, 157
*SAVSYS, 73, 130, 232
*SECADM, 64, 73, 233
*SERVICE, 73–74
*SPLCTL, 73, 157–158, 231
assignments, reviewing, 290
excess, removing, 280
security level 20, 20
security level 30, 21
security level 40, 21
security level 50, 21
See also Authorities
*SPLCTL special authority
defined, 73
drawbacks, 73

operators and, 231
output queue and, 157–159
See also Authorities
Spoofing, 204
Spooled files, managing, 161
State, assignment, 22
Storage Area Network (SAN), 171
STRPASTHR (Start Pass-Through) command, 40
STRSST (Start System Service Tools) command, 93, 94
STRTCP (Start TCP/IP) command, 176, 186
STRTCPSVR (Start TCP/IP Servers) command, 176
System administrator role, 236–237
*ALLOBJ special authority and, 236–237
job functions, 230
required capabilities, 230
System i Access data transfers, 6, 214
System i Access for Web, 188–189
configuration files, 189
illustrated, 188
roles and, 224
security, 188
System i Access for Windows, 185–187
application administration, 187
data transfer, 186
defined, 185
host servers, 186
Microsoft policies, 187
remote command, 186
roles and, 224
security challenges, 185–186
selective install, 187
user view from desktop, 186
System i Security: Protecting i5/OS Data with Encryption, 172
System i Security Reference, 150, 157, 247, 253, 254, 281
System i Security Service Tools, 110
System i Systems Management: Recovering Your System, 297
System security, 19–62
System Service Tools (SST), 63, 93
device profile configuration, 106
panel access, 97
System test, 221
System values
audit-related, 56–62
compliance, 286
general, 27–44
in Internet security, 201–202
locking down, 62
password-related, 44–56
QALWOBJRST, 28, 29–30
QALWUSRDMN, 28, 30
QAUDCTL, 56, 57
QAUDENDACN, 56, 57, 244
QAUDFRCLVL, 56, 58, 244
QAUDLVL, 26, 56, 58–60, 109, 245
QAUDLVL2, 56, 61
QAUTOCFG, 28, 31, 173
QAUTOVRT, 28, 31, 174
QCRTAUT, 28, 32, 119–120
QCRTOBJAUD, 56, 61–62, 252
QDSCJOBITV, 28, 35–36
QDSPGNINF, 28, 33
QFRCCVNRST, 33–35
QINACTITV, 28, 35–36, 202
QINACTMSGQ, 28, 35–36
QLMTDEVSSN, 28, 36–37
QLMTSECOFR, 28, 37–38
QMAXSGNACN, 28, 38–39
QMAXSIGN, 28, 38
QPWDCHGBLK, 55
QPWDEXPITV, 45, 46–47
QPWDEXPWRN, 45, 56
QPWDLMTAJC, 45, 47
QPWDLMTCHR, 45, 47–48
QPWDLMTREP, 45, 48
QPWDLVL, 45–46, 202
QPWDMAXLEN, 45, 48–49
QPWDMINLEN, 45, 48–49
QPWDPOSDIF, 45, 49
QPWDRQDDGT, 45, 50
QPWDRQDDIF, 45, 50–51
QPWDRULES, 45, 53–55
QPWDVLDPGM, 45, 51
QRETSVRSEC, 28, 39–40, 90–91, 184
QRMTSIGN, 28, 40
QSCANFS, 28, 41–42, 151
QSCANFSCTL, 28, 41–42, 151
QSECURITY, 19–27
QSSLCSL, 44
QSSLCSLCTL, 44
QSSLPCL, 44
QUSEADPAUT, 28, 42–43, 275
QVFYOBJRST, 28, 43–44
security-related, 27–62
System-wide auditing, 245–247
Systems Network Architecture (SNA), 174

T

TCP/IP
defined, 175
Distributed Relational Database Architecture (DRDA), 183
Domain Name Service (DNS), 182
File Transfer Protocol (FTP), 178–180
internal addresses, 177
IP packet filtering, 177–178
Network Address Translation (NAT), 178
port security, 177
Post Office Protocol (POP), 181
Remote Execution (REXEC) facility, 182
security, 175–185
servers, starting, 176
Simple Mail Transfer Protocol (SMTP), 180
Simple Network Management Protocol (SNMP), 182–183
Simple Network Time Protocol (SNTP), 183
utilities, 179
Testing applications, 221–222
Threats, evaluating, 5
Trigger programs, 218

U

*UPDATE authority, 81, 114
USEADPAUT (Use adopted authority) parameter, 42, 128, 129, 275
User account management, reviewing, 291–292
User auditing, 247–48, 253
User Function Registration APIs, 133
User Identification (UID)
APIs, 215–216
number, 79
User profile attributes, 64–79
AUT (Authority), 79
in compliance evaluation, 286–287
CURLIB (Current library), 74
DSPSGNINF (Display sign-on information), 76
GRPAUT (Group authority), 76, 78, 79
GRPAUTTYP (Group authority type), 76, 78, 79

GRPPRF (Group profile), 76, 77, 78, 79
INLMNU (Initial menu), 74
INLPGM (Initial program to call), 74
LCLPWDMGT (Local password management), 68
list of, 64–65
LMTCPB (Limit capabilities), 75–76, 77
LMTDEVSSN (Limit device sessions), 76
OWNER (Owner), 76, 78
PASSWORD (User password), 67–68, 77
PWDCHGBLK (Block password change), 69
PWDEXP (Set password to expired), 68
PWDEXPITV (Password expiration interval), 69, 76, 77
SPCAUT (Special authority), 70, 72–74, 77
STATUS (Profile status), 70
SUPGRPPRF (Supplemental groups), 76, 78
TEXT, 77
USRCLS (User class), 70–71
USRPRF (User profile), 65–67, 77

User profiles, 63–83
assigning groups to, 77–79
capabilities, limiting, 75–76
compliance, 286–288
copying, 83
creating with Navigator for i, 81–83
creation/maintenance authority, 64
current library, 74
dates, 287
defined, 63–64
examination commands, 287
inactive, 287
initial menu, 74
initial program, 74
in Internet security, 202–203
names, 65–67
object authority, 288–290
passwords, 67
predefined, 64
private authorities and, 81
status, 70
tools, 80, 81
UID, 79

USRPRF (User profile) parameter, 65–67, 77, 128, 274

V

Validation
job initiation, 23
parameter-passing, 24
program, 51–52

Validation list (*VLDL) objects, 83, 221
See also Validation list users

Validation list users, 83–91
authentication, 83
defined, 63, 84
PRTVLDL command, 84–90
security implications, 90–91
uses, 83

Vendors
*ALLOBJ special authority and, 237
security for, 237–238
security questions for, 220
support, 237–238

Virtual private networks (VPNs), 14, 193
client authentication, 14
connections, 165

Virus scanning, 150–151

W

Web applications, 153
configuring, 205
securing, 204–206

WebSphere applications, 217

Wireless networks
compliance requirements, 195
security, 194–195
in security policy, 14
threat, 194

Work with Service Tools Security Log panel, 108

Work with Service Tools User IDs panel, 94

Work with System Security panel, 106–108
"Allow a service tools user ID" option, 107–108
"Allow new digital certificates" option, 107
"Allow system value security changes" option, 107
illustrated, 107
options, 106
reaching, 106

Workstations, 37

WRKAUT (Work with Authority) command, 141, 142

WRKFCNUSG (Work with Function Usage) command, 134, 135

WRKLNK (Work with Object Links) command, 141, 181

WRKNAMSMTP (Work with Names for SMTP) command, 180

WRKOBJPGP (Work with Objects by Primary Group) command, 117

WRKREGINF (Work with Registration Information) command, 195

WRKSPLF (Work with Spooled Files) command, 275

WRKSYSVAL (Work with System Values) command, 189